Krebs cycle
克雷伯氏循環

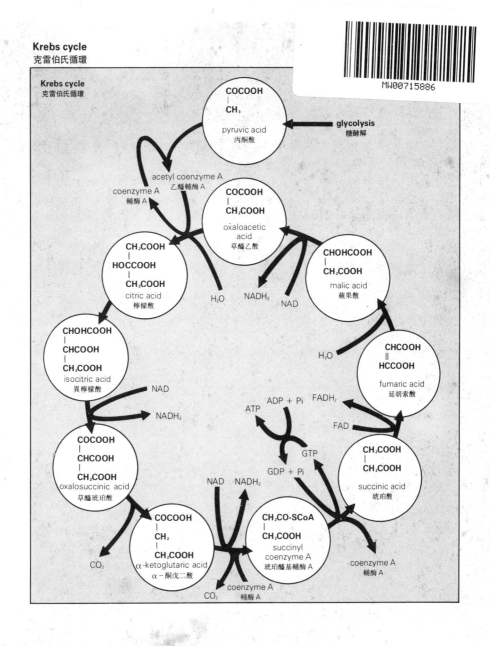

前言

　　朗文英漢科學系列圖解詞典包括生物、化學、科學、植物、地質和物理等冊。這是一套內容既有聯繫而又各自獨立成冊的系列詞典，《朗文英漢生物圖解詞典》為其中的一冊。

　　本書收詞 1800 多個，包括細胞學、分類原則、生物類別、器官組織、生理學、遺傳學、進化以及生態學等全部生物學基本原理的詞彙。

　　這些詞按詞義分科目編排，英漢雙解對照，釋義簡明，概念準確。每一科目的的上、下各個詞條，內容互有關聯。釋義深入淺出，易於理解；其中又標出頁碼和箭嘴號，引導讀者查找相關詞條，提供更多資料作比較，以加深理解，掌握更多詞彙。

　　詞典中收入近 400 幅印刷精美的彩色插圖和圖表，顯示生物的形態、主要種群間的關係以及生物各器官之間的關係，以助讀者更好理解釋義，但釋義並不依賴這些插圖。

　　此外，本詞典還收入維生素、營養素、國際單位制三個實用附錄。詞典後部附索引，按英文字母順序排列，自成一個英漢生物學詞彙素，方便讀者檢索。

　　本詞典適合具高中至大學一、二年級程度的學生以及需要深入了解生物學術語的非生物學專業的讀者查閱使用。

朗文出版（遠東）有限公司
一九九一年四月

朗文英漢生物圖解詞典

LONGMAN ENGLISH-CHINESE
ILLUSTRATED DICTIONARY OF
BIOLOGY

Longman 朗文　　YORK PRESS

English edition © Librairie du Liban 1985
(Original title: Longman Illustrated Dictionary of Biology)
Bilingual edition © Librairie du Liban & Longman Asia Limited 1991, 1993, 1994, 1995

Published by
Longman Asia Limited
18/F., Cornwall House
Taikoo Place
979 King's Road
Quarry Bay
Hong Kong
Tel: 2811 8168
Fax: 2565 7440
Telex: 73051 LGHK HX

朗文出版亞洲有限公司
香港鰂魚涌英皇道 979 號
太古坊康和大廈十八樓
電話：2811 8168
圖文傳真：2565 7440
電傳：73051 LGHK HX

Fourth impression 1995
一九九一年初版
一九九五年第四次印刷

ISBN 962 359 039 3

Produced by Longman Asia Limited
Printed in Hong Kong
SWT/04

The
publisher's
policy is to use
**paper manufactured
from sustainable forests**

Contents 目錄

How to use the dictionary 本詞典的用法

This dictionary contains over 1800 words used in the biological sciences. These are arranged in groups under the main headings listed on pp. 3-4. The entries are grouped according to the meaning of the words to help the reader to obtain a broad understanding of the subject.

At the top of each page the subject is shown in bold type and the part of the subject in lighter type. For example, on pp. 18 and 19:

18 · THE CELL/CARBOHYDRATES

THE CELL/CARBOHYDRATES · **19**

In the definitions the words used have been limited so far as possible to about 1500 words in common use. These words are those listed in the 'defining vocabulary' in the *New Method English Dictionary* (fifth edition) by M. West and J. G. Endicott (Longman 1976). Words closely related to these words are also used: for example, *characteristics*, defined under *character* in West's *Dictionary*.

本詞典共收1800多個用於生物學的詞彙。這些詞彙均按照類別排列在第 3－4 頁所列出的那些主要標題之下。所有的條目都根據各詞的意義加以分組，以幫助讀者對所需查找的科目獲得概括的了解。

在每一頁左或右上方，皆用黑體字印出有關科目名稱，並以秀麗體印出該科目下的分段。例如，在第18頁和第19頁上：

18· 　**細胞** / 碳水化合物

細胞 / 碳水化合物 　·19

釋義部分所使用的詞盡可能限於常用的1500個詞左右，這些詞列於 M·韋斯特和 J·G·恩迪科特合編的《新法英語詞典》（第五版，朗文公司1976年版）中的釋義詞彙表內。本詞典也使用一些和這些詞密切相關的詞，例如使用韋斯特詞典中在"character(特性)"條下解釋的"characteristics(特徵)"這個詞。

---

1. To find the meaning of a word　　　1· 查明詞的意義

Look for the word in the alphabetical index at the end of the book, then turn to the page number listed.

In the index you may find words with a letter or number at the end. These only occur where the same word appears twice in the dictionary: [a] indicates a word which is defined as it relates to animals and [p] a word defined as it relates to plants. For example, **cone**

cone[a] is part of the retina in the eye;

cone[p] is a reproductive structure in some plants.

The numbers also indicate a word which is defined twice in different contexts. For example, **translocation**

translocation[1] is the transport of materials in plants;

translocation[2] is a kind of chromosome mutation.

The description of the word may contain some words with arrows in brackets (parentheses) after them. This shows that the words with arrows are defined near by.

(↑) means that the related word appears above or on the facing page;

(↓) means that the related word appears below or on the facing page.

先於本書末尾的字順索引中，找出所欲查的詞，然後翻到該詞旁邊所註明的頁碼。

在該索引中，你可能會看到在末尾帶有一個字母或數字的一些單詞。這些情況僅存在於同一個詞在本詞典中出現兩次的場合：[a]表示所解釋的詞與動物學有關；[p]則表示所解釋的詞與植物學有關。例如，cone

Cone[a] 視錐——是眼睛視網膜的一部份；

Cone[p] 球果——是某些植物中的一種生殖結構。

數字則表示在不同的上下文中作出過兩種不同解釋的詞。例如，translocation

Translocation[1] 轉移作用——是植物中物質的運輸；

Translocation[2] 易位——是一種染色體突變。

對單詞所作出的解釋之中，可能還會含有一些單詞，其後附有帶括號(圓括號)的箭頭。這表示，帶有箭頭的詞，其解釋就在附近。

（↑）表示有關的詞出現在本條之前或對面頁上；

（↓）表示有關的詞將出現在本條之後或對面頁上。

A word with a page number in brackets after it is defined elsewhere in the dictionary on the page indicated. Looking up the words referred to may help in understanding the meaning of the word that is being defined.

The explanation of each word usually depends on knowing the meaning of a word or words above it. For example, on p. 178 the meaning of *spore mother cell*, *microsporangium*, and the words that follow depends on the meaning of the word *spore*, which appears above them. Once the earlier words are understood those that follow become easier to understand. The illustrations have been designed to help the reader understand the definitions but the definitions are not dependent on the illustrations.

2. To find related words

Look in the index for the word you are starting from and turn to the page number shown. Because this dictionary is arranged by ideas, related words will be found in a set on that page or one near by. The illustrations will also help to show how words relate to one another.

For example, words relating to principles of classification are on pp. 40–41. On p. 40 *classification* is followed by words used to describe taxonomy and the binomial system and illustrations showing the different taxa involved in the classification of a species and the binomial system; p. 41 continues to explain and illustrate classification, explaining natural and artificial classifications and illustrating the relationships between the major groups of organisms.

3. As an aid to studying or revising

The dictionary can be used for studying or revising a topic. For example, to revise your knowledge of gas exchange, you would look up *gas exchange* in the alphabetical index. Turning to the page indicated, p. 112, you would find *respiration*, *respiratory quotient*, *breathing*, *gas exchange*, and so on; on p. 113 you would find *air*, *gill*, *gill filament*, and so on. Turning over to p. 114 you would find *counter current exchange system* etc.

In this way, by starting with one word in a topic you can revise all the words that are important to this topic.

如果一個詞其後附有帶括號的頁碼，就表示對該詞的解釋是在本詞典中的另一處，即所註明該頁碼的那一頁上。查閱這些有關的詞，有助於更好理解原先所查那個詞的詞義。

每一個單詞的解釋通常取決於理解該詞前面的一個詞或幾個詞的詞義。例如，在第178頁上，"孢子母細胞"、"小孢子囊"以及隨後的一些詞的詞義，都取決於理解在它們前面所出現的"孢子"這個詞的詞義。理解了前面出現的那些詞，其後出現的詞也就易於理解。本詞典設計的插圖，有助於讀者理解釋義，但釋義並不依賴插圖。

2. 查找相關的詞

在索引中，查找出你作爲起頭的那個詞，然後翻到所註明的頁碼。由於本詞典是按照概念排列的，所以，在該頁或其附近的一頁上，即可找到一組相關單詞。插圖也將有助於表明各單詞之間的相互關係。

例如，與分類原則有關的單詞均列在第40－41頁上。在第40頁上，"分類法"這一單詞的後面，列有用於闡明分類學及雙名法的一些單詞，插圖則表示涉及一個種的分類之不同分類單位和雙名法。第41頁繼續對分類法進行解釋，並用插圖加以說明，解釋自然分類和人爲分類，用插圖說明生物主要種羣之間的關係。

3. 作爲學習和複習的輔助工具

本詞典可作學習和複習課題之用。例如，爲複習與氣體交換有關的知識，可查閱字順索引中的"氣體交換"，再按註明的頁碼翻到第112頁，即可找到"呼吸（作用）"、"呼吸商"、"呼吸"、"氣體交換"等等；在第113頁上，可找到"空氣"、"鰓"、"鰓絲"等等。翻到第114頁，則可找到"逆流交換系統"等。

這樣，由某一課題中的一個單詞開始查找，即可複習與該課題有關的所有重要的單詞。

4. To find a word to fit a required meaning

It is almost impossible to find a word to fit a meaning in most dictionaries, but it is easy with this book. For example, if you had forgotten the word for the outer whorl of the perianth of a flower, all you would have to do would be to look up *perianth* in the alphabetical index and turn to the page indicated, p. 179. There you would find the word *calyx* with a diagram to illustrate its meaning.

4. 查找適當的詞，以表達確切的意義

在大多數詞典中，要查找適當的詞來確切表達某一意義幾乎是不可能的，但使用本詞典，就可輕易做到這一點。譬如，你忘記了用於表達"花被外輪"的那個單詞，你僅需在字順索引中查出 perianth(花被)，然後再按註明的頁碼翻到第179頁，即可找到"花萼"這一單詞以及用以說明其含義的插圖。

5. Abbreviations used in the definitions

abbr	abbreviated as	縮寫為
adj	adjective	形容詞
e.g.	*exempli gratia* (for example)	例如
etc	*et cetera* (and so on)	等等
i.e.	*id est* (that is to say)	也就是，即
n	noun	名詞

p.	page	頁
pl.	plural	複數
pp.	pages	頁(複數)
sing.	singular	單數
v	verb	動詞
=	the same as	與……同樣

5. 在釋義中採用的各種縮寫

cell theory an idea, developed in 1839, by Theodore Schwann, which states that all living organisms are made up of individual cells and that it is in these cells and by their division that processes such as growth and reproduction (p. 173) take place.

cell (*n*) the basic unit of a plant or animal. It is an individual, usually microscopic (↓) mass of living matter or protoplasm (p. 10). An animal cell consists of a nucleus (p. 13), which contains the chromosomes (p. 13), the cytoplasm (p. 10) which is usually a viscous fluid or gel surrounded by a very thin skin, the plasma membrane (p. 13). A plant cell is similar except that it is surrounded by a cellulose (p. 19) cell wall (↓) and has a fluid-filled vacuole (p. 11).

cell wall the non-living external layer of a cell in plants. It is comparatively rigid but slightly elastic and provides support for the cell. There may be a primary cell wall (p. 14) composed of cellulose (p. 19) and calcium pectate and, in older plants, a secondary cell wall (p. 14) made of layers of cellulose containing other substances, such as the woody lignin (p. 19).

organelle (*n*) any part of a cell, such as the nucleus (p. 13) or flagellum (p. 12), that has a particular and specialized function.

prokaryote (*n*) a cell in which the chromosomes (p. 13) are free in the cytoplasm (p. 10) and not enclosed in a membrane (p. 14): there is no nucleus. Bacteria (p. 42) and blue-green algae (p. 43) are prokaryotes.

細胞學說　1839年由特奧多爾・施萬提出的一種概念。這種概念認為所有生物都是由一個個細胞構成的，而正是在這些細胞中，經由它們的分裂，才使諸如生長和繁殖(第173頁)的過程得以進行。

細胞(名)　植物或動物的基本單元。細胞是生活物質或原生質(第10頁)的單體，通常為顯微(↓)體。動物細胞是由細胞核(第13頁)和細胞質(第10頁)組成。細胞核含有染色體(第13頁)，細胞質通常是由一層很薄的皮膜即質膜(第13頁)包住的一種黏性液體或凝膠。植物細胞除了四周有一層纖維素(第19頁)細胞壁(↓)以及有一個充滿液體的液泡(第11頁)之外，其餘皆與動物細胞相同。

細胞壁　植物細胞的無生命外層。它較為堅硬，但是有彈性，能支撐細胞。植物中可能有一種由纖維素(第19頁)和果膠鈣組成的初生細胞壁(第14頁)；而在一些較老的植物中，則可能有一種由包含一些其他物質，如木質素(第19頁)的多層纖維素組成的次生細胞壁(第14頁)。

細胞器(名)　細胞中具有特殊和專門功能的任何一部份，例如細胞核(第13頁)或鞭毛(第12頁)

原核細胞(名)　一種細胞。在這種細胞中，染色體(第13頁)游離於細胞質(第10頁)中，未被膜(第14頁)所包裹，且沒有核。細菌(第42頁)和藍藻(第43頁)均屬原核生物。

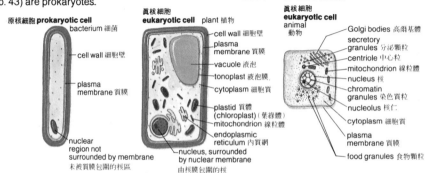

原核細胞 **prokaryotic cell**
bacterium 細菌
cell wall 細胞壁
plasma membrane 質膜
nuclear region not surrounded by membrane 未被質膜包圍的核區

眞核細胞 **eukaryotic cell** plant 植物
cell wall 細胞壁
plasma membrane 質膜
vacuole 液泡
tonoplast 液泡膜
cytoplasm 細胞質
plastid 質體 (chloroplast) (葉綠體)
mitochondrion 線粒體
endoplasmic reticulum 內質網
nucleus, surrounded by nuclear membrane 由核膜包圍的核

眞核細胞 **eukaryotic cell** animal 動物
Golgi bodies 高爾基體
secretory granules 分泌顆粒
centriole 中心粒
mitochondrion 線粒體
nucleus 核
chromatin granules 染色質粒
nucleolus 核仁
cytoplasm 細胞質
plasma membrane 質膜
food granules 食物顆粒

optical microscope 光學顯微鏡

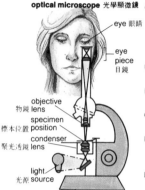

eye 眼睛

eye piece 目鏡

objective lens 物鏡

specimen position 標本位置

condenser lens 聚光透鏡

light source 光源

electron microscope 電子顯微鏡

insulator ①

electron gun ②

condenser lenses ③

specimen chamber ④

objective lens ⑤

specimen position ⑥

projector lenses ⑦

eye ⑧

binocular viewer 雙筒觀測鏡

phosphorescent screen 螢光屏

camera chamber 攝影室

eukaryote (*n*) a cell in which the nucleus (p. 13) is separated from the cytoplasm (p. 10) by a nuclear membrane (p. 13). All organisms, except bacteria (p. 42) and blue-green algae (p. 43) are composed of eukaryotic cells.

unicellular (*adj*) of an organism consisting of one cell only.

multicellular (*adj*) of an organism composed of many cells.

cytology (*n*) the study or science of cells and their activities.

microscopy (*n*) the study, using a microscope (↓), of organisms too small to be seen with the naked eye.

microscope (*n*) an instrument used to give a magnified image of an object that is too small to be seen with the naked eye.

optical microscope a microscope (↑) in which light is passed through the object to be enlarged and passed to the eye through an objective lens system and an eyepiece. This instrument can magnify an object by a maximum of about 1500 times. For larger magnifications, an electron microscope (↓) must be used.

electron microscope a microscope (↑) which can be used to magnify objects by greater than 1500 times and to as much as 500 000 times by using electrons, which have a smaller wavelength than light, to examine the object.

ultrastructure (*n*) the structure of an object which can only be resolved using an electron microscope (↑).

sectioning (*n*) the cutting of an extremely thin slice of tissue (p. 83) which can then be examined using a microscope (↑). The tissue is first frozen or embedded in a material such as paraffin wax before it is cut. **section** (*n*).

microtome (*n*) an instrument used to cut very thin slices of a material.

staining (*n*) a method of examining particular structures inside cells by making parts of the cells opaque to light or electrons using chemicals. Certain kinds of staining materials will stain different structures e.g. iodine stains starch (p. 18).

真核細胞（名） 一種細胞。在這種細胞中，由核膜（第13頁）將細胞核（第13頁）與細胞質（第10頁）分開。所有生物，除細菌（第42頁）和藍藻（第43頁）之外，都是由真核細胞構成的。

單細胞的（形） 指僅由一個細胞構成的生物而言。

多細胞的（形） 指由許多細胞構成的生物而言。

細胞學（名） 研究細胞及其活動的科學。

顯微鏡檢術（名） 利用顯微鏡（↓）對小得用肉眼看不見的生物進行研究的技術。

顯微鏡（名） 對小得用肉眼看不見的物體進行圖像放大的一種儀器。

光學顯微鏡 一種顯微鏡（↑）。使用這種顯微鏡時，光綫透過待放大的物體，然後再通過物鏡系統和目鏡，傳遞給肉眼。這種儀器對物體進行放大的最大值，可達約1,500倍。要獲得更大的放大倍數，則需使用電子顯微鏡（↓）。

電子顯微鏡 一種顯微鏡（↑）。這種顯微鏡使用波長小於光的電子儀器，對物體進行檢查。它可將物體放大到1,500倍以上，乃至放大到500,000倍。

超微結構（名） 使用電子顯微鏡（↑）方可分辨的物體的結構。

切片（名） 指將組織（第83頁）切成極薄的切片。這種極薄的切片即可利用顯微鏡（↑）進行觀察。在切片之前，先將組織加以冰凍和包埋在石蠟之類材料中。（名詞 section 也作"切片"解釋）。

切片機（名） 將材料切成極薄切片的一種器械。

染色（名） 利用化學藥品使細胞的某些部份不能透過光或電子，借以觀察細胞內部特殊結構的一種方法。某些種類的染色物質可使不同的結構染色，例如碘可使澱粉（第18頁）染色。

① 絕緣體　　④ 標本室　　⑦ 投影透鏡
② 電子槍　　⑤ 物鏡　　　⑧ 眼睛
③ 聚光透鏡　⑥ 標本位置

centrifugation (*n*) a method of separating substances of different densities by accelerating them, usually in a rotating container (centrifuge), for quite long periods. Cells may be broken open and suspended in a liquid before centrifugation so that, after centrifugation, the solid particles, the sediment, will fall to the bottom of the container while a supernatant fluid will be left behind above the sediment.

dialysis (*n*) a method of separating small molecules from larger molecules in a mixed solution (p. 118) by separating the solution from water by a membrane (p. 14) through which the small molecules will diffuse (p. 119) leaving behind the larger molecules which are too big to pass through the membrane.

chromatography (*n*) a method of separating mixtures of substances, such as amino acids (p. 21), by making a solution (p. 118) of the substances and allowing the substances to be absorbed (p. 81) and flow through a medium such as paper. The different substances will travel at different rates and so be separated.

chromatogram (*n*) the column or strip of solid on which substances have been separated by chromatography (↑).

electrophoresis (*n*) a method of separating mixtures of substances by suspending them in water and subjecting them to an electrical charge. Different substances will move in different directions and at different rates in response to the charge.

protoplasm (*n*) the contents of a cell.

cytoplasm (*n*) all the protoplasm (↑), or material, inside a cell other than the nucleus (p. 13), which can be thought of as alive. It is usually a viscous fluid or gel containing other organelles (p. 8), such as the Golgi body (↓). **cytoplasmic** (*adj*).

ribosome (*n*) a particle of protein (p. 21) and RNA (p. 24) which is contained in the cytoplasm (↑). Under the control of the DNA (p. 24) in the nucleus (p. 13), protein is produced on the ribosomes by linking together amino acids (p.21) Ribosomes often occur in groups or chains.

離心法(名)　通常使用旋轉容器(離心機)將不同比重的物質進行較長時間加速,從而使它們相互分離的一種方法。在離心之前,可將細胞破裂,懸浮在液體中,這樣,在離心之後,固體粒子即沉澱物就會沉到容器的底部,而上清液則留在沉澱物之上。

透析(作用)(名)　利用膜(第14頁)能使溶液與水分離的手段而使混合溶液(第118頁)中的小分子與大分子實現分離的一種方法。小分子可擴散(第119頁)通過膜,大分子由於太大,以致於不能透過膜,而留了下來。

色譜法(名)　分離如氨基酸(第21頁)之類物質混合物的一種方法。
其做法是:將這些物質作成溶液(第118頁),使它們被吸附(第81頁)並流過一種介質例如紙。這些不同的物質將以不同的速率在介質上行進,從而可將它們分離。

色譜圖(名)　已利用色譜法(↑)將物質分離於其上的固體柱或條。

電泳(名)　分離物質的混合物的一種方法。其做法是:將物質的混合物懸浮在水中,使其經受電荷的作用。這些不同的物質將隨電荷的作用而朝着不同的方向並以不同的速率移動。

原生質(名)　細胞的內含物。

細胞質(名)　細胞內除了細胞核(第13頁)之外的所有原生質(↑)或物質。細胞質可被認為是活的物質。它通常是含有高爾基體(↓)之類的其他細胞器(第8頁)的一種黏性液體或凝膠。(形容詞形式為 cytoplasmic)

核糖體、核蛋白體(名)　細胞質(↑)內的一種含蛋白質(第21頁)和核糖核酸(第24頁)的顆粒。在細胞核(第13頁)內的脫氧核糖核酸(第24頁)的控制下,在核糖體上可將氨基酸(第21頁)結合在一起合成蛋白質。核糖體常以成羣或成串的形式存在。

dialysis 透析

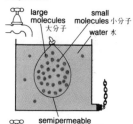

large molecules 大分子
small molecules 小分子
water 水
semipermeable membrane 半透性膜

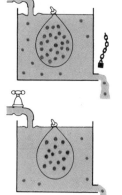

paper chromatography 紙色譜法

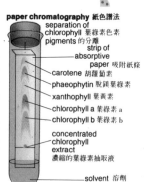

separation of chlorophyll 葉綠素色素 pigments 的分離
strip of absorptive paper 吸附紙條
carotene 胡蘿蔔素
phaeophytin 脫鎂葉綠素
xanthophyll 葉黃素
chlorophyll a 葉綠素 a
chlorophyll b 葉綠素 b
concentrated chlorophyll extract 濃縮的葉綠素抽取液
solvent 溶劑

endoplasmic reticulum 內質網

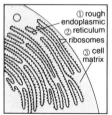

① rough endoplasmic ② reticulum ribosomes ③ cell matrix

smooth ④ endoplasmic reticulum

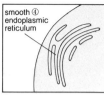

① 粗糙型內質網　② 核糖體
③ 細胞基質　　　④ 光滑型內質網

mitochondrion 線粒體

cristae 嵴　　　matrix 基質

smooth outer membrane
光滑的外膜

vacuole 液泡

cell sap
細胞液

vacuole
液泡

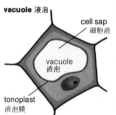

tonoplast
液泡膜

endoplasmic reticulum a meshwork of parallel, interconnected cavities within the matrix (p. 88) of a cell. These are bounded by unit membranes (p. 14) which are continuous with the nuclear membrane (p. 13). **ER** (*abbr*).

ER = endoplasmic reticulum (↑).

rough ER ER (↑) which is covered on the cytoplasm (↑) side with ribosomes (↑).

smooth ER ER (↑) with no ribosomes (↑).

Golgi body a group or groups of flattened cavities within the cytoplasm (↑) of a cell bounded by membranes (p: 14) and connected with the ER (↑). It is similar to smooth ER (↑) but may be used for linking carbohydrates (p. 17) to proteins (p. 21), and it is associated with secretion (p. 106).

mitochondria (*n.pl.*) rod-shaped bodies in the cytoplasm (↑) of a cell. They are bounded by two unit membranes (p. 14) of which the inner is folded inwards into crests or cristae. Cell respiration (p. 30) and energy production take place in these bodies and there are more of them in cells that use a lot of energy.

lysosomes (*n.pl.*) spherical bodies that occur in the cytoplasm (↑) of cells. They are bounded by membranes (p. 14) and contain enzymes (p. 28) which may be released to destroy unwanted organelles (p. 8) or even whole cells.

microtubule (*n*) a fibrous (p. 143) structure made of protein (p. 21) found in the cytoplasm (↑). They may occur singly or in bundles. Their function may be cellular transport, e.g. the spindle (p. 37) fibres in nuclear division (p. 35).

microfilament (*n*) a very fine, thread-like structure made of protein (p. 21) which occurs in the cytoplasm (↑) of most cells.

fibril (*n*) a small fibre (p. 143) or thread-like structure.

vacuole (*n*) a droplet of fluid bounded by a membrane (p. 14) or tonoplast (↓) and contained within the cells of plants and animals except bacteria (p. 42) and blue-green algae (p. 43).

tonoplast (*n*) the inner plasma membrane (p. 13) of a cell in plants which separates the vacuole (↑) from the cytoplasm (↑).

protoplast (*n*) the protoplasmic (↑) material between the tonoplast (↑) and the plasma membrane (p. 13).

內質網　細胞基質（第88頁）內相互平衡及連接的空腔網絡。這些空腔以與核膜（第13頁）相連接的單位膜（第14頁）爲界。（英文縮寫爲 ER）

ER　(Endoplasmic reticulum) 內質網（↑）。

粗糙型 ER　在細胞質（↑）側充滿着核糖體（↑）的內質網（↑）。

光滑型 ER　無核糖體（↑）的內質網。

高爾基體　細胞的細胞質（↑）內的一組或多組扁平空腔，它們以膜（第14頁）爲界，並與內質網（↑）相連。高爾基體與光滑型的內質網（↑）相似，但可供碳水化合物（第17頁）與蛋白質（第21頁）結合之用，同時還與分泌作用（第106頁）有關。

線粒體（名、複）　細胞的細胞質（↑）中的棒狀體。它們被兩層單位膜（第14頁）包覆，內膜向裡褶疊成育突或嵴。細胞呼吸（第30頁）及能量產生就發生在這些棒狀體內。細胞中有較多耗用大量能量的棒狀體。

溶酶體（名、複）　細胞的細胞質（↑）中含的球體。它們被膜（第14頁）包覆，含有多種酶（第28頁）。這些酶可以釋放出來，破壞無用的細胞器（第8頁），甚至整個細胞。

微管（名）　細胞質（↑）中由蛋白質（第21頁）構成的一種纖維（第143頁）結構。它們可單獨地或成束地存在。其功能可能是與細胞的遷移有關，例如紡錘（第37頁）絲在核分裂（第35頁）中所起的功能。

微絲（名）　大多數細胞的細胞質（↑）中含的一種極細的蛋白質（第21頁）構成的絲狀結構。

原纖維（名）　一種小的纖維（第143頁）或絲狀結構。

液泡（名）　由膜（第14頁）或液泡膜（↓）包覆的液態小滴。它存在於除細菌（第42頁）和藍藻（第43頁）以外的植物和動物的細胞中。

液泡膜（名）　植物細胞的內質膜（第13頁），它將液泡（↑）與細胞質（↑）分開。

原生質體（名）　液泡膜（↑）和質膜（第13頁）之間的原生質（↑）物質。

cell sap fluid contained in a plant vacuole (p. 11).

plastid (*n*) in plants, except bacteria (p. 42), blue-green algae (p. 43), and fungi (p. 46), a membrane (p. 14) bounded body in the cytoplasm (p. 10) which contains DNA (p. 24), pigments (p. 126), and food reserves.

chloroplast (*n*) in plants only, the plastid (↑) containing chlorophyll (↓) and the site of photosynthesis (p. 93). It is always bounded by a double unit membrane (p. 14).

細胞液　植物液泡（第11頁）中所含的液體。

質體（名）　除細菌（第42頁）、藍藻（第43頁）和眞菌（第46頁）之外的植物細胞質（第10頁）中的膜（第14頁）包覆體。其內含有脫氧核糖核酸（第24頁）、色素（第126頁）及貯存食物。

葉綠體（名）　僅存於植物中，係含有葉綠素（↓）的質體（↑），也是進行光合作用（第93頁）的部位。葉綠體總是由雙層單位膜（第14頁）所包覆。

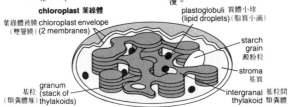

chloroplast 葉綠體

葉綠體被膜 chloroplast envelope（雙層膜）(2 membranes)

plastoglobuli 質體小球 (lipid droplets)（脂質小滴）

starch grain 澱粉粒

stroma 基質

granum 基粒 (stack of thylakoids)（類囊體堆）

intergranal 基粒間 thylakoid 類囊體

chlorophyll (*n*) a green pigment (p. 126) found in the chloroplasts (↑) of plants which is important in photosynthesis (p. 93). There are two forms of chlorophyll; chlorophyll *a* and chlorophyll *b*.

leucoplast (*n*) a colourless plastid (↑) e.g. starch (p. 18) grains.

stroma (*n*) the matrix (p. 88) within a chloroplast (↑) containing starch (p. 18) grains and enzymes (p. 28).

grana (*n.pl.*) disc-shaped, flattened vesicles (↓) in the stroma (↑) of a chloroplast (↑) holding the chorophyll (↑). **granum** (*sing.*).

vesicle (*n*) a thin-walled drop-like structure or a cavity containing fluid.

lamella (*n*) a thin plate-like structure. **lamellae** (*pl.*).

cilium (*n*) in animals and a few plants, a fine thread which projects from the surface of a cell and moves the fluid surrounding it by a beating or rowing action. **cilia** (*pl.*).

flagellum (*n*) a fine, long thread which projects from the surface of a cell and moves with an undulating action. In bacteria, the flagellum provides locomotion (p. 143) by a whip-like action during part of its life history. It is longer than a cilium (↑). **flagella** (*pl.*).

葉綠素（名）　植物葉綠體（↑）中的綠色色素（第126頁）。它在光合作用（第93頁）中具有重要的作用。葉綠素有兩種：葉綠素 a 和葉綠素 b。

白色體（名）　一種無色的質體（↑），例如澱粉（第18頁）粒。

基質（名）　葉綠體（↑）內含有澱粉（第18頁）粒和酶（第28頁）的基質（第88頁）。

基粒（名、複）　保有葉綠素（↑）的葉綠體（↑）基質（↑）中呈圓盤狀的扁平泡囊（↓）。（單數形式爲 granum）

泡囊（名）　含有液體的一種薄壁、滴狀結構或空腔。

片層（名）　一種薄的片狀結構。（複數形式爲 lamellae）

纖毛（名）　動物和少數植物中，突出在細胞表面，藉拍打或劃動可使其周圍液體流動的一種細絲。（複數形式爲 cilia）

鞭毛（名）　突出在細胞表面能作波浪式運動的一種細長絲。在細菌的部份生活史中，鞭毛是藉鞭打動作來爲運動（第143頁）創造條件的。鞭毛比纖毛（↑）長。（複數形式爲 flagella）。

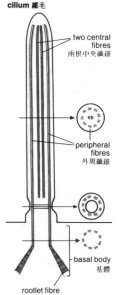

cilium 纖毛

two central fibres 兩根中央纖維

peripheral fibres 外周纖維

basal body 基體

rootlet fibre 小根纖維

basal body a tiny, rod-shaped body situated at the base of a cilium (↑) or flagellum (↑) composed of nine fibrils (p. 11) arranged in a ring at the edge of the cilium. There are also two central fibrils which do not form part of the basal body.

microvilli (*n.pl.*) finger-like projections from the surface of the plasma membrane (↓) of a cell which improve the absorption (p. 81) powers of the cell by increasing its surface area. *See also* villi (p. 103).

nucleus (*n*) a body present within the cells of eukaryotic (p. 9) organisms which contains the chromosomes (↓) of the organism. **nuclear** (*adj*).

nuclear membrane the firm, double unit membrane (p. 14) surrounding the nucleus (↑) and separating it from the cytoplasm (p. 10) while allowing the exchange of materials between the nucleus and the cytoplasm through its pores (p. 120).

nucleolus (*n*) a small, round dense body, one or two of which may be present within the nucleus (↑). It is rich in RNA (p. 24) and protein (p. 21) but is not contained within a membrane (p. 14).

chromosome (*n*) a rod- or thread-shaped body occurring within the nucleus (↑) and which is readily stained by various dyes, hence its name. A chromosome is composed of DNA (p. 24) or RNA (p. 24), and protein (p. 21). Each chromosome is in the form of a long helix (p. 25) of DNA. Chromosomes mostly occur in pairs called homologous chromosomes (p. 39). They are composed of thousands of genes (p. 196) which give rise to and control particular characteristics and functions of the organism, such as eye colour, and which are passed on through the offspring by inheritance (p. 196). Each organism has a consistent number of chromosomes, e.g. in human cells, 23 pairs.

chromatin (*n*) a granular compound of nucleic acid (p. 22) and protein (p. 21) in the chromosome (↑) and which is strongly stained by certain dyes.

plasma membrane an extremely thin membrane (p. 14) separating the cell from its surroundings. It allows the transfer of substances between the cell and its surroundings.

nucleus 核

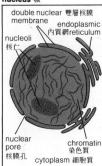

double nuclear 雙層核膜
membrane
　　　endoplasmic
nucleoli　內質網reticulum
核仁
nuclear
pore　　　chromatin
核膜孔　　　染色質
cytoplasm 細胞質

基體　位於纖毛(↑)或鞭毛(↑)基部的一種細小桿狀體。它由在纖毛邊緣成環狀排列的九根原纖維(第11頁)組成。另外還有兩根中央原纖維，它們並不構成基體的組成部份。

微絨毛(名、複)　細胞質膜(↓)表面的指狀突出物。由於它增加了質膜的表面積，因而可改善細胞的吸收(第81頁)能力。參見"絨毛"(第103頁)。

核(名)　真核(第9頁)生物細胞內含的一種物體，含有這類生物的染色體(↓)。(形容詞形式為nuclear)

核膜　圍繞核(↑)並使核與細胞質(第10頁)分開的堅硬雙層單位膜(第14頁)。但通過核膜孔(第120頁)可使核與細胞質之間進行物質交換。

核仁(名)　核(↑)內含的一個或兩個小而圓的致密體。核仁富含核糖核酸(第24頁)和蛋白質(第21頁)，但其外並無膜(第14頁)包覆。

染色體(名)　細胞核(↑)內含的一種桿狀或絲狀體，因其易被各種不同的染料所染色而得名。染色體由脫氧核糖核酸(第24頁)或核糖核酸(第24頁)和蛋白質(第21頁)組成。每一條染色體都是以脫氧核糖核酸的一個長螺旋(第25頁)的形式存在。染色體大都成對出現，被稱為同源染色體(第39頁)。它們由成千上萬個基因(第196頁)組成。基因可以產生和控制生物的特定性狀和功能，例如眼睛的顏色，並可通過遺傳(第196頁)傳給後代。每一種生物的細胞都含有一定數量的染色體，例如人體細胞中有23對染色體。

染色質(名)　染色體(↑)中含有核酸(第22頁)和蛋白質(第21頁)的一種粒狀化合物，它可被某些染料牢固地染色。

質膜　將細胞與其周圍環境分開的一層極薄的膜(第14頁)。它可使物質在細胞與其周圍環境之間進行轉移。

plasmalemma (*n*) the plasma membrane (p. 13) or cell membrane (↓).

unit membrane the common structure, divided into three layers, of the plasma membrane (p. 13), or other membranes, such as the endoplasmic reticulum (p. 11). It comprises a monomolecular film (↓) and a bimolecular leaflet (↓).

monomolecular film a layer, one molecule thick, of protein (p. 21) which occurs either side of the bimolecular leaflet (↓) and forms part of the organization of the unit membrane (↑). Stained and under an electron microscope (p. 9), it appears as a dark stratum (↓).

bimolecular leaflet a layer, two molecules thick, of lipid (p. 20) which is found between two monomolecular films (↑) and forms part of the organization of the unit membrane (↑). Stained and under an electron microscope (p. 9) it appears as a light stratum (↓).

stratum (*n*) a layer. **strata** (*pl.*).

phagocytosis (*n*) the process in which a cell flows around particles in its surroundings and takes them into the cytoplasm (p. 10) to form a vacuole (p. 11).

pinocytosis (*n*) a process in which a cell folds back within itself and surrounds a tiny drop of fluid in its surroundings and takes it into the cytoplasm (p. 10) to form a vesicle (p. 12).

middle lamella in plants, the material which is laid down between adjacent cell walls (p. 8) and sticks the cells together. It is laid down as new cells form.

primary cell wall the first cell wall (p. 8) of a young cell which is laid down as the new cell forms. *See also* secondary cell wall (↓).

secondary cell wall a cell wall (p. 8) which is laid down inside the primary cell wall (↑). It surrounds some of the cells in older plants.

pit (*n*) a small area of the secondary cell wall (↑) which has remained almost unthickened or absent during the formation of the secondary wall. It allows substances to pass between the cells. The pits in one cell correspond in position with the pits in a neighbouring cell.

原生質膜(名) 質膜(第13頁)或細胞膜(↓)。

單位質膜(名) 質膜(第13頁)和內質網(第11頁)之類的一些其他膜的共同結構。它分爲三層。單位膜由一單分子膜(↓)和一雙分子層(↓)所構成。

單分子膜 一個分子厚的蛋白質(第21頁)層。它存在於雙分子層片(↓)的兩側,並成爲單位質膜(↑)結構的組成部份。將它染色後,在電子顯微鏡(第9頁)下觀察,可見到它作爲暗層(↓)出現。

雙分子層 二個分子厚的脂(第20頁)層。它存在於兩個單分子膜(↑)之間,並成爲單位質膜(↑)結構的組成部份。將它染色後,在電子顯微鏡(第9頁)下觀察,可見到它作爲亮層(↓)出現。

層(名) 薄層。(複數形式爲 strata)

吞噬作用(名) 細胞圍繞其周圍的微粒流動,將它們吸收進細胞質(第10頁),形成液泡(第11頁)的過程。

胞飲作用(名) 細胞自身扭彎,包圍其周圍的液態小滴,將其吸收進細胞質(第10頁),形成泡囊(第12頁)的過程。

中膠層 植物中存在於兩層相鄰細胞壁(第8頁)之間,可使細胞黏連在一起的物質。它於新細胞形成時產生。

初生細胞壁 新生細胞最初生成的細胞壁(第8頁)。它於新細胞形成時產生。參見"次生細胞壁"(↓)。

次生細胞壁 初生細胞壁(↑)內側生成的一層細胞壁(第8頁)。它包圍着較老植物中的一些細胞。

紋孔(名) 次生細胞壁(↑)的一個小區。在次生壁形成過程中,它幾乎一直保持不增厚或不存在狀態。它可使物質在細胞之間通過。一個細胞中的紋孔與相鄰細胞的紋孔位置是相對應的。

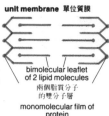

unit membrane 單位質膜

bimolecular leaflet of 2 lipid molecules
兩個脂質分子的雙分子層

monomolecular film of protein
蛋白質的單分子膜

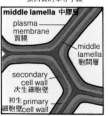

middle lamella 中膠層

plasma membrane 質膜

middle lamella 胞間層

secondary cell wall 次生細胞壁

primary cell wall 初生細胞壁

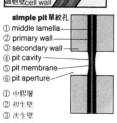

simple pit 單紋孔

① middle lamella
② primary wall
③ secondary wall
④ pit cavity
⑤ pit membrane
⑥ pit aperture

① 中膠層
② 初生壁
③ 次生壁
④ 紋孔腔
⑤ 紋孔膜
⑥ 紋孔口

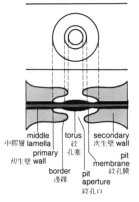

bordered pit 具緣紋孔

middle lamella 中膠層
primary wall 初生壁
torus 紋孔塞
border 邊緣
pit aperture 紋孔口
secondary wall 次生壁
pit membrane 紋孔膜

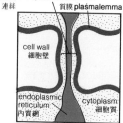

plasmodesmata 胞間連絲
胞質和 plasmodesma comprising
網管組 cytoplasm and tube of
的胞間 endoplasmic reticulum
連絲

質膜 plasmalemma

cell wall
細胞壁

endoplasmic
reticulum
內質網

cytoplasm
細胞質

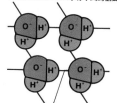

**hydrogen bond between
water molecules** 水分子間的氫鍵

hydrogen bond (attraction
between positive hydrogen
atom and negative oxygen
atom)
氫鍵（帶正電的氫原子和
帶負電的氫原子之間的吸引）

plasmodesmata (*n.pl.*) fine threads of cytoplasm (p. 10) which connect the cytoplasm of neighbouring cells, and may be grouped through the membranes (↑) of pits (↑). Plasmodesmata run through narrow pores (p. 120) in the cellulose (p. 19) cell wall (p. 8). **plasmodesma** (*sing.*).

biochemistry (*n*) the study or science of the chemical substances and their reactions in animals and plants.

organic compound any substance which is a compound of carbon, except for the oxides and carbonates of carbon, and from which all living things are made. Oxygen and carbon are the main components of organic compounds.

inorganic compound a compound which, except for the oxides and carbonates, does not contain carbon, and which is not an organic compound (↑). Salt is an example of an inorganic compound.

hydrogen bond a bond which holds one molecule of water to another molecule making water more stable than it otherwise would be. A molecule of water consists of two hydrogen atoms bonded to one oxygen atom by sharing electrons. The resulting molecule is weakly polar with hydrogen atoms being positively charged and the oxygen negatively charged. These polar molecules are weakly attracted to one another.

acid (*n*) a substance that releases hydrogen (H^+) ions in a watery solution (p. 118) or accepts electrons in chemical reactions. An acid can be an inorganic compound (↑), such as hydrochloric acid, HCl or an organic compound (↑) such as ethanoic acid, CH_3COOH. The acidity of a solution can be measured on the pH scale ($-\log H^+$ concentration). **acidic** (*adj*).

base[1] (*n*) a substance that releases hydroxyl (OH^-) ions in a watery solution (p. 118) or gives up electrons in chemical reactions, e.g. sodium hydroxide, NaOH. **basic** (*adj*).

pH *see* acid (↑).

buffer (*n*) a substance which helps a solution (p. 118) to resist a change in pH (↑) when an acid (↑) or base (↑) is added to the solution. Many biological fluids function as buffers.

胞間連絲（名、複） 連接相鄰細胞的細胞質，並可聚集通過紋孔（↑）膜（↑）的細胞質（第10頁）的細絲。胞間連絲貫穿纖維素（第19頁）細胞壁（第 8 頁）的窄孔（第120頁）。（單數形式為 plasmodesma）

生物化學（名） 研究動物和植物中的化學物質及其反應的科學。

有機化合物 除碳的氧化物和碳酸鹽以外的任何含碳的化合物。一切有生命的東西都是由有機化合物組成的。氧和碳是有機化合物的主要成份。

無機化合物 除碳的氧化物和碳酸鹽以外的一種不含碳的、非有機化合物（↑）的化合物。鹽是無機化合物的一個例子。

氫鍵 將一個水分子與另一個水分子締合，使水比在其他情況下更為穩定的一種鍵。一個水分子由兩個氫原子和一個氧原子通過共價電子鍵合而成。這樣形成的分子，由於氫原子帶正電和氧原子帶負電，而變得極性較弱。這些極性分子相互之間的吸引力較弱。

酸（名） 在水溶液（第118頁）中釋放氫離子（H^+），或者在化學反應中接受電子的一種物質。酸可以是一種無機化合物（↑），例如鹽酸（HCl），或者是一種有機化合物（↑），例如乙酸（CH_3COOH）。溶液的酸度可用 pH 值（$-\log H^+$ 濃度）來度量。（形容詞形式為 acidic）

鹼（名） 在水溶液（第118頁）中釋放羥離子（OH^-），或者在化學反應中放出電子的一種物質，例如氫氧化鈉（NaOH）。（形容詞形式為 basic）

氫離子濃度、pH 值 見酸（↑）。

緩沖劑（名） 有助於溶液（第118頁）在其中添加酸（↑）或鹼（↑）時能阻止其氫離子濃度（↑）發生變化的一種物質。許多生物流體都可起緩沖劑的作用。

condensation example of a condensation reaction 縮合一種縮合反應的例子

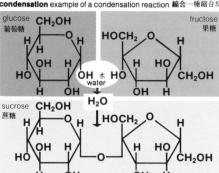

condensation (*n*) a reaction whereby two simple organic compounds (p. 15), such as glucose (↓) and fructose (↓), combine to form another compound, such as sucrose (p. 18) and a molecule of water.

hydrolysis (*n*) a reaction in which water combines with an organic compound (p. 15), such as sucrose (p. 18), to form two new organic compounds, such as glucose (↓) and fructose (↓). The reverse of condensation (↑).

縮合反應（名） 一種化學反應。藉這種反應，兩種簡單的有機化合物（第15頁）如葡萄糖（↓）和果糖（↓）的分子相結合，生成另一種化合物如蔗糖（第18頁）的分子和一個水分子。

水解作用（名） 一種化學反應。在這種反應中，水與一種有機化合物（第15頁）如蔗糖（第18頁）相結合，生成兩種新的有機化合物如葡萄糖（↓）和果糖（↓）。水解是縮合（↑）反應的逆反應。

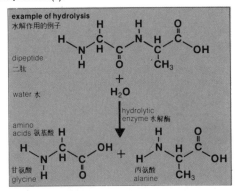

molecular biology the study or science of the structure and activities of the molecules which make up animals and plants.

分子生物學 研究組成動物和植物的各種分子結構和活動性的科學。

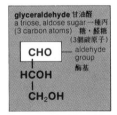

glyceraldehyde 甘油醛
a triose, aldose sugar 一種丙
(3 carbon atoms) 糖，醛糖
（3個碳原子）
CHO — aldehyde group 醛基
HCOH
CH₂OH

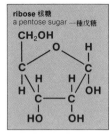

ribose 核糖
a pentose sugar 一種戊糖

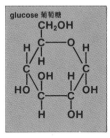

glucose 葡萄糖

fructose 果糖

carbohydrate (*n*) an organic compound (p. 15) containing the elements carbon, hydrogen, and oxygen with the general formula $(CH_2O)_n$. Carbohydrates are essential in the metabolism (p. 26) of all living things.

monosaccharide (*n*) a carbohydrate (↑) composed of small molecules. Monosaccharides are the building blocks from which disaccharides (p. 18) and polysaccharides (p. 18) are built. Common monosaccharides found in cells contain from three to seven carbon atoms. A monosaccharide is the simplest sugar and, if further broken down, ceases to be a sugar.

sugar (*n*) the simplest carbohydrate (↑), a mono-, di- or polysaccharide (p. 18).

triose sugar a monosaccharide (↑) in which n for the general formula of the carbohydrate (↑) is 3. Glyceraldehyde is a triose sugar with the formula $C_3H_6O_3$.

pentose sugar a monosaccharide (↑) in which n for the general formula of the carbohydrate (↑) is 5. Ribose (p. 22) is a pentose sugar with the formula $C_5H_{10}O_5$.

hexose sugar a monosaccharide (↑) in which n for the general formula of the carbohydrate (↑) is 6. Glucose (↓) is a hexose sugar with the formula $C_6H_{12}O_6$. The atoms of hexose sugars may be arranged differently to give different types of sugars e.g. glucose and fructose (↓).

glucose (*n*) a hexose sugar (↑) which is widely found in animals and plants. Glucose provides a major source of energy in living things by being oxidized (p. 32) during respiration (p. 112) into carbon dioxide and water, releasing energy. In plants it is the product of photosynthesis (p. 93) and is stored as starch (p. 18) while in animals it is produced by the digestion (p. 98) of disaccharides (p. 18) and polysaccharides (p. 18) and is stored as glycogen (p. 19). Glucose combines with fructose (↓) to form sucrose (p. 18) by condensation (↓).

fructose (*n*) a hexose sugar (↑) which is widely found in plants. It combines with glucose (↑) to form sucrose (p. 18) by condensation (↑).

碳水化合物／醣類（名）含碳、氫、氧元素並具通式 $(CH_2O)_n$ 的一種有機化合物（第15頁）。碳水化合物在所有生物的新陳代謝（第26頁）中是必不可少的。

單糖／單醣（名）小分子組成的碳水化合物（↑）。單糖是組成貳糖（第18頁）和多糖（第18頁）的結構單元。存在於細胞中的普通單糖含有3個至7個碳原子。單糖是最簡單的糖；如果將其進一步分解，則產物再不是糖。

糖（名）最簡單的碳水化合物（↑），可以是單糖、貳糖或多糖（第18頁）。

丙糖 一種單糖（↑）。其碳水化合物（↑）通式中的 n 為3。甘油醛是化學式為 $C_3H_6O_3$ 的一種丙糖。

戊糖 一種單糖（↑）。其碳水化合物（↑）通式中的 n 為5。核糖（第22頁）是化學式為 $C_5H_{10}O_5$ 的一種戊糖。

己糖 一種單糖（↑）。其碳水化合物（↑）通式的 n 為6。葡萄糖（↓）是化學式為 $C_6H_{12}O_6$ 的一種己糖。己糖中原子的不同排列可產生不同類型的糖，例如葡萄糖和果糖（↓）。

葡萄糖（名）一種己糖（↑）。它廣泛存在於動物和植物中。葡萄糖在呼吸（第112頁）過程中，被氧化（第32頁）成二氧化碳和水，並放出能量，因而構成了生物中的一種主要能源。在植物中，葡萄糖是光合作用（第93頁）的產物，以澱粉（第18頁）的形式儲存；而在動物中，葡萄糖由貳糖（第18頁）和多糖（第18頁）的消化（第98頁）所產生，以糖原（第19頁）的形式儲存。葡萄糖與果糖（↓）經縮合（↑）而生成蔗糖（第18頁）。

果糖（名）一種己糖（↑）。它廣泛存在於植物中。果糖與葡萄糖（↑）經縮合（↑）而生成蔗糖（第18頁）。

galactose (*n*) a hexose sugar (p. 17) which is a constituent of lactose (↓) and is found in many plant polysaccharides (↓) as well as in animal protein (p. 21)-polysaccharide combinations.

disaccharide (*n*) a carbohydrate (p. 17) which results from the combination of two monosaccharides (p. 17) by condensation (p. 16), e.g. maltose (↓) and sucrose (↓).

半乳糖(名) 一種己糖(第17頁)。它是乳糖(↓)的一種成份,存在於許多植物多糖(↓)中以及動物蛋白質(第21頁)—多糖複合物中。

貳糖／貳醣(名) 兩個單糖(第17頁)縮合(第16頁)生成的一種碳水化合物(第17頁),例如麥芽糖(↓)和蔗糖(↓)。

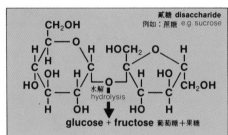

貳糖 disaccharide
例如:蔗糖 e.g. sucrose

glucose + fructose 葡萄糖＋果糖

maltose (*n*) a disaccharide (↑) which is formed from the condensation (p. 16) of two molecules of glucose (p. 17). It is a product of the breakdown of starch (↓) during germination (p. 168) in plants, and digestion (p. 98) in animals. Also known as **malt sugar.**

sucrose (*n*) a disaccharide (↑) which is a compound of one molecule of glucose (p. 17) and one molecule of fructose (p. 17). It is widespread in plants but not in animals. Also known as **cane sugar**.

lactose (*n*) a disaccharide (↑) which is a compound of one molecule of glucose (p. 17) and one molecule of galactose (↑). It occurs in the milk of mammals (p. 80). Also known as **milk sugar**.

polysaccharide (*n*) a carbohydrate (p. 17) which results from the combination of more than two monosaccharides (p. 17) by condensation (p. 16). A polysaccharide has the general formula $(C_6H_{10}O_5)_n$.

starch (*n*) a polysaccharide (↑) which forms one of the main food reserves of green plants. It is found in the leucoplasts (p. 12). It stains blue-black with iodine.

麥芽糖(名) 由兩個分子的葡萄糖(第17頁)經縮合(第16頁)而生成的一種貳糖(↑)。麥芽糖是在植物發芽(第168頁)過程中以及在動物消化(第98頁)過程中澱粉(↓)分解的一種產物。另一個英文名稱是 malt sugar。

蔗糖(名) 一種貳糖(↑)。它是一個分子的葡萄糖(第17頁)和一個分子的果糖(第17頁)生成的一種化合物。許多植物都含有蔗糖,但動物就沒有。另一個英文名稱是 cane sugar。

乳糖(名) 一種貳糖(↑)。它是一個分子的葡萄糖(第17頁)和一個分子的半乳糖(↑)生成的一種化合物。哺乳動物(第80頁)的乳汁中含有乳糖。另一個英文名稱是 milk sugar。

多糖／多醣(名) 兩個以上的單糖(第17頁)縮合(第16頁)而生成的一種碳水化合物(第17頁)。多糖的通式為 $(C_6H_{10}O_5)_n$。

澱粉(名) 構成綠色植物主要食物貯備之一的一種多糖(↑)。它存在於白色體(第12頁)之中。碘可使其染上藍黑色。

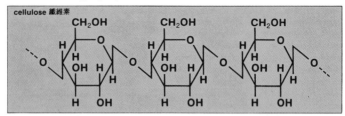

polysaccharide 多糖
e.g. starch (amylopectin) 圖例：澱粉（支鏈澱粉）

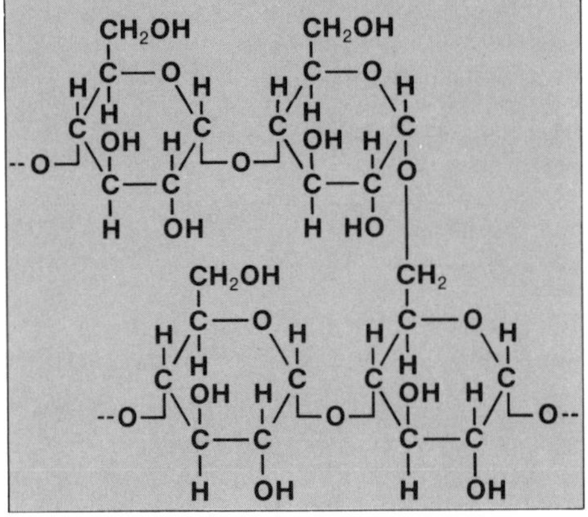

植物細胞壁表面所見的
纖維素微纖維（×24,000）
**cellulose microfibrils in
surface view of plant cell
wall** (× 24,000)

glycogen (*n*) a polysaccharide (↑) stored by animals and by fungi (p. 46). It is made up of many glucose (p. 17) molecules. In vertebrates (p. 74) it is present in large quantities in the liver (p. 103) and muscles (p. 143).

cellulose (*n*) a long-chain polysaccharide (↑) made up of units of glucose (p. 17). It is used for structural support and is the main component of the cell wall (p. 8) in plants.

糖原／醣原（名） 動物和真菌（第46頁）體內貯存的一種多糖（↑）。它由許多葡萄糖（第17頁）分子所組成。脊椎動物（第74頁）的肝藏（第103頁）和肌肉（第143頁）含大量的糖原。

纖維素（名） 由許多葡萄糖（第17頁）單元構成的長鏈多糖（↑）。在植物中，它支承着結構，也是細胞壁（第 8 頁）的主要成份。

cellulose 纖維素

lignin (*n*) a complex organic compound (p. 15) whose structure is not fully understood. With cellulose (↑) it forms the chief components of wood in trees. It is laid down in the cell walls (p. 8) of sclerenchyma (p. 84), xylem (p. 84) vessels, and tracheids (p. 84). It stains red with acidified phloroglucinol. **lignified** (*adj*).

木質素（名） 一種複雜的有機化合物（第15頁）。其結構至今還未完全為人們所瞭解。它與纖維素（↑）都是構成樹木中木材部份的要素。在厚壁組織（第84頁）、木質部（第84頁）、導管和管胞（第84頁）等的細胞壁（第 8 頁）中，都含有木質素。用酸化間苯三酚可使其染成紅色。（形容詞形式爲 lignified）

lipid (*n*) any of a number of organic compounds (p. 15) found in plants and animals with very different structures but which are all insoluble in water and soluble in substances like ethoxyethane (ether) and trichloromethane (chloroform). It is formed by the condensation (p. 16) of glycerol (↓) and fatty acids (↓). Lipids have a variety of functions including storage, protection, insulation, waterproofing, and as a source of energy.

fat (*n*) a lipid (↑) formed from the alcohol glycerol (↓) and one or more fatty acids (↓). It is solid at room temperature.

oil (*n*) a lipid (↑) formed from the alcohol glycerol (↓) and one or more fatty acids (↓). It is liquid at room temperature.

glycerol (*n*) an alcohol with the formula $C_3H_8O_3$ which is formed by the hydrolysis (p. 16) of a fat. It is a sweet, sticky, odourless, colourless liquid. Its modern name is propane-1,2,3,-triol.

fatty acid an organic acid (p. 15) with the general formula $(R(CH_2)_nCOOH)$ which can be united with glycerol (↑) by condensation (p. 16) to give a lipid (↑). In living organisms, fatty acids usually have unbranched chains and an even number of carbon atoms.

triglyceride (*n*) the major component of animal and plant lipids (↑). It is derived from glycerol (↑) which has three reactive hydroxyl groups, by condensation (p. 16) with three fatty acids (↑).

phospholipid (*n*) a lipid (↑) which contains a phosphate group as an essential part of the molecule. It is derived from glycerol (↑) attached to two fatty acids (↑), a phosphate group, and a nitrogenous base. Phospholipids are essential components of cell membranes (p. 14).

saturated (*adj*) of a carbon chain, such as that in a fatty acid (↑), in which each carbon atom is attached by single bonds to carbons, hydrogen atoms, or other groups. It is unreactive.

unsaturated (*adj*) of a carbon chain, such as that in a fatty acid (↑), in which carbon atoms are attached to other groups with at least one double or triple bond. An unsaturated fatty acid is reactive and may be essential to maintain a vital structure or function in an organism.

脂類（名） 植物和動物都含有多種有機化合物（第15頁），此乃其中的一種。這些有機化合物的結構有很大的差異，但都不溶於水，而溶於像乙氧基乙烷（乙醚）和三氯甲烷（氯仿）這樣的一些物質之中。脂類是由甘油（↓）和脂肪酸（↓）經縮合（第16頁）而生成的。它具有各種不同的功能，包括貯存、保護、保溫、防水和作爲能源。

脂肪（名） 由酒精甘油（↓）和一種或多種脂肪酸（↓）生成的一種脂類（↑）。它在室溫下爲固體。

油（名） 由酒精甘油（↓）和一種或多種脂肪酸（↓）生成的一種脂類（↑）。它在室溫下爲液體。

甘油（名） 化學式爲 $C_3H_8O_3$ 的一種醇。它是由脂肪水解（第16頁）生成的。甘油是一種具有甜味、黏性、無氣味、無色的液體。其新名稱爲丙烷—1, 2, 3, —三醇。

脂肪酸 通式爲 $(R(CH_2)_nCOOH)$ 的一種有機酸（第15頁）。它可與甘油（↑）縮合（第16頁），生成脂類（↑）。在生物體內，脂肪酸通常具有無分支的鏈和偶數的碳原子。

甘油三酯（名） 動物和植物脂類（↑）的主要成份。它是從具有三個活性羥基的甘油（↑）與三個脂肪酸（↑）縮合（第16頁）得到的。

磷脂（名） 以磷酸基作爲分子主要成份的一種脂類（↑）。它是由甘油（↑）與兩個脂肪酸（↑）、一個磷酸基和一個含氮鹼的結合得到的。磷脂是細胞膜（第14頁）的主要成份。

飽和的（形） 指碳鏈而言，例如脂肪酸（↑）中的碳鏈。在這種碳鏈中，每個碳原子都由單鍵與其他碳原子、氫原子或別的基團相連接。它是化學惰性的。

不飽和的（形） 指碳鏈而言，例如脂肪酸（↑）中的碳鏈。在這種碳鏈中，各碳原子至少有一個雙鍵或三鍵與其他一些基團相連接。不飽和脂肪酸是具反應性的，可能對於維持生物體的生命結構和功能是必不可少的。

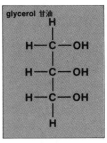

glycerol 甘油

unsaturated 不飽和的

unsaturated 不飽和的
carbon atoms 碳原子

蛋白質的一級、二級、三級和
四級結構
**primary, secondary, tertiary
and quaternary structure of
proteins**

primary 一級
structure 結構　　secondary
　　　　　　　　structure
　　　　　　　　二級結構

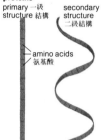

amino acids
氨基酸

tertiary structure 三級結構

quaternary
structure
四級結構

氨基酸之間的肽鍵
**peptide bond between
amino acids**

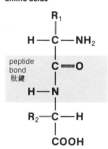

peptide
bond
肽鍵

R₁ and **R₂** are side groups
R₁和R₂是側基

steroid (*n*) a complex, saturated (↑) hydrocarbon in which the carbon atoms are arranged in a system of rings. All steroids are chemically similar but may have very different functions in organisms. The most common steroid in animals is cholesterol.

protein (*n*) a very complex organic compound (p. 15) made up of large numbers of amino acids (↓). Proteins make up a large part of the dry weight of all living organisms.

amino acid an organic compound (p. 15) with an amino group of atoms (-NH₂) and acidic (p. 15) carboxyl (-COOH) groups of atoms on the molecule. The general formula is RCHNH₂COOH with R representing a hydrogen or carbon chain. There are more than twenty naturally occurring amino acids with different R groups. Hundreds of thousands of amino acids are linked together to form a protein (↑). *See also* the diagram, amino acids and the genetic code on p. 204.

dipeptide (*n*) an organic compound (p. 15) which results from linking together two amino acids (↑) by condensation (p. 16).

polypeptide (*n*) an organic compound (p. 15) which results from linking together many amino acids (↑) by condensation (p. 16). In turn, polypeptides may be linked together to form proteins (↑).

peptide bond the link which joins one amino acid to the carboxyl (-COOH) group of another, resulting in the formation of a dipeptide (↑) or polypeptide (↑). A peptide bond can only be broken by the action of a hot acid (p. 15) or alkali.

conjugated protein a protein (↑) which occurs in combination with a non-protein or prosthetic group (p. 30). Haemoglobin (p. 126) is an example of a conjugated protein.

globular protein a protein (↑) which, because of the positive and negative charge on it, forms a complex three-dimensional structure as the opposite charges are attracted together and form weak bonds. A hormone (p. 130) is an example of a globular protein.

類固醇（名）　一種複雜的飽和（↑）烴。其中，碳原子排列成環狀系統。所有類固醇在化學性質上是相類似的，但在生物體內可能具有很不相同的功能。動物中最爲普通的類固醇是膽固醇。

蛋白質（名）　由許多氨基酸（↓）組成的一種非常複雜的有機化合物（第15頁）。蛋白質佔據所有生物體乾重的大部份。

氨基酸　分子中具有一個氨基（−NH₂）和酸性（第15頁）羧基（−COOH）的一種有機化合物（第15頁）。其通式爲 RCHNH₂COOH，R 代表一個氫原子或碳鏈。具有不同 R 基的天然存在的氨基酸有20多種。無數個氨基酸連接在一起即形成蛋白質（↑）。參見第204頁的"氨基酸"和"遺傳密碼"示意圖。

二肽（名）　兩個氨基酸（↑）經縮合（第16頁）相連接而產生的一種有機化合物（第15頁）。

多肽（名）　許多氨基酸（↑）經縮合（第16頁）相連接而產生的一種有機化合物（第15頁）。同樣，多肽也可連接在一起生成蛋白質（↑）。

肽鍵　將一個氨基酸與另一個氨基酸中的羧基（−COOH）連接而生成二肽（↑）或多肽（↑）的鍵。肽鍵只能在熱酸（第15頁）或鹼的作用下斷裂。

結合蛋白質　與非蛋白質或輔基（第30頁）相結合而產生的一種蛋白質（↑）。血紅蛋白（第126頁）是結合蛋白質的一個例子。

球狀蛋白　一種蛋白質（↑）。這種蛋白質由於其上帶有正電荷和負電荷，故當相反的電荷相互吸引在一起，形成弱鍵時，即形成一種複雜的三維結構。激素（第130頁）是球狀蛋白的一個例子。

fibrous protein a protein (p. 21) which occurs as long parallel chains with cross links. Fibrous proteins are insoluble and are used for support and other structural purposes. Keratin in hair, hooves, feathers etc is an example of a fibrous protein.

colloid (*n*) a substance, such as starch (p. 18), that will not dissolve or be suspended in a liquid but which is dispersed in it.

nucleic acid a large, long-chain molecule composed of chains of nucleotides (↓) and found in all living organisms. The carrier of genetic (p. 196) information.

nucleotide (*n*) an organic compound (p. 15) formed from ribose (↓), phosphoric acid (↓), and a nitrogen base (↓).

ribose (*n*) a monosaccharide (p. 17) or pentose sugar (p. 17) which forms an essential part of a nucleotide (↑).

deoxyribose (*n*) a monosaccharide (p. 17) with one less oxygen than ribose (↑).

phosphoric acid an inorganic compound (p. 15) with the formula H_3PO_4 which forms an essential part of nucleotides (↑). The phosphate molecule from phosphoric acid forms a bridge between two pentose (p. 17) molecules.

base[2] (*n*) a substance, such as a purine (↓) or pyrimidine (↓), containing nitrogen, which is attached to the main sugar-phosphate chain in a nucleic acid (↑).

cytosine (*n*) a nitrogen base (↑) derived from pyrimidine (↓) and found in both ribonucleic acid (p. 24) and deoxyribonucleic acid (p. 24).

uracil (*n*) a nitrogen base (↑) derived from pyrimidine (↓) and found only in ribonucleic acid (p. 24).

adenine (*n*) a nitrogen base (↑) derived from purine (↓) and found in both ribonucleic acid (p. 24) and deoxyribonucleic acid (p. 24).

guanine (*n*) a nitrogen base (↑) derived from purine (↓) and found in both ribonucleic acid (p. 24) and deoxyribonucleic acid (p. 24).

thymine (*n*) a nitrogen base (↑) derived from pyrimidine (↓) and found only in deoxyribonucleic acid (p. 24).

纖維狀蛋白　作爲長的平行鏈交聯存在的一種蛋白質(第21頁)。纖維狀蛋白是不溶性的，起支持作用以及結構作用。毛髮、腳蹄和羽毛等中的角蛋白就是纖維狀蛋白的一個例子。

膠體(名)　指澱粉(第18頁)之類的一種物質。在液體中，它不會溶解或懸浮，但可分散於其內。

核酸　存在於所有生物中的一種由核苷酸(↓)鏈組成的長鏈大分子。它是遺傳(第196頁)信息的載體。

核苷酸(名)　由核糖(↓)、磷酸(↓)和含氮鹼基(↓)所組成的一種有機化合物(第15頁)。

核糖(名)　一種單糖(第17頁)或戊糖(第17頁)。它是構成核苷酸(↑)的一種基本成份。

脫氧核糖(名)　比核糖(↑)少一個氧的一種單糖(第17頁)。

磷酸　化學式爲 H_3PO_4 的一種無機化合物(第15頁)。它是構成核苷酸(↑)的一種基本成份。來自磷酸的磷酸基分子在兩個戊糖(第17頁)分子之間形成一個橋。

氮鹼基　一種含氮物質，例如嘌呤(↓)或嘧啶(↓)。在核酸(↑)中，氮鹼基連接在糖—磷酸基主鏈上。

胞嘧啶(名)　一種含氮鹼基(↑)。它由嘧啶(↓)衍生而來，存在於核糖核酸(第24頁)和脫氧核糖核酸(第24頁)兩者之中。

尿嘧啶(名)　一種含氮鹼基(↑)。它由嘧啶(↓)衍生而來，僅存在於核糖核酸(第24頁)之中。

腺嘌呤(名)　一種含氮鹼基(↑)。它由嘌呤(↑)衍生而來，存在於核糖核酸(第24頁)和脫氧核糖核酸(第24頁)兩者之中。

鳥嘌呤(名)　一種含氮鹼基(↑)。它由嘌呤(↓)衍生而來，存在於核糖核酸(第24頁)和脫氧核糖核酸(第24頁)兩者之中。

胸腺嘧啶(名)　一種含氮鹼基(↑)。它由嘧啶(↓)衍生而來，僅存在於脫氧核糖核酸(第24頁)之中。

nucleotide basic structure
核苷酸基本結構

phosphate 磷酸

ribose 核糖

nitrogen base 含氮鹼基

the common bases in the nucleotides of DNA and RNA　DNA 和 RNA 的核苷酸中所具有的共同鹼基

	purines 嘌呤	**pyrimidines** 嘧啶
DNA only 僅 DNA		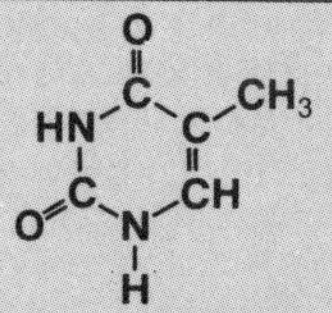 thymine 胸腺嘧啶
DNA and RNA DNA 和 RNA	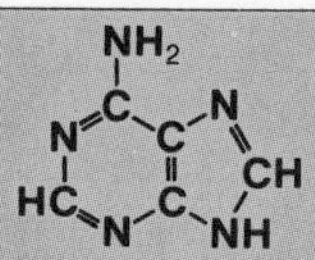 adenine 腺嘌呤 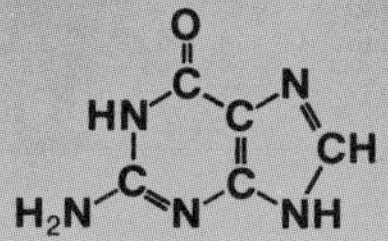 guanine 鳥嘌呤	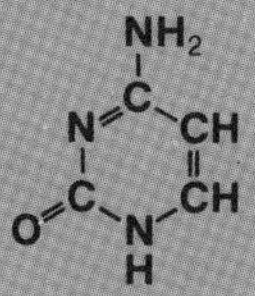 cytosine 胞嘧啶
RNA only 僅 RNA		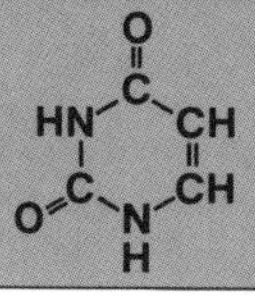 uracil 尿嘧啶

pyrimidine (*n*) an organic compound (p. 15) with the basic formula $C_4H_4N_2$ and with a cyclic structure from which important nitrogen bases (↑) are derived.

pyrimidine base any of the several compounds related to pyrimidine (↑) and present in nucleic acids (↑).

purine (*n*) an organic compound (p. 15) with the basic formula $C_5H_4N_5$, with a double cyclic structure, from which important nitrogen bases (↑) are derived.

purine base any of several compounds related to purine (↑) and present in nucleic acids (↑).

嘧啶（名）　基本化學式為 $C_4H_4N_2$，並具有環狀結構的一種有機化合物(第15頁)。由它可衍生出一些重要的含氮鹼(↑)。

嘧啶氮鹼　與嘧啶(↑)有關的幾種化合物中的任何一種。它們存在於核酸(↑)之中。

嘌呤（名）　基本化學式為 $C_5H_4N_5$，並具有雙環結構的一種有機化合物(第15頁)。由它可衍生出一些重要的含氮鹼(↑)。

嘌呤氮鹼　與嘌呤(↑)有關的幾種化合物中的任何一種。它們存在於核酸(↑)之中。

basic molecular shape of nitrogen base 含氮鹼的基本分子形狀

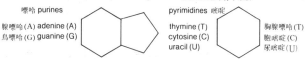

嘌呤 purines
腺嘌呤 (A) adenine (A)
鳥嘌呤 (G) guanine (G)

pyrimidines 嘧啶
thymine (T)　胸腺嘌呤 (T)
cytosine (C)　胞嘧啶 (C)
uracil (U)　尿嘧啶 (U)

RNA ribonucleic acid. A nucleic acid (p. 22) consisting of a large number of nucleotides (p. 22) arranged to form a single strand. The base (p. 22) in each nucleotide is one of cytosine (p. 22), uracil (p. 22), adenine (p. 22), or guanine (p. 22). The sugar is ribose (p. 22). RNA is found in the nucleus (p. 13) of a cell and in the cytoplasm (p. 10). It usually occurs as ribosomes (p. 10) but also as *transfer RNA* and *messenger RNA*. Strands of RNA are produced in the nucleus from DNA (↓), passed to the cytoplasm, and then a ribosome is joined to the RNA. The ribosome moves along the strand of RNA and produces a polypeptide (p. 21) whose structure is controlled by the RNA. *See also* transcription and translation p. 205.

DNA deoyxribonucleic acid. A nucleic acid (p. 22) consisting of a large number of nucleotides (p. 22) arranged to form a single strand. Usually, two strands are coiled round each other to form a double helix (↓). The base (p. 22) in each nucleotide consists of one of cytosine (p. 22), adenine (p. 22), guanine (p. 22), or thymine (p. 22). The sugar is deoxyribose (p. 22). DNA is found in the chromosomes (p. 13) of prokaryotes (p. 8) and eukaryotes (p. 9) and in the mitochondria (p. 11) of eukaryotes. It is the material of inheritance (p. 196) in almost all living organisms and is able to copy itself during nuclear divisions (p. 35).

RNA　RNA即核糖核酸。由排列成單鏈的許多核苷酸(第22頁)組成的一種核酸(第22頁)。每一核苷酸中的氮鹼基(第22頁)爲下列四種鹼基之一：胞嘧啶(第22頁)、尿嘧啶(第22頁)、腺嘌呤(第22頁)和鳥嘌呤(第22頁)。每一核苷酸中的糖都是核糖(第22頁)。RNA 存在於細胞核(第13頁)和細胞質(第10頁)之中。它通常以核糖體(第10頁)的形式出現，但也以轉移 RNA 和信使 RNA 的形式出現。RNA 鏈在細胞核中根據 DNA（↓）製造，然後被傳送至細胞質，此時，核糖體便與其結合。核糖體沿着 RNA 鏈移動至產生多肽(第21頁)。多肽的結構則是由 RNA 控制的。參見第205頁"轉錄"和"轉譯"。

DNA　DNA 即脫氧核糖核酸。由排列成單鏈的許多苷酸(第22頁)組成的一種核酸(第22頁)。通常，兩股鏈互相纏繞形成雙螺旋（↓）。每一核苷酸中的氮鹼基(第22頁)爲下列四種鹼基之一：胞嘧啶(第22頁)、腺嘌呤(第22頁)、鳥嘌呤(第22頁)和胸腺嘧啶(第22頁)。每一核苷酸中的糖是脫氧核糖(第22頁)。DNA 存在於原核生物細胞(第 8 頁)和眞核生物細胞(第 9 頁)的染色體(第13頁)中和眞核生物細胞的線粒體(第11頁)中。DNA 是幾乎所有生物的遺傳(第196頁)物質，在核分裂(第35頁)期間，能自我複製。

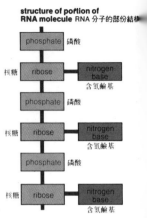

structure of portion of RNA molecule RNA 分子的部份結構

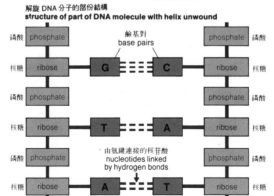

解旋 DNA 分子的部份結構
structure of part of DNA molecule with helix unwound

鹼基對 base pairs

由氫鍵連接的核苷酸 nucleotides linked by hydrogen bonds

DNA 雙螺旋示意圖
**diagram of the DNA
double helix**

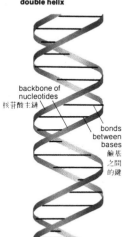

backbone of
nucleotides
核苷酸主鏈

bonds
between
bases
鹼基
之間
的鍵

polynucleotide chain a chain of linked nucleotides (p. 22) which makes up a nucleic acid (p. 22).

Watson-Crick hypothesis a hypothesis (p. 235) based on X-ray crystallography which suggests that DNA (↑) is a double helix (↓) of two coiled chains of alternating phosphate and sugar groups with the sugars linked by pairs of bases (p. 22).

double helix the arrangement of two helical (↓) polynucleotide chains (↑) in DNA (↑).

helix a helix is the curve that results from drawing a straight line on a plane which is then wrapped round a circular cylinder. The two helixes of DNA (↑) intertwine to form a double helix (↑) and are linked by nitrogen bases (p. 22). **helical** (*adj*).

base pairing the links holding the double helix (↑) of DNA (↑) together, each link consisting of a purine (p. 22) linked to a pyrimidine (p. 22) by hydrogen bonds (p. 15).

vitamin (*n*) the name given to a variety of organic compounds (p. 15) which are required by organisms for metabolism (p. 26) and which cannot usually be synthesized by the organism in sufficient quantities to replace that which is broken down during metabolism. *See* p. 238.

Benedict's test a method to determine the presence of monosaccharides (p. 17) and some disaccharides (p. 18) by adding a solution (p. 118) of copper sulphate, sodium citrate, and sodium carbonate to a solution of the sugar which produces a red precipitate (p. 26) when boiled because the sugar reduced the copper sulphate to copper (I) oxide. Sucrose (p. 18) and other non-reducing sugars do not reduce copper sulphate but it can be detected by hydrolysing (p. 16) it first into its component reducing sugars.

Fehling's test this is similar to Benedict's test (↑) but the reagent (p. 26) used is a solution (p. 118) containing copper sulphate, sodium potassium tartrate, and sodium hydroxide.

iodine test a method to determine the presence and distribution of starch (p. 18) in cells by cutting a thin section (p. 9) of the material and mounting it in iodine dissolved in potassium iodide. The starch grains turn blue-black.

多核苷酸鏈 組成核酸(第22頁)的多個核苷酸(第22頁)連接而成的鏈。

華特森—克立克假說 一種以 X 射綫結晶學爲基礎而提出的假說(第235頁)。該假說認爲，DNA(↑)是一種由相互纏繞在一起的兩股鏈所構成的雙螺旋(↓)，每一股鏈上的磷酸基和糖基交替出現，其中糖基皆由配對氮鹼基(第22頁)連接在一起。

雙螺旋 DNA(↑)中的兩條螺旋形的(↓)多核苷酸鏈(↑)的排列方式。

螺旋 在平面上劃一條直綫，然後將其捲附在一個圓柱體上，由此而形成的曲綫即爲螺旋。DNA (↑) 的兩股螺旋相互纏繞形成雙螺旋(↑)，並由含氮鹼基(第22頁)加以連接。(形容詞形式爲 helical)

鹼基配對 將 DNA(↑)雙螺旋(↑)連接在一起的結合形式。每一結合均包含嘌呤(第22頁)與嘧啶(第22頁)之間的氫鍵(第15頁)維繫方式。

維生素(名) 一類有機化合物(第15頁)的總稱，這些有機化合物爲生物體的新陳代謝(第26頁)所必需，通常不能被生物體大量合成，以補充其在新陳代謝過程中所分解的量。見第238頁。

本立德試驗 對單糖(第17頁)和某些貳糖(第18頁)的存在進行測定的一種方法。將硫酸銅、檸檬酸鈉和碳酸鈉的溶液(第118頁)添加到糖溶液中，經加熱煮沸時，由於糖能使硫酸銅還原成氧化銅(I)，因此，該溶液便會產生紅色沉澱(第26頁)。蔗糖(第18頁)和其他一些非還原糖不能使硫酸銅還原，但如果先將它們水解(第16頁)成其還原糖成份，則還是可以對它們進行測定的。

費林氏試驗 這項試驗與本立德試驗(↑)相類似，但所用的試劑(第26頁)是含硫酸銅、酒石酸鈉鉀和氫氧化鈉溶液(第118頁)。

碘液試驗 對細胞內澱粉(第18頁)的存在及其分佈情況進行測定的一種方法。將待測定材料切成薄的切片(第9頁)，然後將它放入溶解於碘化鉀溶液的碘中，則澱粉顆粒會變成藍黑色。

emulsion test a method of testing for the presence of a lipid (p. 20) by dissolving the substance in alcohol (usually ethanol) and adding an equal volume of water. A cloudy white precipitate (↓) indicates a lipid.

alcohol/water test = emulsion test (↑).

Sudan III test a method of testing for a lipid (p. 20) which stains red with Sudan III solution.

greasemark test a method of testing for a lipid (p. 20) by taking a drop of the substance to be tested and placing it on a filter paper. When it is dry, only a lipid leaves a translucent mark when held up to the light.

translucent (adj) of a material that lets light pass through but through which objects cannot be seen clearly.

Millon's test a method of testing for protein (p. 21) by adding a few drops of Millon's reagent (↓) to a suspension of the protein and boiling it. The protein stains brick red.

Biuret test a method of testing for protein (p. 21) by adding an equal volume of 2 per cent sodium hydroxide solution (Biuret A) followed by 0.5 per cent copper sulphate solution (Buiret B). The protein stains purple.

emulsion (n) a colloidal (p. 22) suspension of one liquid in another.

suspension (n) a mixture in which the particles of one or more substances are distributed in a fluid.

fluid (n) a substance which flows, i.e. a liquid or a gas.

precipitate (n) an insoluble solid formed by a reaction which occurs in solution (p. 118).

reagent (n) a substance or solution (p. 118) used to produce a characteristic reaction in a chemical test.

metabolism (n) a general name for the chemical reactions which take place within the cells of all living organisms.

metabolite (n) any of the substances, inorganic (p. 15) or organic (p. 15) such as water or carbon dioxide, amino acids (p. 21) or vitamins (p. 25) which take part in metabolism (↑).

metabolic pathway a series of small steps in which metabolism (↑) proceeds.

乳劑試驗 檢驗脂類(第20頁)存在的一種方法。將待試驗的物質溶解於醇(通常是乙醇)中,然後添加等體積的水,如出現白色霧狀沉澱(↓),即表明有脂類存在。

醇／水試驗 同乳劑試驗(↑)。

蘇丹Ⅲ試驗 檢驗脂類(第20頁)的一種方法。使用蘇丹Ⅲ溶液,可使脂類染成紅色。

脂迹試驗 檢驗脂類(第20頁)的一種方法。將一滴待試驗的物質滴在一張濾紙上,待其乾燥後舉起濾紙對着光線觀察,這時,只有脂類才會留下一個半透明的痕迹。

半透明的(形) 指物質而言,該物質允許光線透過,但卻不能透過該物質清楚地看到其他物體。

米隆試驗 檢驗蛋白質(第21頁)的一種方法。將幾滴米隆試劑(↓)添加到蛋白質的懸浮液中,並將其煮沸,則蛋白質就會染成磚紅色。

雙縮脲試驗 檢驗蛋白質(第21頁)的一種方法。在待試驗物質中,添加等體積的2%的氫氧化鈉溶液(雙縮脲 A),然後添加0.5%的硫酸銅溶液(雙縮脲 B),則蛋白質就會染成紫色。

乳劑、乳濁液(名) 一種液體在另一種液體中的膠體(第22頁)懸浮液。

懸浮液(名) 一種物質或多種物質的顆粒分散於流體中所形成的混合物。

流體(名) 流動的物質,即液體或氣體。

沉澱物(名) 由溶液(第118頁)中進行的反應所生成的不溶性固體。

試劑(名) 在化學試驗中,用來產生特性反應的物質或溶液(第118頁)。

新陳代謝(名) 一切生物的細胞內所發生的化學反應的總稱。

代謝物(名) 參與新陳代謝(↑)的任何無機(第15頁)或有機(第15頁)物質,例如水或二氧化碳、氨基酸(第21頁)或維生素(第25頁)。

代謝途徑 新陳代謝(↑)進行中的一系列小步驟。

metabolism chemical reations in a plant cell 新陳代謝 植物細胞中之化學反應

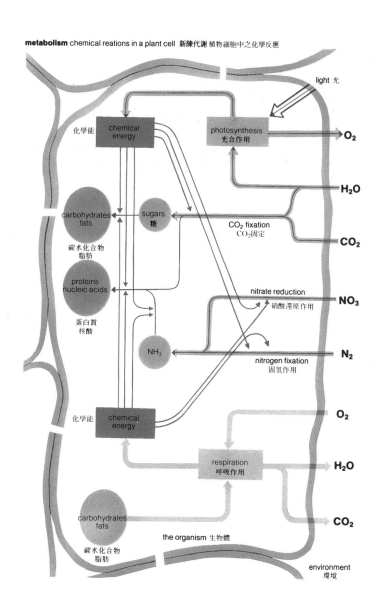

the function of enzymes in catalysis of reactions 酶在催化反應中的功能

synthesis 合成

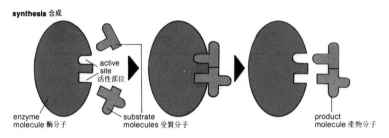

enzyme molecule 酶分子   active site 活性部位   substrate molecules 受質分子   product molecule 產物分子

enzyme (*n*) a protein (p. 21) which increases the rate at which the chemical processes of metabolism (p. 26) take place without being used up by the reaction which it affects. Enzymes are present in all living cells. They are easily destroyed by high temperatures (denatured) and require certain conditions before they will act. The rate of an enzyme-catalyzed (↓) reaction depends upon the concentration of substrate (↓) and enzyme temperature and pH (p. 15). Enzymes increase the rate of reactions by lowering the activation energy.

intracellular (*adj*) within a cell. For example, most enzyme (↑) activity is intracellular i.e. takes place within the cell that produces the enzymes.

extracellular (*adj*) outside a cell. For example, digestive (p. 98) enzymes (↑), which are extracellular in their activity may be secreted (p. 106) into the gut (p. 98) of an animal from other cells where they are produced.

in vivo 'in life' (*adj*) of all the processes which take place within the living organism itself.

in vitro 'in glass' (*adj*) of processes, such as the culture of cell tissues (p. 83) which are carried out experimentally outside the living organism, and originally derived from experiments carried out on parts of an organism in a test tube.

catalyst (*n*) any substance, such as an enzyme (↑), which increases the rate at which a chemical reaction takes place but which is not consumed by the reaction. **catalyze** (*v*).

酶（名）　一種蛋白質（第21頁）。這種蛋白質可加速新陳代謝（第26頁）中所發生的種種化學反應過程，而其本身則未被它所影響的反應消耗掉。酶存在於一切活細胞內。它容易被高溫破壞（變性）。酶在發生作用之前，需要某些條件。酶催化（↓）反應速率取決於受質（↓）的濃度，酶的溫度和 pH 值（第15頁）。酶依靠降低活化能來增加反應速率。

細胞內的（形）　指在細胞之內而言。例如，酶（↑）的大部份活性是存在於細胞內的，即發生於產生酶的細胞之內的。

細胞外的（形）　指在細胞之外而言。例如，消化（第98頁）酶（↑）的活性是存在於細胞外的。消化酶可以從其產生的其他一些細胞分泌（第106頁）到動物的消化道（第98頁）中。

in vivo（形）　即英語的 "in life"（在活體內的）。指發生於生物體自身之內的所有過程而言。

in vitro（形）　即英語的 "in glass"（活在體外的或在試管內的）。指一些細胞組織（第83頁）的培養之類過程而言，這些細胞組織是在生物體外被進行實驗的，它們最初是從於試管中對生物體組織部份進行的那些實驗而得來的。

催化劑（名）　指酶（↑）之類的任何物質。催化劑可增加化學反應進行的速率，但其本身則不被反應所消耗。（動詞形式爲 catalyze）

breakdown 分解

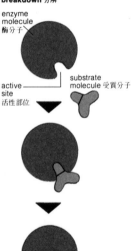

enzyme molecule 酶分子   active site 活性部位   substrate molecule 受質分子

product molecules 產物分子

enzymes control of reaction rate through substrate concentration 酶通過受質濃度控制反應速率

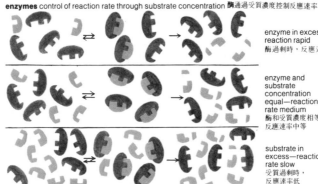

enzyme in excess—
reaction rapid
酶過剩時，反應速率高

enzyme and
substrate
concentration
equal—reaction
rate medium
酶和受質濃度相等時，
反應速率中等

substrate in
excess—reaction
rate slow
受質過剩時，
反應速率低

substrate (*n*) the substance on which something acts e.g. most enzymes (↑) only work on one substrate each and become attached to the substrate molecules.

active site that part of the enzyme (↑) molecule to which specific substrate (↑) molecules become attached.

enzyme-substrate complex the combination of the enzyme (↑) molecule with the substrate (↑) molecule.

lock and key hypothesis a hypothesis (p. 235) which explains the properties of enzymes (↑) by supposing that the particular shape of an enzyme protein (p. 21) corresponds with the shape of particular molecules like a lock and key so that one enzyme will only act as a catalyst (↑) for one specific kind of molecule.

inhibitor (*n*) a substance which slows down or stops a reaction which is controlled by an enzyme (↑). **inhibition** (*n*). **inhibit** (*v*).

competitive inhibition inhibition (↑) when the substrate (↑) and the inhibitor compete for the enzyme (↑). Also known as reversible (p. 30) inhibition.

non-competitive inhibition inhibition (↑) when the inhibitor combines permanently with the enzyme (↑) so that the substrate (↑) is excluded. Also known as non-reversible (p. 30) inhibition.

non-competitive inhibition
非競爭性抑制

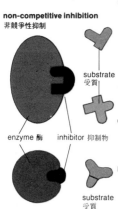

substrate
受質

enzyme 酶   inhibitor 抑制物

substrate
受質

受質、底物（名） 某物作用於其上的物質。例如，大多數酶（↑）各自僅對一種受質起作用，並附着在受質分子上。

活性部位 酶（↑）分子附着特殊受質（↑）分子的那個部位。

酶—受質複合體 酶（↑）分子與受質（↑）分子的結合物。

鎖鑰假說 一種解釋酶（↑）的特性的假說（第235頁）。其依據是，假設一種酶蛋白（第21頁）的特有形狀與一些特殊分子的形狀就像鎖和鑰匙一樣吻合，那麼，一種酶就僅對一種特殊類型的分子起催化劑（↑）的作用。

抑制物（名） 可減緩或阻止由酶（↑）控制的反應的一種物質。（名詞形式為 inhibition，動詞形式為 inhibit）

競爭性抑制 當受質（↑）和抑制物相互競爭酶（↑）時所產生的抑制（↑）。也稱為可逆（第30頁）抑制。

非競爭性抑制 當抑制物與酶（↑）永久地結合，從而排斥受質（↑）所產生的抑制（↑）。也稱為非可逆（第30頁）抑制。

reversible (*adj*) of a reaction or process that is not permanent i.e. it can work in the opposite direction. **reverse** (*v*).

non-reversible (*adj*) not reversible (↑).

cofactor (*n*) an additional inorganic compound (p. 15) which must be present in a reaction before the enzyme (p. 28) will catalyze (p. 28) it.

co-enzyme (*n*) an additional, non-protein (p. 21) organic compound (p. 15) which must be present in a reaction before the enzyme (p. 28) will catalyze (p. 28) it.

prosthetic group a non-protein (p. 21) organic compound (p. 15) which forms an essential part of the enzyme (p. 28) and which must be present in a reaction before the enzyme will catalyze (p. 28) it.

hydrolase (*n*) an enzyme (p. 28) which catalyzes (p. 28) hydrolysis (p. 16) reactions.

carbohydrase (*n*) an enzyme (p. 28) which catalyzes (p. 28) digestion (p. 98) reactions and aids in the breakdown of carbohydrates (p. 17).

oxidase (*n*) an enzyme (p. 28) group which catalyzes (p. 28) oxidation (p. 32) reactions.

dehydrogenase (*n*) an enzyme (p. 28) group which catalyzes (p. 28) reactions in which hydrogen atoms are removed from a sugar.

carboxylase (*n*) an enzyme (p. 28) group which catalyzes (p. 28) reactions in which carboxyl (COOH) groups are added to a substrate (p. 29).

transferase (*n*) an enzyme (p. 28) group which catalyzes (p. 28) reactions in which a group is transferred from one substrate (p. 29) to another.

isomerase (*n*) an enzyme (p. 28) group which catalyzes (p. 28) reactions in which the atoms of molecules are rearranged.

cell respiration the breakdown by oxidation (p. 32) of sugars yielding carbon dioxide, water and energy.

endergonic (*adj*) of a reaction which absorbs (p. 81) energy.

exergonic (*adj*) of a reaction which releases energy.

electron (*n*) a very small, negatively charged particle in an atom which may be raised to higher energy levels and then released during cell respiration (↑).

可逆的（形） 指不是不可變的反應或過程而言,即它可可朝相反的方向起作用。(動詞形式為 reverse)

非可逆的(形) 不可逆的(↑)。

輔助因子(名) 一種附加的無機化合物(第15頁)。在酶(第28頁)催化(第28頁)反應之前,它必須存在於反應物之中。

輔酶(名) 一種附加的非蛋白質(第21頁)有機化合物(第15頁)。在酶(第28頁)催化(第28頁)反應之前,它必須存在於反應物之中。

輔基 一種非蛋白質(第21頁)有機化合物(第15頁)。它構成酶(第28頁)的一種基本成份;在酶(第28頁)催化(第28頁)反應之前,必須存在於反應物之中。

水解酶(名) 一種可催化(第28頁)水解(第16頁)反應的酶(第28頁)。

糖酶(名) 可催化(第28頁)消化(第98頁)反應並有助於碳水化合物(第17頁)分解的一種酶(第28頁)。

氧化酶(名) 一組可催化(第28頁)氧化(第32頁)反應的酶(第28頁)。

脫氫酶(名) 可對從糖中脫去氫原子的這類反應進行催化(第28頁)的一組酶(第28頁)。

羧化酶(名) 可對將羧基(COOH)添加到受質(第29頁)上的這類反應進行催化(第28頁)的一組酶(第28頁)。

轉移酶(名) 可對將一種基團從一種受質(第29頁)轉移到另一種受質上的這類反應進行催化(第28頁)的一組酶(第28頁)。

異構酶(名) 可對分子中的原子進行重排的這類反應進行催化(第28頁)的一組酶(第28頁)。

細胞呼吸 經糖氧化(第32頁)所發生的分解,結果產生二氧化碳、水和能量。

吸能的(形) 指反應能吸收(第81頁)能量的。

放能的(形) 指釋放能量的反應而言。

電子(名) 原子中極小的帶負電荷的粒子。在細胞呼吸(↑)過程中,它可被提到較高的能級,然後加以釋放。

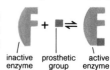

prosthetic group 輔基
role in enzyme reaction 在酶反應中的作用

inactive enzyme 非活性酶　prosthetic group 輔基　active enzyme 活性酶

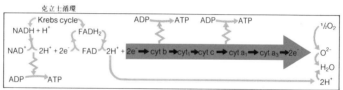

克立士循環

electron carrier system 呼吸中的電子載體系統
in respiration

electron (hydrogen) carrier system a system which operates during cell respiration (↑) in which electrons (↑) (initially released as part of a hydrogen atom which splits into an electron and a proton) are collected by an electron acceptor (↓) and passed to another electron acceptor at lower energy levels. The energy released in the process is used to convert ADP (p. 33) to ATP (p. 33).

electron acceptor a molecule which functions as a coenzyme (↑) with a dehydrogenase (↑) that catalyzes (p. 28) the removal of hydrogen during cell respiration (↑). It accepts electrons (↑) and passes them on to electron acceptors at lower energy levels.

NAD nicotinamide adenine dinucleotide. One of the most important coenzymes (↑) or electron acceptors (↑) concerned with cell respiration (↑).

NADP nicotinamide adenine dinucleotide phosphate. An important coenzyme (↑) or electron acceptor (↑) similar to NAD (↑).

電子(氫)載體系統　在細胞呼吸(↑)過程中運轉的一種系統。這種系統中，電子(↑)(最初作爲能分裂成一個電子和一個質子的氫原子的一部份釋放出來)被一個電子受體(↓)接收，然後傳遞給處於較低能級的另一個電子受體。在這個過程中所釋放的能量用於將腺苷二磷酸(第33頁)轉變成腺苷三磷酸(第33頁)。

電子受體　起脫氫酶(↑)輔酶(↑)作用的一種分子。脫氫酶在細胞呼吸(↑)過程中可催化(第28頁)受質脫氫。電子受體可接受電子(↑)，然後將其傳遞給處於較低能級的電子受體。

NAD　即烟醯胺腺嘌呤二核苷酸。參與細胞呼吸(↑)的最重要的輔酶(↑)或電子受體(↑)之一。

NAOP　即烟醯胺腺嘌呤二核苷酸磷酸。類似於NAD(↑)的一種重要輔酶(↑)或電子受體(↑)。

nicotinamide adenine dinucleotide (NAD)
adding a further phosphate group at **p** gives NADP
烟醯胺腺嘌呤二核苷酸（NAD）
在 P 上再加上一個磷基
即產生 NADP

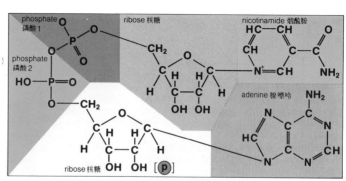

oxidation (*n*) a reaction in which a substance (1) loses electrons (p. 30); (2) has oxygen added to it; or (3) has hydrogen removed from it. **oxidize** (*v*).

reduction (*n*) a reaction in which a substance (1) gains electrons (p. 30); (2) has oxygen removed from it; or (3) has hydrogen added to it. **reduce** (*v*).

cytochrome (*n*) one of a system of coenzymes (p. 30) involved in cell respiration (p. 30) having prosthetic groups (p. 30) which contain iron. Cytochromes are involved in the production of ATP (↓) by oxidative phosphorylation (p. 34).

flavoprotein (*n*) FP. An important coenzyme (p. 30) involved in cell respiration (p. 30).

vitamin B the collective name for a group of vitamins (p. 25) which play an important role in cell respiration (p. 30) by functioning as coenzymes (p. 30).

aerobic (*adj*) of a reaction, for example, respiration (p. 112) which can only take place in the presence of free, gaseous oxygen. In aerobic respiration, organic compounds (p. 15) are converted to carbon dioxide and water with the release of energy. Organisms that use aerobic respiration are called aerobes.

anaerobic (*adj*) of a reaction, for example, respiration (p. 112) which takes place in the absence of free gaseous oxygen. In anaerobic respiration organic compounds (p. 15) such as sugars are broken down into other compounds such as carbon dioxide and ethanol with a lower release of energy. Organisms that use anaerobic respiration are called anaerobes.

basal metabolism the smallest (or minimum) amount of energy needed by the body to stay alive. It varies with the age, sex and health of the organism.

BMR basal metabolic rate = basal metabolism (↑).

metabolic rate in cell respiration (p. 30) the rate at which oxygen is used up and carbon dioxide is produced.

calorific value the amount of heat produced, measured in calories, when a given amount of food is completely burned. *See also* joule (p. 97).

氧化(作用)(名) 一種反應。在這種反應中，一種物質至少發生下列作用之一：(1)失去電子(第30頁)；(2)加入氧；(3)脫去氫。(動詞形式爲 oxidize)

還原作用(名) 一種反應。在這種反應中，一種物質至少發生下列作用之一：(1)獲得電子；(2)失去氧；(3)獲得氫。(動詞形式爲 reduce)

細胞色素(名) 參與細胞呼吸(第30頁)、具有含鐵輔基(第30頁)的一組輔酶(第30頁)之一。細胞色素經氧化磷酸作用(第34頁)參與腺苷三磷酸(↓)的生成。

黃素蛋白(名) 英文縮寫爲 FP。一種參與細胞呼吸(第30頁)的重要輔酶(第30頁)。

維生素 B 一族維生素(第25頁)的總稱。該族維生素具有輔酶(第30頁)的功能，在細胞呼吸(第30頁)中起重要作用。

需氧的(形) 指反應而言，例如僅在有游離氧態氧存在的情況下才能進行的呼吸(第112頁)。在需氧呼吸中，有機化合物(第15頁)被轉變成二氧化碳和水，並放出能量。進行需氧呼吸的生物被稱爲需氧生物。

缺氧的(形) 指反應而言，例如在無游離氧態氧存在的情況下所進行的呼吸(第112頁)。在缺氧呼吸中，糖之類有機化合物(第15頁)被分解成其他化合物諸如二氧化碳和乙醇，並放出少量的能量。進行缺氧呼吸的生物被稱爲厭氧生物。

基礎代謝 維持生命所需的最低(或最少)能量。它因生物體的年齡、性別和健康情況而異。

BMR 基礎代謝率之縮寫。與基礎代謝(↑)同。

代謝率 在細胞呼吸(第30頁)中，氧氣消耗和二化碳產生的速率。

熱值、卡值 一定量的食物完全燃燒時所產生的熱量(以卡計)。參見"焦耳"(第97頁)。

aerobic respiration 需氧呼吸

carbohydrate 碳水化合物
糖
酵解 glycolysis

water 水

oxygen 氧

carbon dioxide 二氧化碳

Krebs cycle and electron transfer chain

克士循環和電子傳遞鏈

mitochondrion 線粒體

ATP

ADP + Pi

ATP adenosine triphosphate. An organic compound (p. 15) composed of adenine (p. 22) ribose (p. 22), and three inorganic phosphate groups. It is a nucleotide (p. 22) and is responsible for storing energy temporarily during cell respiration (p. 30). It is formed by the addition of a third phosphate group to ADP (↓) which stores the energy that is released when required in other metabolic (p. 26) processes.

ADP adenosine diphosphate. The organic compound (p. 15) which accepts a phosphate group to form ATP (↑).

ATP　即腺苷三磷酸。由腺嘌呤（第22頁）、核糖（第22頁）和三個無機磷酸基組成的一種有機化合物（第15頁）。它是一種核苷酸（第22頁），在細胞呼吸（第30頁）過程中，負責臨時貯存能。它是在 ADP（腺苷二磷酸）（↓）上添加了第三個磷酸基而生成的，其內貯存着能量。當在其他一些代謝（第26頁）過程中需要能量時，它便將之釋放出來。

ADP　ADP 即腺苷二磷酸。一種接受一個磷酸基即生成ATP（↑）的有機化合物（第15頁）。

ADP, ATP and their reactions

ADP、ATP 及其反應

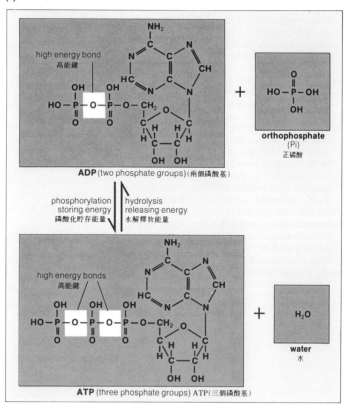

ADP (two phosphate groups)（兩個磷酸基）

ATP (three phosphate groups) ATP（三個磷酸基）

phosphate bond a bond which links the phosphate groups in ATP (p. 33) and which is often misleadingly referred to as a high energy bond. Energy is stored throughout the ATP molecule but is released as the phosphate bonds are broken and other bonds are formed.

oxidative phosphorylation the process in which ATP (p. 33) is produced from ADP (p. 33) in the presence of oxygen during aerobic (p. 32) cell respiration (p. 30).

glycolysis (*n*) the first part of cellular respiration (p. 30) in which glucose (p. 17) is converted into pyruvic acid (↓) in the cytoplasm (p. 10) of all living organisms. It uses a complex system of enzymes (p. 28) and coenzymes (p. 30). It produces energy for short periods in the form of ATP (p. 33) when there is a shortage of oxygen.

pyruvic acid an organic compound (p. 15) which is formed as the end product of glycolysis (↑). For every molecule of glucose (p. 17) two molecules of pyruvic acid are formed.

Kreb's cycle a part of cellular respiration (p. 30) in which pyruvic acid (↑) in the presence of oxygen and via a complex cycle of enzyme- (p. 28) controlled reactions produces energy in the form of ATP (p. 33) and intermediates which give rise to other substances such as fatty acids (p. 20) and amino acids (p. 21). It takes place in the mitochondria (p. 11).

fermentation (*n*) a process in which pyruvic acid (↑) in the absence of oxygen uses up hydrogen atoms and so produces NAD (p. 31) allowing it to be used again in glycolysis (↑).

lactic acid fermentation fermentation (↑) from which lactic acid is produced. In higher animals this takes place especially in the muscles (p. 143) where there is an oxygen debt (p. 117).

磷酸鍵　ATP（第33頁）中連接磷酸基的鍵。這種鍵常常被誤稱爲是高能鍵。整個 ATP 分子中貯存着能量，但這種能量只有磷酸鍵斷裂生成其他鍵時才被釋放出來。

氧化磷酸化(作用)　在需氧（第32頁）細胞呼吸（第30頁）過程中，在有氧的情況下，由 ADP（第33頁）生成 ATP（第33頁）的過程。

糖酵解、糖解作用（名）　細胞呼吸（第30頁）的第一階段。在這一階段中，所有生物細胞質（第10頁）內的葡萄糖（第17頁）被轉化成丙酮酸（↓）。糖解利用一套複雜的酶（第28頁）和輔酶（第30頁）的系統。當氧氣短缺時，通過糖解作用，即可在短時間內，以 ATP（第33頁）的形式產生能量。

丙酮酸　作爲糖酵解（↑）最終產物所生成的一種有機化合物（第15頁）。對於每一個葡萄糖的分子（第17頁）來說，可生成兩個丙酮酸的分子。

克立士循環　細胞呼吸（第30頁）的一個階段。在這一階段中，丙酮酸（↑）在有氧的情況下，經過一個由酶（第28頁）控制的一些反應所組成的複雜循環，以 ATP（第33頁）的形式產生能量，同時生成一些中間產物。這些中間產物還能產生其他一些物質，例如脂肪酸（第20頁）和氨基酸（第21頁）。克立士循環在線粒體（第11頁）中進行。

發酵(作用)（名）　丙酮酸（↑）在缺氧情況下，消耗氫原子，產生 NAD（第31頁），使其重新用於糖酵解（↑）的一種過程。

乳酸發酵　產生乳酸的發酵作用（↑）。在高等動物中，這種發酵尤其發生於有氧償（第117頁）的那些肌肉（第143頁）之中。

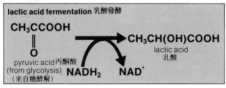

lactic acid fermentation 乳酸發酵

CH_3CCOOH
‖
O
pyruvic acid 丙酮酸
(from glycolysis)
（來自糖酵解） $NADH_2$

$CH_3CH(OH)COOH$
lactic acid
乳酸

NAD^+

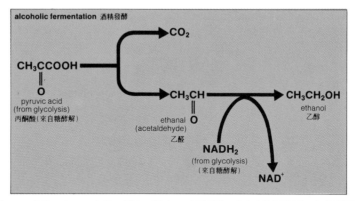

alcoholic fermentation 酒精發酵

$$CH_3CCOOH \longrightarrow$$
$$\|$$
$$O$$
pyruvic acid
(from glycolysis)
丙酮酸（來自糖酵解）

$$\longrightarrow CO_2$$

$$\longrightarrow CH_3CH \longrightarrow CH_3CH_2OH$$
$$\|$$
ethanal O
(acetaldehyde)
乙醛
ethanol
乙醇

$$NADH_2$$
(from glycolysis)
（來自糖酵解）

$$NAD^+$$

alcoholic fermentation fermentation (↑) in which ethanol (alcohol) and carbon dioxide are produced. This process is made use of in the brewing and wine-making industries in which yeasts (p. 49) decompose sugars to provide energy for their reproduction (p. 173) and growth.

nuclear division the process in which the nucleus (p. 13) of a cell divides into two in the development of new cells and new tissue (p. 83) so that growth may occur or damaged cells be replaced. There are two types: mitosis (p. 37) and meiosis (p. 38).

centriole (n) a structure similar to a basal body (p. 13). Centrioles are found outside the nuclear membrane (p. 13) and divide at mitosis (p. 37) forming the two ends of the spindle (p. 37).

chromatid (n) one of a pair of thread-like structures which together appear as chromosomes (p. 13) and which shorten and thicken during the prophase (p. 37) of nuclear division (↑).

centromere (n) a region, somewhere along the chromosome (p. 13), where force is exerted during the separation of the chromatids (↑) in mitosis (p. 37) and meiosis (p. 38).

chromomere (n) one of a number of granules of chromatin (p. 13) which occur along a dividing chromosome (p. 13) probably as a result of the coiling and uncoiling within the chromatids (↑). It appears as a 'bump' or constriction.

酒精發酵 產生乙醇（酒精）和二氧化碳的發酵作用（↑）。在這種發酵作用中，酵母（第49頁）將糖分解，爲其生長和繁殖（第173頁）提供所需的能量，釀造和製酒工業所利用的正是這一作用。

核分裂 在形成新細胞和新組織（第83頁）時，一個細胞的核（第13頁）分裂成兩個的過程。這樣，生長才得以進行，受損的細胞才得以替換。核分裂有兩種類型：有絲分裂（第37頁）和減數分裂（第38頁）。

中心粒（名） 類似於基體（第13頁）的一種結構。中心粒存在於核膜（第13頁）的外面；在有絲分裂（第37頁）時，分裂形成紡錘體（第37頁）的兩端。

染色單體（名） 共同作爲染色體（第13頁）出現的一對絲狀結構之一。在核分裂（↑）的前期（第37頁），染色單體縮短、變粗。

着絲粒（名） 染色單體（↑）在有絲分裂（第37頁）和減數分裂（第38頁）中發生分離時，位於染色體（第13頁）長度方向上某處的一個着力部位。

染色粒（名） 染色質（第13頁）中的諸多顆粒之一。這些顆粒很可能是作爲染色單體（↑）內部繞螺旋和解螺旋的結果，沿着分裂的染色體（第13頁）而出現的。染色粒表現爲一個"凸緣"或縊痕。

somatic cell any cell in a living organism other than a germ cell (↓) and which contains the characteristic number of chromosomes (p. 13), normally diploid (↓), for the organism.

germ cell a cell that gives rise to a gamete (p. 175). A cell in a living organism, other than a somatic cell (↑), and which takes part in the reproduction (p. 173) of the organism. It contains only half of the characteristic number of chromosomes (p. 13) of the organism i.e. it is haploid (↓).

體細胞 生物體中除生殖細胞(↓)以外的任何細胞。體細胞具有生物體特有的染色體(第13頁)數，通常爲二倍體(↓)。

生殖細胞 即可產生配子(第175頁)的細胞。生物體中除體細胞(↑)以外的任何細胞。生殖細胞參與生物體的生殖(第173頁)。它只具生物體特有的染色體(第13頁)數的一半，即它是單倍體(↓)。

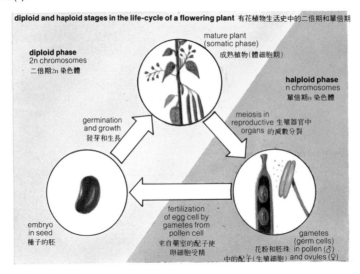

diploid and haploid stages in the life-cycle of a flowering plant 有花植物生活史中的二倍期和單倍期

diploid phase
2n chromosomes
二倍期2n 染色體

mature plant
(somatic phase)
成熟植物(體細胞期)

halploid phase
n chromosomes
單倍期n 染色體

germination
and growth
發芽和生長

meiosis in
reproductive 生殖器官中
organs 的減數分裂

embryo
in seed
種子的胚

fertilization
of egg cell by
gametes from
pollen cell
來自藥室的配子使
卵細胞受精

gametes
(germ cells)
in pollen (♂)
花粉和胚珠 and ovules (♀)
中的配子(生殖細胞)

haploid (*adj*) of a cell which has only unpaired chromosomes (p. 13); half the diploid (↓) number of chromosomes which are not paired in the haploid state. Germ cells (↑) of most animals and plants are haploid. *See also* polyploidy etc p. 207.

diploid (*adj*) of a cell which has chromosomes (p. 13) which occur in homologous (p. 39) pairs Somatic cells (↑) of most higher plants and animals are described as diploid. Double the haploid (↑) number.

單倍體的(形) 指下列的這種細胞而言，這種細胞僅有不成對的染色體(第13頁)，而且，在單倍體狀態下，不成對的染色體數爲二倍(↓)數的一半。大多數動物和植物的生殖細胞(↑)都爲單倍體。參見第207頁的"多倍性"等。

二倍體的(形) 指下列的這種細胞而言，這種細胞的染色體(第13頁)以同源(第39頁)配對形式出現。大多數高等植物和動物的體細胞(↑)被認爲是二倍體，其染色體數爲單倍(↑)數的二倍。

mitosis 有絲分裂
(only two pairs of homologous chromosomes shown for (為清晰 clarity) 起見，僅顯示兩對同源染色體)

chromosomes 染色體
nuclear membrane 核膜
nucleolus 核仁

prophase chromosomes become visible in the nucleus, each one duplicated into two chromatids, joined by a centromere 前期核中的染色體可以 看見，每一染色體複製成由一著 絲粒連結的兩個染色單體

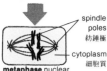

spindle poles 紡錘極
cytoplasm 細胞質

metaphase nuclear membrane and nucleolus have disintegrated. Spindle fibres form. Chromosomes shorter and thicker, arranged midway between the spindle poles. 中期核膜和核仁已經解體。 紡錘絲形成。染色體縮短，變粗， 排列在兩個紡錘極的中間。

anaphase chromatids separate at centromeres. Sister chromatids drawn to opposite poles of the spindle. 後期染色單體在着絲粒處分離。 姐妹染色單體被牽引至紡錘體 兩個相反的極

telophase nuclear membrane and nucleoli reform. Chromosomes begin to lose their compact structure 末期核膜和核仁重新形成。 染色體開始失去其致密結構

Interphase chromosomes no longer visible. 間期染色體已看不見。

mitosis (*n*) the usual process of nuclear division (p. 35) into two daughter nuclei (p. 13) during vegetative growth. During mitosis each chromosome (p. 13) duplicates itself, each one of the duplicates going into separate daughter nuclei. The daughter cells are identical to each other and to the parent cell.

spindle (*n*) fibrous (p. 143) material which forms from the centrioles (p. 35) during mitosis (↑) and meiosis (p. 38). It takes part in the distribution of chromatids (p. 35) to the daughter cells. The chromosomes (p. 13) are arranged at its equator (↓) during metaphase (↓).

pole (*n*) one of the two points on the spindle (↑) which is the site of the formation of the spindle fibres (p. 143) from the centrioles (p. 35).

equator (*n*) part of the spindle (↑) midway between the poles (↑) to which the chromosomes (p. 13) become attached by the spindle attachment.

interphase (*n*) a stage in the cell cycle when the cell is preparing for nuclear division (p. 35). At this stage, the DNA (p. 24) is replicating to produce enough for the daughter cells.

prophase (*n*) the first main stage in nuclear division (p. 35) in which the chromosomes (p. 13) become visible and then the chromatids (p. 35) appear while the nucleolus (p. 13) and nuclear membrane (p. 13) begin to dissolve.

metaphase (*n*) a main stage in nuclear division (p. 35) at which the nuclear membrane (p. 13) has disappeared and the chromosomes (p. 13) lie on the equator (↑) of the spindle (↑). Then the chromatids (p. 35) start to move apart.

anaphase (*n*) a main stage in nuclear division (p. 35) in which the centromeres (p. 35) divide and the chromatids (p. 35) move to opposite poles (↑) by the contraction of the spindle (↑).

telophase (*n*) a main stage in nuclear division (p. 35) at which the chromatids (p. 35) arrive at the poles (↑) and the cytoplasm (p. 10) may divide to form two separate daughter cells in interphase (↑). The spindle (↑) fibres (p. 143) dissolve while the nucleolus (p. 13) and nuclear membrane (p. 13) in each daughter cell reform and the chromosomes (p.13) regain their thread-like form.

有絲分裂（名）　在營養生長期間，形成兩個子核 （第13頁）的核分裂（第35頁）的通常過程。在有 絲分裂時，每一個染色體（第13頁）都進行自我 複製，隨後，各個重複染色體分別進入已分裂 開的子核內，這些子細胞彼此相同，與親代細 胞也相同。

紡錘體（名）　在有絲分裂（↑）和減數分裂（第38頁） 期間，由中心粒（第35頁）形成的纖維狀（第143 頁）物質。紡錘體參與染色單體（第35頁）分配 給子細胞的活動。在中期（↓），染色體（第13 頁）排列在紡錘體的赤道（↓）上。

紡錘極（名）　紡錘體（↑）的兩個末端，是由中心粒 （第35頁）形成紡錘絲（第143頁）的部位。

赤道（名）　位於兩個紡錘極（↑）中間的紡錘體（↑） 部份。染色體（第13頁）由紡錘絲連接到紡錘極 上。

間期（名）　細胞在準備進行核分裂（第35頁）時的細 胞周期中的一個階段。在這一階段，DNA （第24頁）進行複製，為子細胞提供足夠數量的 DNA。

前期（名）　核分裂（第35頁）中的第一個主要階段。 在這一階段，染色體（第13頁）可以看見，以 後，染色單體（第35頁）出現，而核仁（第13頁） 和核膜（第13頁）開始消失。

中期（名）　核分裂（第35頁）中的一個主要階段。在 這一階段，核膜（第13頁）已經消失，染色體 （第13頁）位於紡錘體（↑）的赤道（↑）上，此 後，染色單體（第35頁）開始離開。

後期（名）　核分裂（第35頁）中的一個主要階段。在 這一階段，着絲粒（第35頁）分裂，染色單體 （第35頁）由於紡錘體（↑）的收縮而向相反的兩 極（↑）移動。

末期（名）　核分裂（第35頁）中的一個主要階段。在 這一階段，染色單體（第35頁）到達兩極（↑）， 細胞質（第10頁）分裂成間期（↑）的兩個獨立的 子細胞。紡錘（↑）絲（第143頁）消失，而每一 子細胞中的核仁（第13頁）和核膜（第13頁）重新 形成；染色體（第13頁）又恢復其絲狀形態。

meiosis (*n*) nuclear division (p. 35) of a special kind which begins in a diploid (p. 36) cell and takes place in two stages. Each stage is similar to mitosis (p. 37) but the chromosomes (p. 13) are duplicated only once before the first division so that each of the four resulting daughter cells is haploid (p. 36). It occurs during the formation of the gametes (p. 175).

減數分裂、成熟分裂（名） 一種特殊的核分裂（第35頁）。這種核分裂以二倍體（第36頁）細胞開始，按兩個階段進行。每一個階段都類似於有絲分裂（第37頁），但染色體（第13頁）在第一次分裂之前僅複製一次，因此，所產生的四個子細胞全都是單倍體（第36頁）。減數分裂發生在配子（第175頁）形成過程中。

meiosis (cytoplasm and membrane not shown) 減數分裂（細胞質和膜未作顯示）

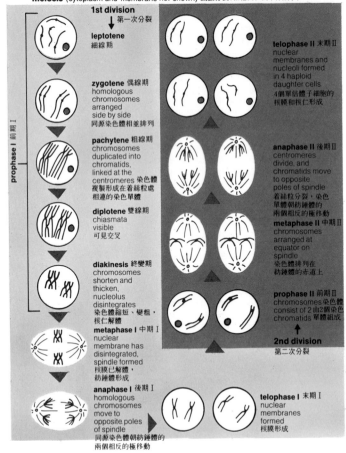

1st division
第一次分裂

leptotene
細線期

zygotene 偶線期
homologous chromosomes arranged side by side
同源染色體相並排列

pachytene 粗線期
chromosomes duplicated into chromatids, linked at the centromeres 染色體複製形成在着絲粒處相連的染色單體

diplotene 雙線期
chiasmata visible
可見交叉

diakinesis 終變期
chromosomes shorten and thicken, nucleolus disintegrates
染色體縮短、變粗，核仁解體

metaphase I 中期 I
nuclear membrane has disintegrated, spindle formed
核膜已解體，紡錘體形成

anaphase I 後期 I
homologous chromosomes move to opposite poles of spindle
同源染色體朝紡錘體的兩個相反的極移動

prophase I 前期 I

telophase I 末期 I
nuclear membranes formed
核膜形成

telophase II 末期 II
nuclear membranes and nucleoli formed in 4 haploid daughter cells
4個單倍體子細胞的核膜和核仁形成

anaphase II 後期 II
centromeres divide, and chromatids move to opposite poles of spindle
着絲粒分裂，色色單體朝紡錘體的兩個相反的極移動

metaphase II 中期 II
chromosomes arranged at equator on spindle
染色體排列在紡錘體的赤道上

prophase II 前期 II
chromosomes 染色體 consist of 2 由2個色色 chromatids 單體組成

2nd division
第二次分裂

differences between
mitosis and meiosis 有絲分裂與減數分裂之間的差異

mitosis	有絲分裂	meiosis	減數分裂
occurs in somatic cells during growth and repair	發生於生長和修複過程中的體細胞內	occurs in the sex organs during gamete formation	發生於配子形成過程中的性器官內
no pairing or separation of homologous chromosomes	無同源染色體配對或分離	pairing and separation of homologous chromosomes	同源染色體配對和分離
no chiasmata formed	不形成交叉	chiasmata formed which may lead to crossing over and recombination	形成交叉並導致交換和重組
one separation of nuclear material i.e. separation of chromatids only	核物質的一次分離，即僅僅是染色單體的分離	two sepatations of nuclear material i.e. separation of homologous chromosomes (1st division) and chromatids (2nd division)	核物質的兩次分離，即同源染色體的分離（第一次分裂）和染色單體的分離（第二次分離）
2 daughter nuclei formed	形成兩個子核	4 daughter nuclei formed	形成四個子核
daughter nuclei identical	子核相同	daughter nuclei not identical	子核不相同
daughter nuclei diploid	子核二倍體	daughter nuclei haploid	子核單倍體

bivalent (*n*) one of the pairs of homologous (↓) chromosomes (p. 13) which associate during the first prophase (p. 37) of meiosis (↑).

chiasmata (*n.pl.*) the points at which homologous (↓) chromosomes (p. 13) remain in contact as the chromatids (p. 35) move apart during the first prophase (p. 37) of meiosis (↑). There may be up to eight chiasmata in a bivalent (↑) pair of chromosomes. **chiasma** (*sing.*).

terminalization (*n*) the process in which the chiasmata (↑) move to the ends of the chromosomes (p. 13) during the prophase (p. 35) of meiosis (↑).

pairs of homologous
chromosomes
成對同源染色體

centromeres
着絲粒

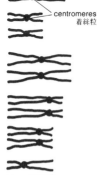

homologous chromosomes two chromosomes (p. 13) which form a pair in which the genes (p. 196) arranged along their length control identical characteristics of the organism, such as eye colour or height.

first meiotic division the first of two major stages of meiosis (p. 37) in which a nuclear division (p. 35) similar to mitosis (p. 37) takes place resulting in the separation of homologous (↑) chromosomes (p. 13).

second meiotic division the second of two major stages in meiosis (p. 37) in which a second nuclear division (p. 35) takes place and the two daughter cells formed from the first meiotic division (↑) each divide into two to result in four haploid (p. 36) daughter cells each containing one of the sister chromatids (p. 35).

二價染色體（名） 在減數分裂（↑）前期（第37頁）I 時聯合成對的同源（↓）染色體（第13頁）。

交叉（名、複） 在減數分裂（↑）前期（第37頁）I，當染色單體（第35頁）分離時，同源（↓）染色體（第13頁）之間仍保持接觸的點。在一對二價（↑）染色體中，有可能出現8個交叉。（單數形式爲 chiasma）

（交叉）移端（作用）（名） 在減數分裂（↑）前期（第35頁），交叉（↑）移至染色體（第13頁）末端的這一過程。

同源染色體 形成一對的兩個染色體（第13頁）。在這些染色體上，沿其長度方向排列的基因（第196頁），控制着生物體的同一性狀，例如眼睛的顏色或身體的高度。

第一次減數分裂 減數分裂（↑）的兩個主要階段中的第一個階段。在這一階段，發生類似於有絲分裂（第37頁）的核分裂（第35頁）引起同源（↑）染色體（第13頁）分離。

第二次減數分裂 減數分裂的兩個主要階段中的第二個階段。在這一階段，發生第二次核分裂（第35頁），由第一次減數分裂（↑）生成的兩個子細胞各自分裂成兩個，產生各含一個姐妹染色體（第35頁）的四個單倍體（第36頁）子細胞。

classification (*n*) the arrangement of all living organisms into an ordered series of named and related groups. **classify** (*v*).

organisms (*n*) any living thing. Organisms can grow and reproduce (p. 175).

taxon (*n*) the general term for any group in a classification (↑) no matter what its rank (↓). **taxa** (*n.pl.*).

taxonomy (*n*) the science of classification (↑).

binomial system a system of naming every known living organism, first devised by the Swedish botanist, Carolus Linnaeus (1707-78), in which the organism is given a two-part scientific name which is usually Latinized. The first word indicates the genus (↓) while the second word indicates the species (↓). While the common names of organisms may only be understood in their place of origin, the scientific name is recognized internationally by scientists. For example, the bird with the English common name, peregrine falcon, is given the scientific name, *Falco peregrinus*.

species (*n*) a group of similar living organisms whose members can interbreed to produce fertile (p. 175) offspring but which cannot breed with other species groups. **specific** (*adj*).

genus (*n*) a group of organisms containing a number of similar species (↑). Of the scientific name, *Falco peregrinus*, *Falco* is the generic name referring to all birds that are classified (↑) as falcons.

分類法（名） 根據一個有序系列的指定類別和相關類別，對所有生物進行的排列。（動詞形式爲 classify）

生物、有機體（名） 任何有生命的物體。生物可以生長和繁殖（第175頁）。

分類單元、分類羣（名） 分類（↑）中對於任何羣體的總稱，不管其等級（↓）如何。（複數形式爲 taxa）

分類學（名） 關於分類（↑）的科學。

雙名法 對每一個已知生物進行命名的一種體系。是由瑞典植物學家 C・林奈 (Carolus Linnaeus) (1707—78) 首先創立。用雙名法時，給生物以一個由兩部份組成的、通常拉丁化的學名。第一個單詞表示其屬（↓），第二個單詞表示其種（↓）。雖然一些生物的普通名稱也許只有在其產地才爲人們所知，但其學名則在國際上被科學家們所承認。例如，具有英語普通名稱 peregrine falcon（游隼）的鳥，給其的學名爲 Falco peregrinus（游隼）。

種（名） 一羣相似的生物，其中的各個成員能進行種間雜交，產生能育的(175頁)後代。但是，它們不能與其他種的羣體繁殖後代。（形容詞形式爲 specific）

屬（名） 一羣包含若干個相似種（↑）的生物。在學名Falco peregrinus（游隼）之中，Falco（隼）是屬名，指納入隼屬分類（↑）的所有的鳥。

classfication of the Peregrine falcon showing the series of ranks and their names		
rank	scientific name of taxonomic groups (taxa)	common name
kingdom	Animalia	animals
phylum	Chordata	vertebrates
class	Aves	birds
order	Falconiformes	birds of prey
family	Falconidae	falcons
genus	*Falco*	true falcons
species	*peregrinus*	Peregrine falcon

游隼的分類(其等級系列和名稱)		
等級	分類羣(分類單元)的學名	普通名
界	動物界	動物
門	脊索動物門	脊椎動物
綱	鳥綱	鳥類
目	鷹隼目	猛禽
科	隼科	隼
屬	隼屬	眞隼
種	廣佈種	游隼

rank (*n*) one of a number of major groups into which living organisms are classified (↑). The largest group which contains organisms that have different body plans from those in any other large group is called a kingdom (↓). Each kingdom may be further divided, on the basis of diversity (p. 213), into a number of phyla, and so on. The principal rank names arranged in order from the largest groups to the most basic are kingdom, phylum, class, order, family, genus (↑) and species (↑).

kingdom (*n*) the highest rank (↑) or taxon (↑). Most simply, all life can be grouped into either the Plant or Animal kingdoms. This, however, is an oversimplification and in this book we divide living organisms into five kingdoms: Monera (p. 42), Protista (p. 44), Fungi (p. 46), Plants, Animals.

artificial classification a classification (↑) in which the organisms are arranged into groups on the basis of apparent analogous (p. 211) similarities which, in fact, have no common ancestry.

natural classification a classification (↑) in which the organisms are arranged into groups on the basis of homologous (p. 211) similarities which demonstrate a common ancestry.

等級(名)　對生物進行分類(↑)所形成的若干個主要分類羣之一。其中最大的分類羣稱爲界(↓)，它包含的生物所具有的體平面不同於其他任何大分類羣的生物。每個界根據多樣性(第213頁)，又可以進一步分爲若干個門，如此等等。按照從最大分類羣到最基本的分類羣的順序排列，各主要等級的名稱分別爲界、門、綱、目、科、屬(↑)和種(↑)。

界(名)　最高的等級(↑)或分類單元(↑)。最爲簡單的分類法是可以將所有的生物或者歸入植物界，或者歸入動物界。然而，這種做法過於簡單化。在本書中，我們將生物分爲五個界：原核生物界(第42頁)、原生生物界(第44頁)、眞菌界(第46頁)、植物界和動物界。

人爲分類　一種分類法(↑)。採用這種分類法對各種生物進行分類時，是以它們表現的同功(第211頁)相似性爲基礎的，而實際上，這些生物表現的同功相似性中，並未包含共同的祖先。

自然分類　一種分類法(↑)。採用這種分類法對各種生物進行分類時，是以它們顯示具有共同祖先的同源(第211頁)相似性爲基礎的。

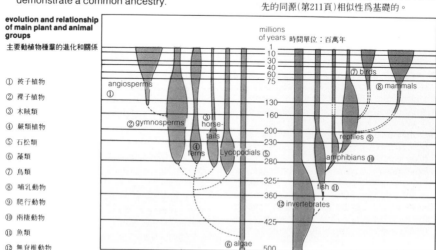

evolution and relationship of main plant and animal groups
主要動植物種羣的進化和關係

① 被子植物
② 裸子植物
③ 木賊類
④ 蕨類植物
⑤ 石松類
⑥ 藻類
⑦ 鳥類
⑧ 哺乳動物
⑨ 爬行動物
⑩ 兩棲動物
⑪ 魚類
⑫ 無脊椎動物

microbiology (*n*) the study or science of very small (microscopic p. 9) or submicroscopic living organisms. It includes bacteriology and virology.

Monera (*n*) the kingdom (p. 41) of prokaryotic (p. 8) organisms which includes the bacteria (↓) and blue-green algae (↓).

bacteria (*n.pl.*) a group of microscopic (p. 9) prokaryotic (p. 8) organisms that may be unicellular (p. 9) or multicellular (p. 9). They lack organelles (p. 8) bounded by membranes (p.14) and contain no large vacuoles (p. 11). Most bacteria are heterotrophic (p. 92) but some are autotrophic (p. 92). Their respiration (p. 112) may be either aerobic (p. 32) or anaerobic (p. 32). Bacteria reproduce (p. 173) mainly by asexual cell division. Heterotrophic bacteria may cause disease. They are important in the decay of plant and animal tissue (p. 83) to release food materials for higher plants and in sewage breakdown. **bacterium** (*sing.*).

bacillus (*n*) a rod-shaped bacterium (↑). **bacilli** (*pl*).

coccus (*n*) a spherical-shaped bacterium (↑). **cocci** (*pl*).

streptococcus (*n*) a coccus (↑) which occurs in chains.

staphylococcus (*n*) a coccus (↑) which occurs in clusters.

spirillum (*n*) a spiral-shaped bacterium (↑). **spirilla** (*pl*).

Gram's stain a stain used in the study of bacteria (↑). Bacteria which take the violet stain are gram-positive while others that do not are gram-negative. Gram-positive bacteria are more readily killed by antibiotics (p. 233).

myxobacterium (*n*) a bacillus (↑) that has a delicate flexible cell wall (p. 8) and is able to glide along solid surfaces.

spirochaete (*n*) a spirillum (↑) which is able to move by flexing its body. Some are parasitic (p. 92) and cause diseases such as syphilis.

rickettsia (*n*) any of the various bacilli (↑) which live as parasites (p. 110) on some arthropods (p. 67) and which can be transmitted to humans causing diseases such as typhus.

微生物學（名） 研究很小的（顯微鏡可見的（第9頁））或亞顯微的生物的科學。微生物學包括細菌學和病毒學。

原核生物界（名） 包括細菌（↓）和藍藻（↓）在內的原核（第8頁）生物之界（第41頁）。

細菌（名、複） 一羣可以是單細胞的（第9頁）或多細胞的（第9頁）顯微鏡可見的（第9頁）原核（第8頁）生物。它們缺乏以膜（第14頁）爲界的細胞器（第8頁），且不含大的液泡（第11頁）。大多數細菌是異養的（第92頁），但有些細菌是自養的（第92頁）。它們的呼吸（第112頁）或許是需氧的（第32頁），或許是缺氧的（第32頁）。細菌主要靠無性的細胞分裂進行繁殖（第173頁）。異養細菌可引起疾病；它們對動植物組織（第83頁）的腐敗，釋放出高等植物所需要的養料，以及對污水分解都起着重要的作用。（單數形式爲 bacterium）

桿菌（名） 一種桿狀細菌（↑）。（複數形式爲 bacilli）

球菌（名） 一種球狀細菌（↑）。（複數形式爲 cocci）

鏈球菌（名） 排列成鏈狀的一種球菌（↑）。

葡萄球菌（名） 排列成串狀的一種球菌（↑）。

螺菌（名） 一種螺旋狀細菌（↑）。（複數形式爲 spirilla）

革蘭氏染劑 一種用於研究細菌（↑）的染色劑。染上該紫色染料的細菌爲革蘭氏陽性細菌，而未染上該紫色染料的其他一些細菌則爲革蘭氏陰性細菌。革蘭氏陽性細菌更容易被抗菌素（第233頁）所殺滅。

黏細菌（名） 具有纖薄韌性的細胞壁（第8頁）、能夠沿着固體表面滑動的一種桿菌（↑）。

螺旋體（名） 能夠靠菌體蠕動而運動的一種螺菌（↑）。某些螺旋體是寄生的（第92頁），可引起像梅毒那樣的一些疾病。

立克次氏體（名） 作爲寄生物（第110頁）寄生在某些節肢動物（第67頁）身體上的各種桿菌（↑）中的任何一種。它可以傳染給人體，引起像斑疹傷寒那樣的一些疾病。

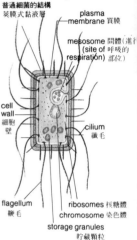

structure of a generalized bacterium
普通細菌的結構

capsule or slime layer 莢膜式黏液層
plasma membrane 質膜
mesosome 間體（進行 (site of 呼吸的 respiration) 部位）
cell wall 細胞壁
cilium 纖毛
flagellum 鞭毛
ribosomes 核糖體
chromosome 染色體
storage granules 貯藏顆粒

bacteria 細菌

myxobacterium 黏細菌

bacillus 桿菌

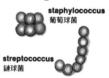

coccus 球菌

staphylococcus 葡萄球菌

streptococcus 鏈球菌

spirillum 螺菌

filamentous blue green algae
絲狀藍藻

actinomycete (*n*) a soil-dwelling gram-positive
(↑) bacterium (↑) with its cells arranged in
filaments (p. 181). It may be used to produce
antibiotics (p. 233) such as streptomycin.

pathogen (*n*) any parasitic (p. 92) bacterium (↑),
virus (↓), or fungus (p. 44) which produces
disease.

toxin (*n*) a poison produced by a living organism,
especially a bacterium (↑), and which may
cause the symptoms of disease which results
from the action of a pathogen (↑). **toxic** (*adj*).

blue-green algae a group of microscopic (p. 9)
prokaryotic (p. 8) organisms known as
Cyanophyta. They contain chlorophyll (p. 12)
and other pigments (p. 126), and are widely
distributed wherever water is present. Some
are able to take in (fix) atmospheric nitrogen
into organic compounds (p. 15).

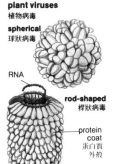

plant viruses
植物病毒
spherical
球狀病毒

RNA

rod-shaped
桿狀病毒

protein
coat
蛋白質
外殼

virus (*n*) a pathogen (↑) which may or may not be
a living organism and which is so small that it
can only be observed with the aid of an electron
microscope (p. 9). It has no normal organelles
(p. 8). A virus will only grow within its host (p. 111)
but occurs as non-living chemicals outside.
Viruses are often named after their hosts to
which they are specific and the symptoms they
cause, for example, tobacco mosaic virus. A
virus consists of an outer protein (p. 21) coat
which surrounds a core of nucleic acids (p. 22).

bacteriophage (*n*) a virus (↑) which infects a
bacterium (↑). It consists of a head, which
contains its DNA (p. 24) or RNA (p. 24),
enclosed by a protein (p. 21) coat, and a tail
which ends in a plate that bears a number of
tail fibres (p. 143). **phage** (*abbr*).

放線菌（名） 其細胞排列成絲狀體（第181頁）的一
種土壤革蘭氏陽性（↑）細菌（↑）。這種細菌可
用來生產鏈黴素之類抗菌素（第233頁）。

病原體（名） 可引起疾病的任何寄生（第92頁）細菌
（↑）、病毒（↓）或眞菌（第44頁）。

毒素（名） 由生物特別是由細菌（↑）產生的一種毒
物。這種毒物可引起由病原體（↑）作用而產生
的一些疾病症狀。（形容詞形式為 toxic）

藍藻 被稱為藍藻門的一羣顯微鏡可見的（第9頁）
原核（第8頁）生物。藍藻含有葉綠素（第12頁）
和其他色素（第126頁），廣泛分佈於有水的地
方。某些藍藻還能呼吸（固定）空氣中的氮，將
其變為有機化合物（第15頁）。

病毒（名） 一種可以是生物，也可以不是生物的病
原體（↑）。這種病原體是如此之小，以致只有
借助電子顯微鏡（第9頁）方可對其進行觀察。
它無正常的細胞器（第8頁）。病毒僅能生長於
其寄主（第111頁）體內，但也可作爲無生命的
化學物質存在於外界。病毒常常是以其特定的
寄主及其所引起的症狀來命名的，例如烟草花
葉病毒。病毒有一個蛋白質（第21頁）外殼，它
包圍着核酸（第22頁）組成的核心部份。

噬菌體（名） 能感染細菌（↑）的一種病毒（↑）。它
由一個頭部和一根尾巴構成。其頭部含有被蛋
白質（第21頁）外殼包圍的 DNA（第24頁）或
RNA（第24頁）；其尾巴的端部爲一長有若干
尾絲（第143頁）的盤狀結構。（英文縮寫爲
phage）

life cycle of a bacteriophage 噬菌體的生活史

bacteriophage
噬菌體

head
頭部

tail
尾巴

tail
fibres
尾絲

1 bacteriophage attacking
a bacterium

bacterium
cell wall
細菌細
胞壁

nucleic
acid
injected
注入的
核酸

1. 噬菌體侵襲細菌

2 parts of new
bacteriophages
synthesized in
bacterial cell

3 bacterium destroyed,
new bacteriophages
released

2. 在細菌細胞中
合成的新的
噬菌體組成部份

3. 細菌溶解，
新噬菌體釋放

Protista (n) a kingdom (p. 41) of unicellular (p. 9) eukaryotic (p. 9) organisms, some of which have been previously allocated to the plant or animal kingdoms, or even both, e.g. Protozoa (↓) and unicellular algae (↓). **protistan** (adj).

binary fission asexual reproduction (p. 173) in which a single parent organism gives rise to two daughter organisms. The nucleus (p. 13) divides by mitosis (p. 37) followed by the division of the cytoplasm (p. 10).

algae (n.pl.) organisms, with a unicellular (p. 9) or simple multicellular (p. 9) body plan, that are able to manufacture their own food material by photosynthesis (p. 93). Unicellular types belong to the Protista (↑) while multicellular types, e.g. seaweeds, are regarded as plants. **alga** (sing.).

phycology (n) the science or study of algae (↑).

Protozoa (n) a division of the Protista (↑) in which the microscopic (p. 9) organisms are unicellular (p. 9), exist as a continuous mass of cytoplasm (p. 10), ingest (p. 98) their food and lack chloroplasts (p. 12) and cell walls (p. 8). Protozoans are widespread and important in natural communities. **protozoan** (adj).

Amoeba (n) a genus (p. 40) of Protozoa (↑). Its members consist of a single motile (p. 173) cell, able to take in food particles by engulfing them using pseudopodia (↓). **amoebae** (pl.).

pseudopodium (n) a temporary protuberance into which the cytoplasm (p. 10) of a protozoan (↑) flows and which enables it to move and feed. **pseudopodia** (pl.).

amoeboid movement the process of locomotion (p. 143) which results from the formation of pseudopodia (↑).

food vacuole a vacuole (p. 11) containing a food particle and a drop of water, engulfed by the pseudopodia (↑) of a protozoan (↑).

ectoplasm (n) the external plasma membrane (p. 13) of a protozoan (↑). A fibrous (p. 143) gel with a less granular structure than the endoplasm (↓) which it surrounds. It takes part in amoeboid movement (↑) and in cell division.

endoplasm (n) the cytoplasm (p. 10) of a protozoan (↑). It is more fluid and granular than ectoplasm (↑)

原生生物界（名） 單細胞（第9頁）眞核（第9頁）生物之界（第41頁）。某些單細胞眞核生物以前一直歸於植物界或動物界，甚至歸於這兩個界，例如原生動物門（↓）和單細胞藻類（↓）。（形容詞形式爲 protistan）

二分裂 一種無性生殖（第173頁）方式。這種生殖方式，可由單個親代生物產生兩個子代生物。親代生物的細胞核（第13頁）藉有絲分裂（第37頁）方式分裂，隨後細胞質（第10頁）分裂。

藻類（名、複） 具單細胞（第9頁）或簡單多細胞（第9頁）體平面的生物。這些生物能藉光合作用（第93頁）製造自身所需資料。單細胞類型藻類屬於原生生物界（↑）；而多細胞類型藻類，如海藻，則被視爲植物。（單數形式爲 alga）

藻類學（名） 研究藻類（↑）的科學。

原生動物門（名） 原生生物界（↑）一門。屬該門的都是在顯微鏡（第9頁）下可見的單細胞（第9頁）生物。它們以細胞質（第10頁）的連續團塊形態存在，沒有葉綠體（第12頁），亦無細胞壁（第8頁），食物依靠自身直接攝取（第98頁）。其分佈廣泛，在自然羣社中佔重要地位。（形容詞形式爲 protozoan）

變形蟲屬、阿米巴屬（名） 原生動物門（↑）的一個屬（第40頁）。其成員均係單個能動的（第173頁）細胞，並能利用僞足（↓）吞噬食物顆粒，將其攝入體內。（複數形式爲 amoebae）

僞足（名） 原生動物（↑）的細胞質（第10頁）流動所形成的一種暫時性突出物。它使原生動物能運動和攝食。（複數形式爲 pseudopodia）

變形運動 由僞足（↑）形成而產生的運動（第143頁）過程。

食物泡 含有一顆食物微粒和一滴水的液泡（第11頁）。該微粒和水是由原生動物（↑）的僞足（↑）所吞噬的。

外質（名） 原生動物（↑）的外質膜（第13頁）。其纖維（第143頁）凝膠的粒狀結構不像由它所包圍着的內質（↓）那樣顯著。外質參與變形運動（↑）和細胞分裂。

內質（名） 原生動物（↑）的細胞質（第10頁）。與外質（↑）相比，內質更富於流動性，其粒狀結構也更爲顯著。

binary fission in a bacterium
細菌的二分裂

pair of chromosomes
成對染色體

replicate chromosome 複製染色體

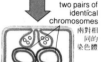

two pairs of identical chromosomes
兩對相同的染色體

two cells identical 與親代細胞
to parent cell 相同的兩個細胞

Amoeba 變形蟲

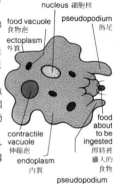

nucleus 細胞核
food vacuole 食物泡
ectoplasm 外質
pseudopodium 僞足
contractile vacuole 伸縮泡
endoplasm 內質
food about to be ingested 即將被攝入的食物
pseudopodium 僞足

the movement of a cilium
纖毛的運動

forward stroke
前進動作

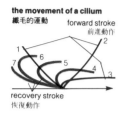

recovery stroke
恢復動作

① 攝食區　⑤ 伸縮泡
② 胞咽　　⑥ 食物泡
③ 小核　　⑦ 大核
④ 纖毛　　⑧ 溝

gel (*n*) a jelly-like material.

granule (*n*) a small particle. **granular** (*adj*).

Paramecium (*n*) a genus (p. 40) of Protozoa (↑). Although it is unicellular (p. 9), its organization is more complex than that of *Amoeba*. It moves by means of cilia (p. 12), it possesses two kinds of nuclei (p. 13), meganuclei (↓) and micronuclei (↓), and it reproduces (p. 173) asexually by transverse binary fission (↑).

草履蟲 *Paramecium*
④cilia
① region of ingestion
② gullet
③ micronucleus
⑤ contractile vacuole
food ⑥ vacuole
meganucleus ⑦
oral groove ⑧

ciliate movement the process of locomotion (p. 143) which involves the beating of stiffened cilia (p. 12) against the water. On the recovery stroke the cilia relax so that they do not push against the water in the reverse direction.

eye spot an organelle (p. 8) which is sensitive to light. It occurs in many Protozoa (↑).

oral groove a ciliated (p. 12) groove in *Paramecium* (↑) into which food particles are drawn by the beating of cilia. It leads to the gullet and the region in which the food is ingested (p. 98).

micronucleus (*n*) the smaller of the two nuclei (p. 13) of *Paramecium* which divides by mitosis (p. 37) and supplies gametes (p. 175) during conjugation (↓).

meganucleus (*n*) the larger of the two nuclei (p. 13) of *Paramecium* (↑) which is concerned with making protein (p. 21) for the organism.

conjugation (*n*) a process of sexual reproduction (p. 173) in *Paramecium* (↑) and other Protozoa (↑) in which two cells temporarily come together and exchange gametes (p. 175).

Euglena (*n*) a genus (p. 40) of Protista (↑). It moves by means of a flagellum (p. 12) and reproduces (p. 173) by transverse binary fission (↑). It has no rigid cell wall (p. 8) but an elastic transparent pellicle. It contains chloroplasts (p. 12) by which it produces its own food substances by photosynthesis (p. 93), but is also able to ingest (p. 98) food through a gullet.

Euglena
眼蟲藻
flagellum
鞭毛
gullet
胞咽

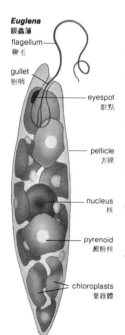

eyespot
眼點

pellicle
表膜

nucleus
核

pyrenoid
澱粉核

chloroplasts
葉綠體

凝膠（名）　一種膠狀物質。

粒、顆粒（名）　一種小的粒子。（形容詞形式為 granular）

草履蟲屬（名）　原生動物門（↑）的一個屬（第40頁）。雖然草履蟲屬是單細胞（第9頁）生物，但其結構比變形蟲屬複雜。草履蟲靠纖毛（第12頁）進行運動。它具有兩種核（第13頁），即大核（↓）和小核（↓）；藉橫向二分裂（↑）進行無性生殖（第173頁）。

纖毛運動　藉剛性纖毛（第12頁）在水中顫動而進行的運動（第143頁）過程。在做恢復動作時，纖毛鬆弛，使水在相反方向不再受到推斥。

眼點　對光敏感的一種細胞器（第8頁）。許多原生動物（↑）都有眼點。

口溝　草履蟲（↑）的有纖毛（第12頁）溝。纖毛顫動將食物顆粒吸入口溝內。口溝通向胞咽和攝食（第98頁）區。

小核（名）　草履蟲（↑）的兩個核（第13頁）中較小的一個核。它在接合（↓）過程中藉有絲分裂（第37頁）方式分裂，並提供配子（第175頁）。

大核（名）　草履蟲（↑）的兩個核中較大的一個核。它與該生物的蛋白質（第21頁）製造有關。

接合（名）　草履蟲（↑）和其他原生動物（↑）中的一種有性生殖（第173頁）過程。在這種過程中，兩個細胞暫時連接在一起，進行配子（第175頁）交換。

眼蟲藻屬、裸藻屬（名）　原生生物界（↑）的一個屬（第40頁）。裸藻靠鞭毛（第12頁）進行運動，藉橫向二分裂（↑）進行生殖（第173頁）。它無硬質的細胞壁（第8頁），但有一層具有彈性的透明表膜。它含有葉綠體（第12頁）可藉光合作用（第93頁）產生其自身的食物，而且也可以通過一個胞咽消化（第98頁）食物。

mycology (*n*) the science or study of fungi (↓).
Fungi (*n*) a kingdom (p. 41) of eukaryotic (p. 9) organisms that are unable to make food material by photosynthesis (p. 93). Instead they take up all their nutrients (p. 92) from their surroundings. They may be microscopic (p. 9) or quite large. They may be unicellular (p. 9) or made up of hyphae (↓). They live either as saprophytes (p. 92) or parasites (p. 110) of plants and animals. Fungi may reproduce (p. 173) sexually and asexually. **Fungus** (*sing*).
hypha (*n*) a branched, haploid (p. 36) filament (p. 181) which is the basic unit of most fungi (↑). It is a tubular structure composed of a cell wall (p. 8) with a lining of cytoplasm (p. 10) and surrounding a vacuole (p. 11). In some fungi, the hyphae (*pl*) may be divided by cross walls or septa. The cell wall is composed mainly of the material chitin (p. 49).
mycelium (*n*) a mass of hyphae (↑) which make up the bulk of a fungus (↑). **mycelia** (*pl*).

眞菌學(名) 研究眞菌(↓)的科學。
眞菌(名) 不能利用光合作用(第93頁)製造食物的眞核(第9頁)生物之界(第41頁)。眞菌從其周圍環境中攝取它們所需要的一切養料(第92頁)。它們之中有的在顯微鏡(第9頁)下方可見到,有的則相當大;有的是單細胞的(第9頁),有的則是由菌絲(↓)構成的。它們作爲植物和動物的腐生菌(第92頁)或寄生菌(第110頁)而存在。眞菌可以進行有性生殖(第173頁)和無性生殖。(單數形式爲 fungus)
菌絲(名) 一種呈分枝狀的單倍(第36頁)絲狀體(第181頁)。這種絲狀體是大多數眞菌(↑)的基本單元。它是一種管狀結構,由一個細胞的細胞壁(第8頁)以及其內的細胞質(第10頁)和被細胞質圍繞着的液泡(第11頁)構成。在某些眞菌中,菌絲可由橫壁或隔膜分開。細胞壁主要由幾丁質(第49頁)組成。
菌絲體(名) 組成眞菌(↑)主體的許多菌絲(↑)的集合體。(複數形式爲 mycelia)

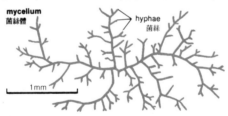

mycellum 菌絲體
hyphae 菌絲
1mm

coenocytic (*adj*) of hyphae (↑) which consist of tubular masses of protoplasm (p. 10) containing many nuclei (p. 13).
dikaryon (*adj*) of a hypha (↑) or mycelium (↑) made up of cells containing two haploid (p. 36) nuclei (p. 13) which divide simultaneously when a new cell is formed.
Phycomycetes (*n.pl.*) a group of Fungi (↑) which possess hyphae (↑) without septa (cross walls). Phycomycetes reproduce (p. 173) sexually by means of zygospores (↓) and asexually by means of zoospores (↓). This group includes the large genus (p. 40) of pin moulds, *Mucor*, and the related genus *Rhizopus*.

多核的、無隔多核的(形) 指菌絲(↑)而言,它由含有許多核(第13頁)的管狀原生質(第10頁)體組成。
雙核的、雙核體的(形) 指由含有兩個單倍體(第36頁)核(第13頁)的一些細胞組成的菌絲(↑)或菌絲體(↑)而言。這些單倍體核的分裂與新細胞形成同時進行。
藻狀菌綱(名、複) 具有菌絲(↑)而無隔膜(橫壁)的一羣眞菌(↑)。藻狀菌綱藉接合孢子(↓)進行有性生殖(第173頁),並藉游動孢子(↓)進行無性生殖。這羣眞菌包括毛霉屬這個大屬(第40頁)和相關屬——根霉屬。

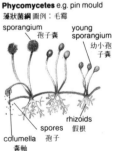

Phycomycetes e.g. pin mould
藻狀菌綱 圖例:毛霉
sporangium 孢子囊
young sporangium 幼小孢子囊
rhizoids 假根
spores 孢子
columella 囊軸

接合孢子的形成階段
zygospores
stages in formation

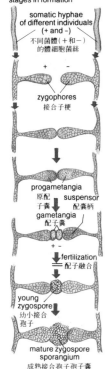

somatic hyphae
of different individuals
(+ and -)
不同菌體（＋和－）
的體細胞菌絲

+ -

zygophores
接合子梗

progametangia
原配 suspensor
子囊 配囊柄
gametangia
配子囊

+ -
fertilization
配子融合

young
zygospore
幼小接合
孢子

mature zygospore
sporangium
成熟接合孢子孢子囊

**asci and ascospores
of Ascomycetes**
子囊菌綱的子囊和
子囊孢子
asci each
with 8
ascospores
每一個子囊含
有8個子囊孢子

ascospores
released
explosively
from asci
從子囊中彈出
的子囊孢子

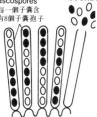

homothallic (*adj*) of the sexual reproduction
(p. 173) of certain fungi (↑) and algae (p. 44) in
which a single thallus (p. 52) produces the
opposite, differently sized gametes (p. 175) to
perform the sexual functions so that the species
is, in effect, hermaphrodite (p. 175).

heterothallic (*adj*) of the sexual reproduction
(p. 173) of certain fungi (↑) and algae (p. 44) in
which reproduction can only take place
between two genetically different thalli (p. 52)
which cannot reproduce independently. In
some fungi, the two thalli may be different in
form so that they are either male or female
while in others there may be no difference in
form but the gametes (p. 175) are different in
size between the two genetically different
strains of the same species (p. 40).

zygospore (*n*) a thick-walled resting spore
(p. 178) produced by a phycomycete (↑) during
sexual reproduction (p. 173) by the fusion of
two gametes (p. 175) called gametangia.

zoospore (*n*) the naked, flagellate (p. 12) spore
(p. 178) produced in a sporangium (p. 178)
during asexual reproduction (p. 173).

Ascomycetes (*n.pl.*) a group of Fungi (↑) which
possess hyphae (↑) with septa. Ascomycetes
reproduce (p. 173) sexually by means of
ascospores (↓) and asexually by means of
conidia (↓). *Penicillium* is an important genus
(p. 40) of Ascomycetes from which antibiotics
(p. 233) are manufactured.

ascus (*n*) a near-cylindrical or spherical cell in
which ascospores (↓) are formed. A number of
asci (*pl.*) may be grouped together into a fruit
body which is visible to the naked eye.

ascospore (*n*) the spore (p. 178) which forms in
the ascus (↑) as a result of the fusion of haploid
(p. 36) nuclei (p. 13) followed by meiosis (p. 38)
to restore the haploid state. Normally, each
ascus contains eight ascospores.

septum (*n*) a wall across a hypha (↑). **septa** (*pl.*).

conidium (*n*) a spore (p. 178) or bud which is
produced during asexual reproduction (p. 173)
from the tips of particular hyphae (↑). **conidia**
(*pl.*). See diagram on p. 48.

同宗配合的(形) 指某些眞菌(↑)和藻類(第44頁)
的有性生殖(第173頁)而言，在這些眞菌和藻
類中，單個葉狀體(第52頁)可產生對生的、大
小不同的配子(第175頁)，以進行有性生殖，
因此，此物種實際上是雌雄同體(第175頁)。

異宗配合的(形) 指某些眞菌(↑)和藻類(第44頁)
的有性生殖(第173頁)而言，在這些眞菌和藻
類中，只能在不能獨立進行生殖的兩個在遺傳
意義上不相同的葉狀體(第52頁)之間發生生
殖。在某些眞菌中，這兩個葉狀體的形態可能
是不相同的，因此，它們或者是雄性的，或者
是雌性的；而在其他一些眞菌中，有一種
(第40頁)的兩個遺傳意義上，不相同的株系之
間，葉狀體的形態上可能沒有差別，但這些配
子(第175頁)的大小是不相同的。

接合孢子(名) 藻狀菌(↑)在有性生殖(第173頁)
過程中，經稱爲配子囊的兩個配子(第175頁)
的融合而產生的一種厚壁休眠孢子(第178
頁)。

游動孢子(名) 在無性生殖(第173頁)過程中，由孢
子囊(第178頁)產生的裸露的、具鞭毛(第12
頁)的孢子(第178頁)。

子囊菌綱(名、複) 具有單隔膜菌絲(↑)的一羣眞
菌(↑)。子囊菌綱藉子囊孢子(↓)進行有性生
殖(第173頁)，並藉分生孢子(↓)進行無性生
殖。青霉屬是子囊菌綱中一個重要的屬(第40
頁)。由它可產生抗菌素(第233頁)。

子囊(名) 一種近似圓柱形或球形的細胞。子囊孢
子(↓)在其內形成。若干子囊可以聚集成肉眼
可見的子實體。

子囊孢子(名) 在子囊(↑)內形成的孢子(第178
頁)。它是由於單倍體(第36頁)核(第13頁)融
合以及後隨使單倍體狀態恢復的減數分裂(第
38頁)而產生的。在正常情況下，每一個子囊
含有8個子囊孢子。

隔膜(名) 橫穿菌絲(↑)的壁。(複數形式爲
septa)。

分生孢子(名) 在無性生殖(第173頁)過程中，從
特定菌絲(↑)的頂端產生的孢子(第178頁)或
孢芽。(複數形式爲 conidia)。圖見第48頁。

Basidiomycetes (*n.pl.*) a group of Fungi (p.46) which possess hyphae (p.46) with septa. Hyphae are often massed into substantial fruit bodies such as mushrooms (↓) or toadstools (↓). They reproduce (p. 173) sexually by basidiospores (↓). *Agaricus*, including the field mushroom, is a genus (p.40) of this group.

擔子菌綱（名、複） 具有帶隔膜的菌絲（第46頁）的一羣眞菌（第46頁）。菌絲常常聚集成堅實的子實體，例如蘑菇（↓）和毒菌（↓）。它們藉擔子（↓）進行有性生殖（第173頁）。蘑菇屬是擔子菌綱的一個屬（第40頁），野蘑菇即歸於此屬。

conidia e.g. *Penicillium*
分生孢子圖例：青霉菌

chains of conidia
分生孢子鏈

mushrooms and toadstools
蘑菇和毒菌

mushroom (*n*) the common name for the fruit body of Basidiomycetes (↑) belonging to the order Agaricales. The name is usually used for those species (p. 40) that are good to eat.

toadstool (*n*) the common name for the fruit body of Basidiomycetes (↑) belonging to the order Agaricales and which are not referred to as mushrooms (↑). It is not necessarily a poisonous species (p. 40).

basidiospore (*n*) a haploid (p. 36) spore (p. 178) produced following sexual reproduction (p. 173) and meiosis (p. 38) and borne externally on the fruit bodies of Basidiomycetes (↑).

basidium (*n*) a club-shaped or cylindrical cell on which the basidiospores (↑) are borne on short stalks, usually four at a time. **basidial** (*adj*).

sterigmata (*n.pl.*) the stalks on a basidium (↑) on which the basidiospores (↑) are borne. Each basidial cell usually bears four sterigmata.

cap (*n*) the umbrella-shaped structure which crowns the central stem of the larger fungi (p. 46) forming the fruit body and in which the spores (p. 178) are produced.

pileus (*n*) = cap (↑).

蘑菇、蕈（名） 擔子菌綱（↑）蘑菇目的子實體的俗名。此名通常用於那些可食用的種（第40頁）。

毒菌（名） 擔子菌綱（↑）蘑菇目的子實體的俗名。但並不將其稱爲蘑菇（↑）。毒菌並非必然能夠全部劃歸成爲一個有毒的種（第40頁）。

擔孢子（名） 擔子菌綱（↑）在有性生殖（第173頁）和減數分裂（第38頁）之後產生的一種單倍體（第36頁）孢子（第178頁）。它着生在子實體的外部。

擔子（名） 一種杵狀或圓柱狀細胞。其短柄上着生擔孢子（↑），通常一次長有四個擔孢子。（形容詞形式爲 basidial）

擔孢子梗（名、複） 擔子（↑）上的小柄。其上着生擔孢子（↑）。每一個擔子細胞通常長有四個擔孢子梗。

菌帽（名） 一種傘形結構。它存在於形成子實體的較大眞菌（第46頁）的中央莖的頂端。孢子（第178頁）產生於其內。

菌蓋（名） 同菌帽（↑）。

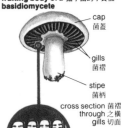

fruiting body of a 擔子菌的子實體
basidiomycete

cap
菌蓋

gills
菌褶

stipe
菌柄

cross section 菌褶
through 之橫
gills 切面

gill 菌褶

basidia
擔子

basidia,
each with
4 basidiospores
每一個擔子上長
有四個擔孢子
sterigmata
擔孢子梗

basidium
擔子柄

basidiospores
擔孢子

gills (*n.pl.*) the fin-like structures which occur on the underside of the cap (↑) of the fruit body of a fungus (p. 46). They bear the spore- (p. 178) producing cells or basidia (↑).

yeast (*n*) unicellular (p. 9) fungi (p. 46) which are very important in brewing and baking and as sources of proteins (p. 21) and minerals. Most yeasts are Ascomycetes (p. 47).

Fungi Imperfecti a loose grouping of fungi (p. 46) which only reproduce (p. 173) asexually.

rust (*n*) a parasitic (p. 92) basidiomycete (↑) fungus (p. 46). Rusts are serious pests of crops and may cause huge losses.

blight (*n*) a disease of plants, such as potatoes, which results from the rapid spread of the hyphae (p. 46) of fungi (p. 46), such as *Phytophthora*, through the leaves of the host (p. 110).

chitin (*n*) a horny material found in the cell walls (p. 8) of many fungi (p. 46) and composed of polysaccharides (p. 18). It is similar to the material that protects the bodies of insects (p. 69).

mycorrhiza (*n*) the symbiotic (p. 228) association (p. 227) which may occur between a fungus (p. 46) and the roots of certain higher plants, especially trees.

lichen (*n*) a symbiotic (p. 228) association (p. 227) of an alga (p. 44) and a fungus (p. 46) to form a slow-growing plant which colonizes (p. 221) such inhospitable environments (p. 218) as rocks in mountainous areas or the trunks of trees.

slime mould widely distributed fungi (p. 46) consisting of masses of protoplasm (p. 10) containing many nuclei (p. 13) and occurring in damp conditions. They reproduce by means of spores (p. 178) and are often classified (p. 40) with fungi. During part of their life history, slime moulds are able to undertake amoeboid movement (p. 44).

菌褶（名、複）　眞菌（第46頁）子實體菌蓋（↑）下側的鰭狀結構。菌褶上長有產生孢子（第178頁）的細胞或擔子（↑）。

酵母（名）　單細胞的（第9頁）眞菌（第46頁）。它們在釀造和焙烤中具有非常重要的作用，也是蛋白質（第21頁）和礦物質的來源。大多數酵母屬於囊菌綱（第47頁）。

半知菌類　僅進行無性生殖（第173頁）的一個鬆散類羣的眞菌（第46頁）。

銹病菌（名）　一種寄生的（第92頁）擔子菌綱的眞菌（第46頁）。銹病菌是農作物的大敵，可造成巨大的損失。

疫病（名）　馬鈴薯之類植物的一種病害。這種病害是由諸如疫霉屬眞菌（第46頁）的菌絲（第46頁）通過寄主（第110頁）的葉子迅速蔓延而引起的。

幾丁質、甲殼質（名）　許多眞菌（第46頁）細胞壁（第8頁）中所含由多糖（第18頁）組成的角質物。幾丁質類似於保護昆蟲（第69頁）身體的物質。

菌根（名）　眞菌（第46頁）與某些高等植物尤其是樹木的根之間存在的共生（第228頁）結合體（第227頁）。

地衣（名）　藻類（第44頁）和眞菌（第46頁）的一種共生（第228頁）結合體（第227頁）。這種結合體是一種生長緩慢的植物。它生長（第221頁）在山區岩石之類荒涼環境（第218頁）之中，也可生長在樹幹之上。

黏菌　由含有多個核（第13頁）的原生質（第10頁）團構成的眞菌（第46頁），生長在潮濕的環境中，分佈極爲廣泛。它們藉孢子（第178頁）進行生殖，通常歸入眞菌類。黏菌在其生活史的部份時間中，能作變形運動（第44頁）。

cellular slime mould 細胞狀黏菌

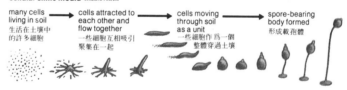

many cells living in soil
生活在土壤中的許多細胞

cells attracted to each other and flow together
一些細胞互相吸引聚集在一起

cells moving through soil as a unit
一些細胞作爲一個整體穿過土壤

spore-bearing body formed
形成載孢體

botany (*n*) the study or science of plants or plant life.

Plantae (*n*) one of the five kingdoms (p. 41) of living organisms containing all plants that are capable of making their own food by photosynthesis (p. 93). It includes the multicellular (p. 9) algae (p. 44), Musci (p. 52), Filicales (p. 56), gymnosperms (p. 57) and angiosperms (p. 57).

Thallophyta (*n*) in the two-kingdom (p. 41) classification (p. 40) of living organisms, a division made up of all those non-animal organisms in which the body is not differentiated into stem, roots, and leaves etc. Reproduction (p. 173) takes place sexually by fusion of gametes (p. 175) and asexually by spores (p. 178). It includes bacteria (p. 42), blue-green algae (p. 43), fungi (p. 46) and lichens (p. 49).

vascular plant a plant which possesses a vascular system (p. 127) to transport water and food materials through the plant and which also provides support for the plant.

植物學(名) 研究植物和植物生命的科學。

植物界(名) 生物的五個界(第41頁)之一。它包含能通過光合作用(第93頁)爲自身製造食物的一切植物。植物界包括多細胞(第 9 頁)藻類(第44頁)、蘚類(第52頁)、眞蕨類(第56頁)、裸子植物(第57頁)和被子植物(第57頁)。

菌藻植物門(名) 在生物的兩界(第41頁)分類法(第40頁)中,由所有那些非動物的生物所組成的一個門。該門中的生物體不分化成莖、根和葉等,藉配子(第175頁)的融合進行有性生殖(第173頁),並藉孢子(第178頁)進行無性生殖。菌藻植物門包括細菌(第42頁)、藍藻(第43頁)、眞菌(第46頁)和地衣(第49頁)。

維管植物 具有維管系統(第127頁)的植物。維管系統可爲植物輸送水份和養料,也能起支持植物的作用。

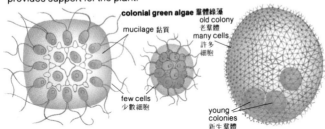

colonial green algae 羣體綠藻
mucilage 黏質
old colony 老羣體
many cells. 許多細胞
few cells 少數細胞
young colonies 新生羣體

Chlamydomonas 衣藻
flagellum 鞭毛
contractile vacuole 伸縮胞
stigma (light detector) 眼點 (感光器)
cell wall 細胞壁
nucleus 核
chloroplast 葉綠體
pyrenoid (food storage area) 澱粉核 (食物貯存區)

Chlorophyta (*n*) a division of mainly multicellular (p. 9) algae (p. 44) in which the plants are mainly freshwater although there are some marine forms. These are the green algae and contain chlorophyll (p. 12) for photosynthesis (p. 93). They store food as starch (p. 18) and fats.

Chlamydomonas (*n*) a unicellular (p. 9) genus (p.40) of Chlorophyta (↑) which is found widely in freshwater ponds. They possess two flagella (p. 12) and a cup-shaped chloroplast (p. 12) contained within the cell wall (p. 8).

綠藻植物門(名) 主要爲多細胞(第 9 頁)藻類(第44頁)的一門。該門的藻類植物雖有一些海生類型,但主要分佈於淡水中。這些藻類植物是綠藻,含有進行光合作用(第93頁)所需的葉綠素(第12頁)。它們以澱粉(第18頁)和脂肪的形式貯存食物。

衣藻屬、單胞藻(名) 綠藻門(↑)中的一個單細胞(第 9 頁)屬(第40頁),廣泛存在於淡水池塘中。它們具有兩根鞭毛(第12頁)和一個包含在細胞壁(第 8 頁)內的杯狀葉綠體(第12頁)。

Spirogyra 水綿

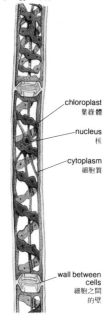

chloroplast
葉綠體

nucleus
核

cytoplasm
細胞質

wall between
cells
細胞之間
的壁

Spirogyra (*n*) a genus (p. 40) of typical filamentous (p. 181) algae of the Chlorophyta (↑). They are found in freshwater and consist of a simple chain of identical cells each containing the characteristic spiral chloroplast (p. 12).

Phaeophyta (*n*) a division of the algae (p. 44) in which the plants are largely marine, such as the large seaweeds of the shore. They contain chlorophyll (p. 12) and the brown pigment (p. 126), fucoxanthin, and they are referred to as the brown algae. They are multicellular (p. 9) and store food as sugars.

Fucus (*n*) a typical genus (p. 40) of the Phaeophyta (↑). They are the common seaweeds of the intertidal zone. Each plant is differentiated into a holdfast for clinging to the rocks, a tough stalk called a stipe, and flat fronds. They are commonly described as wracks.

bladder[P] (*n*) an air-filled sac which occurs in some members of the Phaeophyta (↑). Plants containing these bladders are commonly called bladderwracks.

conceptacle (*n*) one of the many cavities which occur at the tips of the fronds of some members of the Phaeophyta (↑) such as bladderwracks. They open by a pore (p. 120) called an ostiole, and, as well as sex organs, contain masses of sterile hairs called paraphyses.

水綿屬（名） 綠藻門（↑）中典型的絲狀（第181頁）藻類的一個屬（第40頁）。它們分佈於淡水中；皆由一些相同連接的單一鏈構成。每一個細胞都含有特有的螺旋狀葉綠體（第12頁）。

褐藻植物門（名） 藻類（第44頁）的一個門，褐藻門的藻類植物大多數分佈於海水中，例如海岸邊一些的大海藻。它們含有葉綠素（第12頁）和褐色素（第126頁）——墨角藻黃素，被稱爲褐藻，它們爲多細胞（第9頁）藻類植物，以糖的形式貯存食物。

墨角藻屬（名） 褐藻門（↑）中的一個典型的屬（第40頁）。這些藻類植物是潮間帶的普通海藻。每一種藻類植物均分化出一種附着於岩石的固着器，一種稱爲葉柄的堅韌的柄和扁平的藻體。它們通常被稱作墨角藻。

氣囊[P]（名） 褐藻門（↑）某些成員中具有的一種充氣囊。含有這些氣囊的藻類植物，通常被稱爲囊褐藻。

生殖窩（名） 褐藻門（↑）的某些成員如囊褐藻的藻體頂端含有的許多空腔。它們的開口是一個稱爲孔口的小孔（第120頁），其內含有大量稱爲隔絲的不育性絲狀體，以及雌、雄性器官。

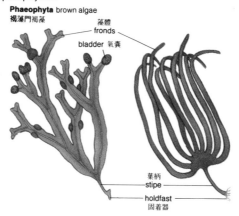

Phaeophyta brown algae
褐藻門褐藻

藻體
fronds

bladder 氣囊

葉柄
stipe

holdfast
固着器

Bryophyta (*n.pl.*) a division of the Plantae (p. 50) including the Hepaticae (↓). Anthocerotae (↓), and the Musci (↓). Bryophytes lack vascular tissues (p. 83) although the stems of some mosses have a central strand of conducting tissue. They are mostly plants of warm places but some are aquatic; others live in desert habitats (p. 217) or cold places and may sometimes be the dominant form of plant life. All bryophytes show a clear alternation of generations (p. 176) with a conspicuous, food-independent gametophyte (p. 177) generation and a short-lived sporophyte (p. 177) generation dependent on the gametophyte. They are small flattened plants with leaves and a stem but no roots and are attached by a rhizoid (↓).

Hepaticae (*n.pl.*) liverworts. A family of the Bryophyta (↑). These are the simplest bryophytes and may be either a flattened, leafless gametophyte (p. 177), a thallose (↓) liverwort, or a creeping, leafy gametophyte known as a leafy liverwort. A typical thallose liverwort is ribbon-like and has Y-shaped branches. They are usually aquatic, living in damp soil or as epiphytes (p. 228).

Anthocerotae (*n.pl.*) hornworts. A family of the Bryophyta (↑). The plant consists of a lobed, green thallus (↓) anchored by a rhizoid (↓) to the substrate which is usually moist soil or mud:

Musci (*n.pl.*) mosses. A family of the Bryophyta (↑). These are the most advanced bryophytes and they either grow in erect, virtually unbranched cushions or in feathery, creeping, branched mats. They are widely distributed throughout the world living in damp conditions, such as in woodland, or they may be aquatic, may even survive in drier conditions, such as on walls or roofs of houses.

thallus (*n*) a general term for the plant body which is not differentiated into root, stem or leaf e.g. liverworts. **thalloid** (*adj*).

rhizoid (*n*) an elongate, single cell, such as in a liverwort, or a multicellular (p. 9) thread, such as in a moss, that anchors the gametophyte (p. 177) to the substrate. It is not a true root.

苔蘚植物門（名、複） 植物界（第50頁）的一個門，其中包括苔綱（↓）、角苔綱（↓）和蘚綱（↓）。雖然某些蘚類的莖內具有中央束狀輸導組織，但苔蘚植物缺乏維管組織（第83頁）。它們大多數是生長在溫暖地帶的植物，但是，有一些是水生植物；另外一些則生長在荒漠的生境（第217頁）或寒冷的地方，有時可能成爲植物生命的優勢型。所有的苔蘚植物都表現出一種明顯的世代交替（第176頁），具有顯著的、養料自給自足的配子體（第177頁）世代和養料依賴於配子體的短壽命孢子體（第177頁）世代。苔蘚植物都是小而平扁的植物，具有葉和莖，沒有根，但附生假根（↓）。

苔綱（名、複） 即苔類植物。苔蘚植物門（↑）的一個綱。苔綱植物是最簡單的苔蘚植物；它們或者是平扁、無葉的配子體（第177頁）——原植體（↓）苔類植物，或者是匍匐多葉的配子體——稱爲多葉苔類植物。典型的原植體苔類植物是帶狀的，具有 Y 型的分支。苔綱植物通常爲水生植物，生長在潮濕土壤中，或者以附生植物（第228頁）的形式出現。

角苔綱（名、複） 即角苔類。苔蘚植物門（↑）的一個綱。其植物體爲分裂的綠色葉狀體（↓），藉假根（↓）固着在通常是潮濕土壤或淤泥的地層上。

蘚綱（名、複） 即蘚類。苔蘚植物門（↑）的一個綱。蘚綱植物是最高級的苔蘚植物；它們或者長成直立的、實際上是無分支的鋪地植物，或者長成羽毛狀的、匍匐的鋪地植物。蘚綱植物廣泛分佈於世界各地；它們生活在潮濕的環境條件下，例如在森林中，也可以生活在水中，甚至還能在較爲乾燥的環境條件下生存，例如長在牆上或房頂上。

葉狀體、原植體（名） 不分化爲根、莖和葉的植物體的總稱，例如苔類植物。（形容詞形式爲 thalloid）

假根（名） 如苔類植物中所具有的一種長形單細胞；或者如蘚類植物中所具有的一種多細胞（第 9 頁）絲狀體。假根可使配子體（第177頁）固着在附着層。它並非是眞正的根。

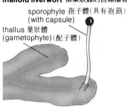

liverwort sporophyte 苔類植物孢子體

capsule containing spores 含有孢子的孢蒴

seta 蒴柄

sporophyte 孢子體

foot embedded in gametophyte tissues 埋入配子體組織的基足

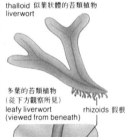

thalloid liverwort 似葉狀體的苔類植物

sporophyte 孢子體（具有孢蒴）(with capsule)

thallus 葉狀體 (gametophyte)（配子體）

rhizoids 假根

thalloid 似葉狀體的苔類植物 liverwort

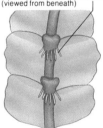

多葉的苔類植物（從下方觀察所見）

leafy liverwort (viewed from beneath) 多葉的苔類植物

rhizoids 假根

protonema 原絲體

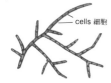
cells 細胞

acrocarpous moss 頂生蘚蘚類

pleurocarpous moss 側生蘚蘚類

foot[P](*n*) the lower part of the sporophyte (p. 171) generation of a bryophyte (↑) which remains embedded in the archegonium (p. 177).

capsule (*n*) (1) the end part of the sporophyte (p. 177) generation of liverworts or mosses which, at maturity, contains the spores (p. 178). (2) a dry fruit, such as that of the poppy, formed from two or more carpels (p. 179) which, during dehiscence (p. 185), open by a variety of slits or pores (p. 120) to release the seeds.

seta[P] (*n*) the stalk of the capsule (↑).

columella (*n*) the central, sterile tissue (p. 83) within the capsule (↑) of liverworts and mosses.

calyptra (*n*) a hood-like structure which covers the capsule (↑) of mosses until mature. It is the remains of the archegonium (p. 177).

operculum[P] (*n*) the lid of a moss capsule (↑) which is shed to reveal the peristome teeth (↓).

peristome teeth a ring of teeth at the tip of the capsule (↑) of mosses which open and close in response to varying levels of moisture – they open when dry and close when moist.

elater (*n*) a spindle-shaped body contained in the capsule (↑) of liverworts. Spiral thickenings change shape with varying moisture levels causing the elaters to flick spores (p. 178) from the capsule.

protonema (*n*) the branching filament (p. 181) that grows from the germinating spore (p. 178) of mosses. It develops buds which grow into the leafy gametophyte (p. 177) generation. Also known as **first thread protonemata** (*pl.*).

基足[P]（名）　處於孢子體（第171頁）世代中的苔蘚植物（↑）的下部。基足始終包埋在頸卵器（第177頁）內。

孢蒴、蒴果、蒴囊（名）　(1)處於孢子體（第177頁）世代中的苔類植物或蘚類植物的端部。成熟時，孢蒴內含有孢子（第178頁）。(2)一種乾果，例如罌粟的乾果，由兩枚或兩枚以上的心皮（第179頁）組成。蒴果開裂（第185頁）時，種子從各種裂縫或孔（第120頁）中釋出。

蒴柄[P]（名）　孢蒴（↑）的柄。

蒴軸（名）　苔類植物或蘚類植物孢蒴（↑）內的中央不育組織（第83頁）。

蒴帽（名）　蘚類植物成熟前覆蓋孢蒴（↑）的一種盔狀結構。蒴帽是頸卵器（第177頁）的殘留物。

蒴蓋[P]（名）　蘚類植物孢蒴（↑）上的蓋狀結構。揭下蒴蓋即顯露出蒴齒（↓）。

蒴齒　蘚類植物孢蒴（↑）頂端的一圈齒。這些齒隨濕度改變而開閉。乾燥時，蒴齒開啟；潮濕時，蒴齒則關閉。

彈絲（名）　苔類植物孢蒴（↑）內所含有的一種棱形體。其螺紋加厚部份可隨濕度變化而變形，從而使彈絲能將孢子（第178頁）從孢蒴中彈出。

原絲體（名）　由蘚類植物的萌發孢子（第178頁）長出的分支絲狀體（第181頁）。原絲體可產生較多的芽，這些芽發育後，即進入多葉的配子體（第177頁）世代。原絲體也稱爲**第一絲狀原絲體**。

paraphyses 隔絲

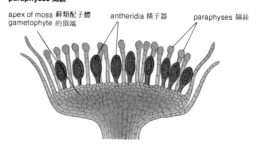
apex of moss 蘚類配子體
gametophyte 的頂端
antheridia 精子器
paraphyses 隔絲

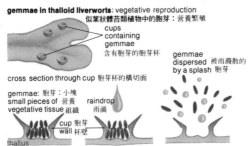

gemmae in thalloid liverworts: vegetative reproduction
似葉狀體苔類植物中的胞芽：營養繁殖

cups containing gemmae 含有胞芽的胞芽杯

cross section through cup 胞芽杯的橫切面

gemmae: 胞芽：小塊
small pieces of 營養
vegetative tissue 組織

raindrop 雨滴

cup 胞芽
wall 杯壁

葉狀體 thallus

gemmae dispersed 被雨濺散的 by a splash 胞芽

liverwort sporophyte discharging spores
可釋放孢子苔類植物孢子體

capsule walls 孢蒴壁

spores 孢子

elaters 彈絲

elater with helical thickenings in cell wall
細胞壁內具有螺紋加厚的彈絲

gemmae (*n.pl.*) minute, lens-shaped bodies produced by liverworts as a means of asexual reproduction (p. 173). **gemma** (*sing.*).

gemmae cup the receptacle or cup-shaped body on the upper surface of the gametophyte (p. 177) generation in liverworts which contains the gemmae (↑).

Pteridophyta (*n.pl.*) a division of the Plantae (p. 50) including the Lycopodiales (↓), Equisetales (↓) and the Filicales (↓). Pteridophytes have a well-developed vascular system (p. 127). They are widely distributed, especially in the tropics, and live mainly on land. There is alternation of generations (p.176) between gametophyte (p.177) and sporophyte (p. 177) phases in which the latter is the most prominent when the plant is differentiated into roots, stems, leaves and rhizomes (p. 174).

vascular cryptogams an alternative name for the Pteridophyta (↑), so called because there is a clear vascular system (p. 127) but no prominent organs of reproduction (p. 173), such as in the angiosperms (p. 57).

homosporous (*adj*) of plants with only a single kind of spore (p. 178) which gives rise to a hermaphrodite (p. 175) generation of gametophytes (p. 177). It occurs in some of the Pteridophyta (↑).

heterosporous (*adj*) of plants with two distinct kinds of spores (p. 178) which give rise to male and female gametophyte (p. 177) generations respectively. It occurs in some Pteridophytes (↑) and is thought to represent an evolutionary (p. 208) step towards the production of seeds.

胞芽（名、複）　苔類植物以一種無性生殖（第173頁）手段而產生的一些微小的透鏡狀體。（單數形式爲 gemma）

胞芽杯　苔類植物配子體（第177頁）上表面處的生殖托或杯狀體。胞芽杯內含有胞芽（↑）。

蕨類植物門（名、複）　植物界（第50頁）的一個門，其中包括石松綱（↓）、木賊綱（↓）和眞蕨綱（↓）。蕨類植物的維管系統（第127頁）十分發達。它們分佈廣泛，尤其是分佈在熱帶地區，主要生長在陸地上。存在着配子體（第177頁）世代和孢子體（第177頁）世代之間的交替（第176頁）。在世代交替中，植物體分化爲根、莖、葉和根狀莖（第174頁），以孢子體世代最爲顯著。

維管隱花植物　蕨類植物（↑）的另一名稱。這是因爲它有明顯的維管系統（第127頁），而沒有像被子植物（第57頁）那樣具有顯著的生殖（第173頁）器官。

具同型孢子的（形）　指只產生單種孢子（第178頁）的一些植物而言。這種孢子可導致配子體（第177頁）雌雄同體（第175頁）世代的發生。它存在於某些蕨類植物（↑）中。

具異型孢子的（形）　指產生兩種不同類孢子（第178頁）的一些植物而言。這兩種孢子可分別導致雄配子體（第177頁）世代和雌配子體世代的發生。它存在於某些蕨類植物（↑）中，被認爲是朝着產生種子方向進化（第208頁）的一個象徵。

維管植物中的孢子同型和孢子異型
homospory and heterospory in vascular plants

homosporous ①
bryophytes ②
some pteridophytes ③ (e.g. ferns)
heterosporous ④
some pteridophytes ⑤ (e.g. club mosses)
gymnosperms ⑥
angiosperms ⑦

① 具同型孢子的　④ 具異型孢子的
② 苔蘚植物　　　⑤ 某些蕨類植物
③ 某些蕨類植物　　　（例如石松類）
　　（例如蕨類）　⑥ 裸子植物
　　　　　　　　⑦ 被子植物

heterosporous plants the production of microspores 具異型孢子的植物 小孢子的產生

♂ gametophyte develops within microspore wall and produces motile ♂ gametes
配子體在小孢子壁內發育，產生游動的配子

anther 花藥

microsporophyll 小孢子葉

microsporangia with microspores (young pollen grains)
具有小孢子的小孢子囊(幼小花粉粒)

cross-section of anther 花藥的橫切面

microsporophyll 小孢子葉

microsporophylls 小孢子葉

microsporangium 小孢子囊

microspore (young pollen grain) 小孢子(幼小花粉粒)

♂ cone 球果

moss 蘚類　　conifer 針葉樹　　angiosperm 被子植物

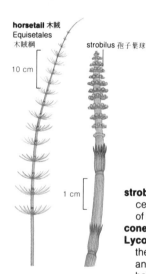

horsetail 木賊
Equisetales
木賊綱

10 cm

1 cm

strobilus 孢子葉球

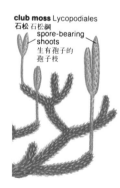

club moss Lycopodiales
石松 石松綱
spore-bearing shoots
生有孢子的孢子枝

strobilus (*n*) the reproductive (p. 173) structure of certain members of the Pteridophyta (↑). It consists of sporophylls (↓) on an axis. **strobili** (*pl.*).

cone[P] = strobilus (↑).

Lycopodiales (*n.pl.*) club mosses. A division of the Pteridophyta (↑). They are an ancient group and even attained tree-like forms. They may be heterosporous (↑) or homosporous (↑) and bear densely packed small leaves on branched stems. They are evergreen.

sporophyll (*n*) a modified leaf that bears a sporangium (p. 178).

Equisetales (*n.pl.*) horsetails. A division of the Pteridophyta (↑). They are an ancient group and even attained tree-like forms. They are characterized by having whorls (p. 83) of small leaves on upright stems with strobili (↑) at the tips. They are homosporous (↑).

microphyll (*n*) a foliage leaf typical of the Lycopodiales (↑) and the Equisetales (↑) which may be very small and has a simple vascular system (p. 127) comprising a single vein running from the base to the apex.

孢子葉球（名）　蕨類植物門（↑）某些成員的生殖（第173頁）結構。孢子葉球由生於莖軸上的孢子葉（↓）構成。（複數形式為 strobili）

球果[P]　同孢子葉球（↑）。

石松綱（名、複）　即石松類。蕨類植物門（↑）的一個綱。石松綱植物是一種古老的植物羣，其形態甚至以喬木狀出現。它們或是具異型孢子的（↑）植物，或是同型孢子的（↑）植物，在分枝的莖幹上長着密集的小葉，為常綠植物。

孢子葉（名）　生有孢子囊（第178頁）的一種變異葉。

木賊綱（名、複）　即木賊類。蕨類植物門（↑）的一個綱。木賊綱植物是一種古老的植物羣，其形態甚至以喬木狀出現。其特徵是在直立的莖幹上長有許多輪（第83頁）狀的小葉，而在這些輪狀小葉的頂端，則生有孢子葉球（↑）。它們是具同型孢子的（↑）植物。

小型葉（名）　石松綱（↑）和木賊綱（↑）的典型的營養葉。這種葉可以長得很小，但有簡單的包括從底部延伸至頂端的單葉脈的維管系統（第127頁）。

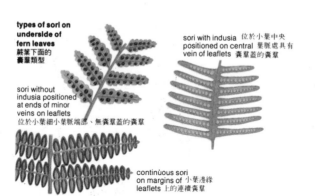

types of sori on underside of fern leaves
蕨葉下面的囊堆類型

sori without indusia positioned at ends of minor veins on leaflets
位於小葉細小葉脈端部、無囊堆蓋的囊堆

sori with indusia 位於小葉中央 positioned on central 葉脈處具有 vein of leaflets 囊堆蓋的囊堆

continuous sori on margins of 小葉邊緣 leaflets 上的連續囊堆

long sori on either side of midrib
在中脈兩側的長列囊堆

ferns 蕨類
Filicales 真蕨綱

simple frond (megaphyll)
單葉（巨型葉）

circinate vernation in young frond
拳卷幼葉捲疊式

compound frond (megaphyll)
複葉（巨型葉）

Filicales (n.pl.) true ferns. A division of the Pteridophyta (p. 54). The plants are characterized by obvious frond-like leaves, often with sporangia (p. 178) on the undersides and rhizomes (p. 174) underground. They are homosporous (p. 54).

megaphyll (n) a large frond-like foliage leaf with a branched system of veins. It is typical of the Filicales (↑).

frond (n) a large well-divided leaf typical of the Filicales (↑).

sorus (n) a reproductive (p. 173) organ made up of a group of sporangia (p. 178) which occur on the undersides of leaves in Filicales (↑).

indusium (n) a flap of tissue (p. 83) covering the sorus (↑). **indusia** (pl.).

annulus (n) an arc or ring of cells in the sporangia (p. 178) of Filicales (↑) which are involved in opening the sporangium on drying to release the spores (p. 178).

circinate vernation the way in which the young fronds (↑) of the Filicales (↑) occur rolled-up.

真蕨綱（名、複） 即真蕨類。蕨類植物門（第54頁）的一個綱。真蕨綱植物的特徵是，具有明顯的似蕨類葉的葉子。在其下面通常生有孢子囊（第178頁），而且，在地下長有根狀莖（第174頁）。它們是具同型孢子的（第54頁）植物。

巨型葉（名） 具有葉脈分支系統的、巨大的、似蕨類葉的營養葉。有巨型葉是真蕨綱（↑）植物的特徵。

蕨葉（名） 真蕨綱（↑）植物的一種典型的、巨大的全裂葉。

孢子囊堆（名） 由一叢孢子囊（第178頁）組成的生殖（第173頁）器官。它存在於真蕨綱（↑）植物葉子的下面。

囊堆蓋（名） 覆蓋囊堆（↑）的一片組織（第83頁）。（複數形式爲 indusia）

環帶（名） 圍繞真蕨綱（↑）植物孢子囊（第178頁）的一列弧形或環形細胞。這些細胞在乾燥狀態下，參與孢子囊的開啟，以釋出孢子（第178頁）。

拳捲幼葉捲疊式 真蕨綱（↑）植物的幼葉（↑）捲疊所採用的方式。

sporangia of ferns 蕨類植物中的孢子囊
fern sorus 蕨的囊堆

sporangia 孢子囊

indusium 囊堆蓋

underside of frond 葉的下面

sporangiophore 孢囊柄

wall of sporangium 孢子囊壁

spores 孢子

annulus 環帶

Spermatophyta (*n.pl.*) seed plants. A division of the Plantae (p. 50) including the Gymnospermae (↓) and the Angiospermae (↓). They are widely distributed and are the dominant plants on land today. The body is highly organized and differentiated into root, stem and leaf, and there is a well-developed vascular system (p. 127). They are heterosporous (p. 54) with a dominant sporophyte (p. 177) generation which is the plant itself. The male gametophyte (p. 177) is the pollen (p. 181) grain while the female gametophyte is the egg which becomes the seed after fertilization (p. 175).

Gymnospermae (*n.pl.*) a division of the Spermatophyta (↑) which includes trees and shrubs in which the seeds are naked and not enclosed in a fruit. Most have cones (p. 55).

Angiospermae (*n.pl.*) flowering plants. A division of the Spermatophyta (↑) which includes the dominant plants on land. They are highly differentiated and have microsporophylls (p. 178) and megasporophylls (p. 179) combined into true flowers as stamens (p. 181) and carpels (p. 179).

Dicotyledonae (*n*) dicotyledons. A class of the Angiospermae (↑) in which the seeds have two cotyledons (p. 168) or seed leaves, the leaves are net veined, the flower parts are usually in multiples of four or five, and the root system includes a tap root (p. 81) with lateral branches. An example of a dicotyledon is the buttercup.

種子植物門（名、複）　即種子植物。植物界（第50頁）的一個門，其中包括裸子植物綱（↓）和被子植物綱（↓）。種子植物分佈廣泛，爲當今陸地上的優勢植物。其植物體結構完善，分化成根、莖和葉，並有相當發達的維管系統（第127頁）。種子植物是具異型孢子的（第54頁）植物，其孢子體（第177頁）世代是佔優勢的，而處於孢子體世代者即該植物的本身。雄配子體（第177頁）是花粉（第181頁）粒，而雌配子體則是卵，受精（第175頁）以後，卵即變成種子。

裸子植物綱（名、複）　種子植物門（↑）的一個綱。它包括各種喬木和灌木，其種子裸露，不包於果實內。大多數裸子植物具有球果（第55頁）。

被子植物綱（名、複）　即有花植物。種子植物門（↑）的一個綱。它包括陸地上的優勢植物。被子植物是高度分化的植物，具有小孢子葉（第178頁）和大孢子葉（第179頁），它們結合在一起形成以雄蕊（第181頁）和心皮（第179頁）形式出現的眞花。

雙子葉植物亞綱（名）　即雙子葉植物。被子植物（↑）綱的一個亞綱。雙子葉植物的種子有兩枚子葉（第168頁），葉上有網狀葉脈；花的各部份通常爲4或5的倍數。根系包括帶有側支的直根（第81頁）。毛茛屬植物爲雙子葉植物的一個例子。

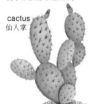

conifer an example of a gymnosperm
針葉樹 裸子植物的一個例子

dictotyledons some examples
雙子葉植物的幾個例子

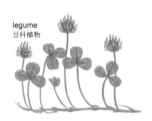

cactus
仙人掌

legume
豆科植物

oak 櫟樹

composite 菊科植物

Monocotyledonae (*n*) monocotyledons. A class of the Angiospermae (p. 57) in which the seeds have one cotyledon (p. 168) or seed leaf, the leaves are usually parallel veined, and the flower parts are usually in multiples of three. Typical monocotyledons are the grasses.

ephemeral (*adj*) of a plant in which the complete life cycle from germination (p. 168) to the production of seed and death is very short so that many generations of the plant may be completed within a single year.

annual (*adj*) of a plant that completes its whole life cycle from the germination (p. 168) of seeds to the production of the next crop of seeds, followed by the death of the plant, within one year.

biennial (*adj*) of a plant that completes its whole life cycle from the germination (p. 168) of seeds to the production of the next crop of seeds, followed by the death of the plant, within two years. During the first year, the plant produces foliage and photosynthesizes (p. 93) to provide an energy store for the reproductive (p. 173) activities of the second year.

perennial (*adj*) of a plant which survives for a number of years and may or may not reproduce (p. 173) within the first year.

單子葉植物亞綱(名)　即單子葉植物。被子植物綱(第57頁)的一個亞綱。單子葉植物的種子有一枚子葉(第168頁)，葉上通常有平行葉脈；花的各部份通常爲3的倍數。禾本科植物是典型的單子葉植物。

短生的(形)　指植物而言，這種植物從發芽(第168頁)到產生種子和死亡的整個生活史十分短暫，所以該植物的許多世代可以在一年之內完成。

一年生的(形)　指植物而言，這種植物的整個生活史，即從種子發芽(第168頁)直到下一茬種子長成，以及繼之而來的植物死亡爲期一年。

二年生的(形)　指植物而言，這種植物的整個生活史，即包括從種子發芽(第168頁)直到下一茬種子長成，以及繼之而來的植物死亡爲期兩年。在第一年，植物長葉和進行光合作用(第93頁)，爲第二年的生殖(第173頁)活動貯存能量。

多年生的(形)　指植物而言，這種植物可生存若干年，在第一年內，或許進行生殖(第173頁)或許不能進行生殖。

monocotyledons 單子葉植物的幾個例子
some examples

grass
禾本科植物

palm
棕櫚

orchid
蘭

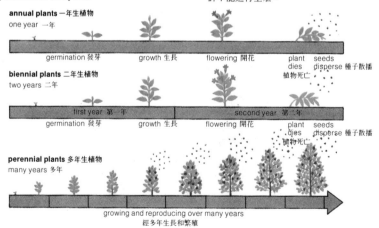

annual plants 一年生植物
one year 一年

germination 發芽　　　growth 生長　　　flowering 開花　　　plant seeds
　　　　　　　　　　　　　　　　　　　　　　　　　　　　　dies disperse 種子散播
　　　　　　　　　　　　　　　　　　　　　　　　　　　　　植物死亡

biennial plants 二年生植物
two years 二年

first year 第一年　　　　　　　　second year 第二年
germination 發芽　　　growth 生長　　　flowering 開花　　　plant seeds
　　　　　　　　　　　　　　　　　　　　　　　　　　　　　dies disperse 種子散播
　　　　　　　　　　　　　　　　　　　　　　　　　　　　　植物死亡

perennial plants 多年生植物
many years 多年

growing and reproducing over many years
經多年生長和繁殖

two types of tree
兩種類型的喬木

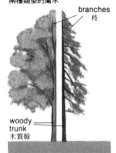

branches
枝

woody
trunk
木質幹

deciduous	evergreen
with	with
sympodial	monopodial
branching	branching
具有合軸	具有總狀
分枝的	分枝的
落葉喬木	常綠喬木

deciduous (*adj*) (1) of a plant that sheds its leaves periodically, in accordance with the season, so that water loss by transpiration (p. 120) is reduced during periods of very dry or cold weather when water is in short supply. (2) in animals, of dentition (p. 104), for example, milk teeth (p. 104) which are shed and replaced by adult teeth.

evergreen (*adj*) of a plant that bears leaves throughout the year and which has adaptations, such as a leathery cuticle (p. 83) or needle-like leaves as in the gymnosperms (p. 57), conifers, to reduce water losses.

herbaceous (*adj*) of a perennial (↑) plant in which the foliage dies back each year while the plant survives as, for example, a bulb (p. 174), corm (p. 174), or tuber (p. 174). Herbaceous plants have no wood in their stems or roots.

tree (*n*) a woody, perennial (↑) plant which usually reaches a height of greater than 4 to 6 metres (13 to 20 feet) and which has a single stem from which branches grow at some distance from the ground level.

sapling (*n*) a young tree.

shrub (*n*) a woody, perennial (↑) plant which is smaller than a tree and from which branches grow quite close to ground level.

climber (*n*) a plant which though rooted in the ground uses other plants to support itself. Climbers use long coiled threadlike tendrils, suckers or adventitious roots (p. 81) to hold on to other plants and sometimes they twist around their stems.

foliage (*n*) all the leaves of a plant together. **foliar** (*adj*).

落葉的；脫落的(形)　(1)指植物而言，這種植物可根據季節定期脫葉，因此，在非常乾燥或寒冷的天氣，當供水不足時，可減少由於蒸騰作用(第120頁)所引起的水份散失；(2)指動物中乳牙(第104頁)之類的出牙而言，乳牙脫落，即由恒牙替代。

常綠的(形)　指植物而言，這種植物一年四季皆長有葉子，對環境有適應性。例如，裸子植物(第57頁)中的針葉樹，具有革質角質層(第83頁)和針狀葉，以減少水份散失。

草本的(形)　指多年生(↑)植物而言，這種植物的葉子每年都枯萎，而植物卻以鱗莖(第174頁)、球莖(第174頁)或塊莖(第174頁)的形式殘存下來。草本植物的莖和根內不含木質。

喬木(名)　一種多年生(↑)木本植物。這種植物通常可長到4至6米(13至20英尺)以上的高度。它有一主幹，在離地面某一高度處從主幹長出分枝。

幼樹(名)　年幼的樹。

灌木(名)　一種多年生(↑)木本植物。這種植物比喬木小，其上長出的分枝十分接近地面。

攀緣植物(名)　一種紮根於地下但卻依賴其他植物支撐自己的植物。攀緣植物利用長而捲曲的線狀捲鬚、吸盤或不定根(第81頁)依附在其他植物上；有時它們圍繞着其他植物的莖幹盤旋而上。

葉子(名)　一種植物所有葉子的總稱。(形容詞形式為 foliar)

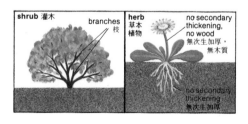

shrub 灌木　　branches
枝

herb
草本
植物

no secondary
thickening,
no wood
無次生加厚，
無木質

no secondary
thickening
無次生加厚

zoology (*n*) the study or science of animals or animal life.

Metazoa (*n*) a term used to describe all those truly multicellular (p. 9) animals as opposed to those animals which belong to the Protozoa (p. 44).

Coelenterata (*n*) a phylum of multicellular (p. 9) invertebrate (p. 75), aquatic and usually marine animals which includes the corals (↓) and jellyfishes. The body is radially symmetrical (↓) and consists of a simple body cavity which opens to the exterior by a mouth which is surrounded by a ring of tentacles (p. 71) that may have stinging cells or nematoblasts and are used for trapping prey and for defence. The body wall consists of an endoderm (p. 166) and an ectoderm (p. 166) separated by a jelly-like mesogloea. Reproduction (p. 173) takes place sexually, and asexually by budding (p. 173).

tissue grade the state of organization of animal cells into different types of tissue (p. 83) for different functions, such as muscular (p. 143) tissue and nervous tissue (p. 91) leading to greater co-ordination of activities such as response and locomotion (p. 143).

symmetrical (*adj*) of structures whose parts are arranged equally and regularly on either side of a line or plane (bilateral symmetry (p. 62)) or round a central point (radial symmetry (↓)).

asymmetrical (*adj*) not symmetrical (↑).

radial symmetry the condition in which the form of an organism is such that its structures radiate from a central point so that, if a cross section is made through any diameter, one half will be a mirror image of the other.

diploblastic (*adj*) of an animal whose body wall is composed of two layers, an endoderm (p. 166) and an ectoderm (p. 166), separated by a jelly-like mesogloea.

enteron (*n*) a sac-like body cavity which functions as a digestive (p. 98) tract or gut (p. 98).

planula larva the small, ciliated (p. 12) larva (p. 165) of a member of the Coelenterata (↑) which results from sexual reproduction (p. 173) and which swims to a suitable site before settling·and growing into a polyp (↓).

動物學（名） 研究動物或動物生命的科學。

後生動物（名） 原生動物門（第44頁）所屬動物的相對詞，用於描述所有那些真正的多個細胞（第9頁）動物。

腔腸動物門（名） 多細胞（第9頁）無脊椎動物（第75頁）的一個門。腔腸動物為水生動物，通常為海洋動物，包括珊瑚（↓）和水母。其身體呈輻射對稱（↓），由單體腔構成。體腔有口與外界相通。口的周圍環生觸手（第71頁）。觸手上長有刺細胞，用於捕食和防衞。體壁由內胚層（第166頁）和外胚層（第166頁）構成，兩胚層之間被膠狀的中膠層隔開。除進行有性生殖（第173頁）外，還有進行出芽（第173頁）的無性生殖。

組織級（名） 動物細胞為發揮不同機能而形成的不同組織（第83頁）類型的組成狀態。例如肌肉（第143頁）組織和神經組織（第91頁）可使諸如反應和運動（第143頁）之類的活動更為協調。

對稱的（形） 指結構而言，其組成部份在一條直線或一個平面的兩側均等而有規則地排列（兩側對稱（第62頁）），或圍繞一中心點均等而有規則地排列（輻射對稱（↓））。

不對稱的（形） 非對稱的（↑）。

輻射對稱 生物體體型結構的一種對稱狀態。其結構是從體內一中心點向外呈輻射狀的，如果沿着任一直徑作一橫切面，則生物體的一半將是另一半的鏡像。

雙胚層的（形） 指動物而言，其體壁是由兩個胚層組成的，其中一個為內胚層（第166頁），另一個為外胚層（第166頁），中間被一個膠狀的中膠層隔開。

消化腔（名） 一種起消化（第98頁）道或腸（第98頁）作用的囊狀體腔。

浮浪幼體 腔腸動物（↑）個體的小而帶有纖毛（第12頁）的幼蟲（第165頁）。它由有性生殖（第173頁）產生；在其游泳到一個適合地點之後，便安居下來，發育成水螅體（↓）個體。

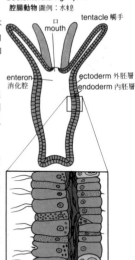

Coelenterata e.g. *Hydra*
腔腸動物 圖例：水螅

口 mouth
tentacle 觸手
enteron 消化腔
ectoderm 外胚層
endoderm 內胚層

endoderm 內胚層
mesogloea 中膠層
ectoderm 外胚層
nematoblast 刺細胞

jellyfish 水母
a scyphozoan
缽水母

Physalia
僧帽水母

浮囊
pneumatophore　sea
level 海水面

gonozoid 生殖個員

small 小指狀
dactylozoid 個員

gastrozoid
營養個員

fishing
dactylozoid
捕食指狀個員

Hydrozoa (*n*) a class of colonial and mainly marine Coelenterata (↑) in which alternation of generations (p. 176) is typical to give free-swimming medusae (↓) that reproduce (p. 173) sexually giving rise to sedentary polyps (↓) which reproduce asexually by budding (p. 173).

polyp (*n*) the sedentary stage in the life cycle of Coelenterata (↑) in which the body is tubular and surrounded by the tentacles (p. 71) at one end while attached to the substrate at the other. It reproduces (p. 173) asexually by budding (p. 173).

medusa (*n*) the free-swimming stage in the life cycle of Coelenterata (↑) in which the body is usually bell-shaped and surrounded by tentacles (p. 71) at one end. It reproduces (p. 173) sexually. **medusae** (*pl.*).

Scyphozoa (*n*) a class of the Coelenterata (↑) which comprises the jellyfishes and which may have no polyp (↑) form. The tentacles (p. 71) surround the mouth and bear stinging hairs.

Anthozoa (*n*) a class of marine Coelenterata (↑), including the sea anemones and corals (↓), in which the medusa (↑) stage is absent. The enteron (↑) is divided by vertical walls or septa and the animals may be colonial or solitary.

coral (*n*) any of the members of the Anthozoa (↑) which are today all colonial and in which the polyp (↑) is contained by a jelly-like, horny, or calcareous (containing CaCO₃) matrix (p. 88).

水螅綱(名)　腔腸動物門(↑)的一個綱。水螅綱動物形成羣體，主要是海生的。其世代交替(第176頁)具有典型性：自由游泳水母體(↓)經由有性生殖(第173頁)產生固着的水螅體(↓)；固着的水螅體經由出芽(第173頁)無性生殖，產生自由游泳的水母體。

水螅體(名)　腔腸動物(↑)處於其生活史中固着階段時的體型。其身體呈圓筒狀，一端被觸手(第71頁)所圍繞，另一端則附着於水中物體的面上。水螅體經由出芽(第173頁)進行無性生殖(第173頁)。

水母體(名)　腔腸動物(↑)處於其生活史中自由游泳階段時的體型。其身體通常呈鐘狀，一端被觸手(第71頁)所圍繞。水母體進行有性生殖(第173頁)。(複數形式爲 medusae)

缽水母綱(名)　腔腸動物門(↑)的一個綱。它包括各種水母；可以沒有水螅體(↑)。觸手(第71頁)圍繞着口，並長有螫毛。

珊瑚蟲綱(名)　生活在海洋中的腔腸動物門(↑)的一個綱。珊瑚蟲綱包括海葵和珊瑚(↓)，沒有水母體(↑)階段。消化腔(↑)被直壁或隔壁分開。該綱動物可形成羣體或單獨生活。

珊瑚(名)　珊瑚蟲綱(↑)中的任何一個成員。珊瑚現今都形成羣體。在珊瑚中，水螅體(↑)被包含在膠狀、角質或鈣質的(含 CaCO₃)基質(第88頁)中。

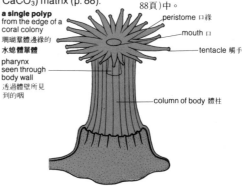

a single polyp
from the edge of a coral colony
珊瑚羣體邊緣的
水螅體單體

pharynx
seen through
body wall
透過體壁所見
到的咽

peristome 口緣

mouth 口

tentacle 觸手

column of body 體柱

Platyhelminthes (*n*) a phylum of multicellular (p. 9) invertebrate (p. 75) animals which includes the flatworms (↓). The body is bilaterally symmetrical (↓) and worm-like, and consists of a single opening to the gut which is often branched. There is no coelom (p. 167) or vascular system (p. 127). The body wall consists of an ectoderm (p. 166), mesoderm (p. 167), and endoderm (p. 166). They are usually hermaphrodite (p. 175).

flatworm (*n*) any of the members of the Platyhelminthes (↑) which have a flattened shape from above downwards that allows the oxygen used in respiration (p. 112) to diffuse into all parts of the body. There are three groups which include the mainly marine flatworms proper, the parasitic (p. 92) tapeworms, and the parasitic flukes.

扁形動物門（名） 多細胞（第 9 頁）無脊椎動物（第75頁）的一個門。它包括各種扁蟲（↓）。扁形動物的身體呈兩側對稱（↓），並成蠕蟲狀；其通常爲分支的消化道僅有一個口。沒有體腔（第167頁）和血管系統（第127頁）。體壁由外胚層（第166頁）、中胚層（第167頁）和內胚層（第166頁）構成。它們通常是雌雄同體（第175頁）的。

扁蟲（名） 扁形動物門（↑）中的任何一個成員。扁蟲身體背腹扁平，使呼吸作用（第112頁）所需的氧能擴散至身體各部份。扁蟲可分爲三類：主要是海生的眞扁蟲、寄生（第92頁）條蟲和寄生吸蟲。

Platyhelminthes e.g. *Planaria*
扁形動物|圖例：眞渦蟲

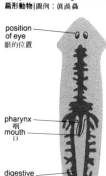

position of eye
眼的位置

pharynx 咽
mouth 口

digestive tract
消化道

Planaria transverse section of body 眞渦蟲 蟲體的橫切面
實質 parenchyma — circular muscles 環肌
內胚層 endoderm — longitudinal muscles 縱肌
— ectoderm 外胚層
gut
消化道 — cilia 纖毛

triploblastic (*adj*) of those animals, such as the Platyhelminthes (↑), in which the body wall consists of three layers, the ectoderm (p. 166), the mesoderm (p. 167), which is formed from cells which have moved from the surface layer, and the endoderm (p. 166).

bilateral symmetry the condition in which one half of the organism, from a section taken down its long axis, is a mirror image of the other half.

acoelomate (*adj*) of those animals without a coelom (p. 167) e.g. Platyhelminthes (↑).

flame cell one of a number of cup-shaped cells that occur in animals, such as the Platyhelminthes (↑), which, by the beating of their cilia (p. 12), draw fluid waste products into their cavity, and then to the exterior.

sucker (*n*) an organ of attachment, for example, in parasitic (p. 92) Platyhelminthes (↑) an adaptation of the pharynx (p. 99) is used to attach the organism to the host (p. 110).

三胚層的（形） 指像扁形動物之類的那些動物而言，這些動物的體壁由三個胚層組成，即外胚層（第166頁）、由表層轉移來的細胞所形成的中胚層（第167頁）和內胚層（第166頁）。

兩側對稱 以沿着生物體的主軸所取的切面爲對稱面，生物體可作互爲鏡像反映的兩部份的對稱狀態。

無體腔動物的（形） 指沒有體腔（第167頁）的那些動物而言，例如扁形動物（↑）。

焰細胞 扁形動物（↑）之類動物體內含有的許多杯狀細胞。焰細胞可藉其纖毛（第12頁）擺動，將液態廢物引入細胞腔內，然後將其從細胞腔的另一端排出。

吸盤（名） 一種吸附器官。例如，在寄生的（第92頁）扁形動物（↑）中，其咽（第99頁）即轉用於將該類扁形動物吸附到寄主（第110頁）體上。

Turbellaria (*n*) a class of the Platyhelminthes (↑) which includes free-living, mainly aquatic flatworms (↑) with a ciliated (p. 12) ectoderm (p. 166).

Planaria (*n*) a genus (p. 40) of the Turbellaria (↑) which includes freshwater forms that have numerous cilia (p. 12) on the underside that assist in locomotion (p. 143), respiration (p. 112) and direct food particles into the mouth.

Trematoda (*n*) a class of the Platyhelminthes (↑) which includes internal parasites (p. 110), such as the flukes, that have a complex life cycle including more than one host (p. 110), usually a vertebrate (p. 74) and an invertebrate (p. 75). They have suckers (↑), a branched gut (p. 98), and a thickened cuticle (p. 145) to resist digestion (p. 98) by the host.

bilharzia (*n*) a disease of humans living especially in Africa which is caused by a liver fluke that spends part of its life in freshwater snails, which are eaten by fish and then by humans. It enters the liver (p. 103) from the gut (p. 98) along the bile duct (p. 101).

渦蟲綱（名） 扁形動物門（↑）的一個綱。它包括具有纖毛（第12頁）外胚層（第166頁）的各種營自由生活，主要為水生的扁蟲（↑）。

真渦蟲屬（名） 渦蟲綱（↑）的一個屬（第40頁）。它包括各種淡水類型的真渦蟲。這些真渦蟲在其下側長有許多纖毛（第12頁），纖毛能協助運動（第143頁）、呼吸（第112頁）和將食物顆粒直接攝入口中。

吸蟲綱（名） 扁形動物門（↑）的一個綱。它包括如吸蟲之類的內寄生物（第110頁）在內。這些內寄生物有複雜的生活史，其寄主（第110頁）一個以上，通常一個寄主為脊椎動物（第74頁），另一個寄主為無脊椎動物（第75頁）。它們具有吸盤（↑）、分支消化道（第98頁）和抗寄主消化作用（第98頁）的加厚角質層（第145頁）。

住血吸蟲病（名） 居住在非洲的人特別易於罹患的一種疾病。這種疾病是由部份時間生活在淡水螺類中的肝吸蟲所引起。這種帶有肝吸蟲的淡水螺類為魚所吃，繼之，魚又為人所食。這樣，肝吸蟲便從人的腸（第98頁）沿着膽管（第101頁）進入肝臟（第103頁）。

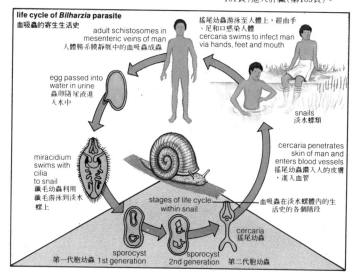

life cycle of *Bilharzia* parasite
血吸蟲的寄生生活史

adult schistosomes in mesenteric veins of man
人體腸系膜靜脈中的血吸蟲成蟲

cercaria swims to infect man via hands, feet and mouth
搖尾幼蟲游泳至人體上，經由手、足和口惑染人體

egg passed into water in urine
蟲卵隨尿液進入水中

snails
淡水螺類

cercaria penetrates skin of man and enters blood vessels
搖尾幼蟲鑽入人的皮膚，進入血管

miracidium swims with cilia to snail
纖毛幼蟲利用纖毛游泳到淡水螺上

stages of life cycle within snail
血吸蟲在淡水螺體內的生活史的各個階段

cercaria
搖尾幼蟲

sporocyst 1st generation
第一代胞幼蟲

sporocyst 2nd generation
第二代胞幼蟲

Cestoda (*n*) a class of the Platyhelminthes (p. 62) which includes the internal parasites (p. 110), the tapeworm, that have a complex life cycle including more than one host (p. 110), both usually vertebrates (p. 74). They are armed with suckers as well as powerful grappling hooks on the head for attachment to the wall of the host's gut (p. 98). The body is divided into sections and has a tough cuticle (p. 145) to prevent digestion (p. 98) by the host.

Nematoda (*n*) a phylum of multicellular (p. 9) invertebrate (p. 75) animals which includes the roundworms (↓). The phylum includes terrestrial (p. 219), aquatic, and parasitic (p. 92) forms which have no cilia (p. 12), and an alimentary canal (p. 98) with a mouth and an anus (p. 103).

roundworm (*n*) any of the members of the Nematoda (↑) which have a characteristic rounded, unsegmented body that tapers at each end. They move by lashing the whole body into s shapes. The sexes are usually separate and the females lay large numbers of eggs. They are able to withstand adverse conditions by secreting (p. 106) a protective coat around the body.

threadworm (*n*) = roundworm (↑).

pseudocoel (*n*) a fluid-filled body cavity between the digestive (p. 98) tract and the other organs of roundworms (↑).

Annelida (*n*) a phylum of multicellular (p. 9) invertebrate (p. 75), mainly free-living, and typically marine aquatic animals which includes the 'true' segmented worms (↓). They have a central nervous system (p. 149), a thin cuticle (p. 145), and bristle-like chaetae (↓) on the body.

條蟲綱（名） 扁形動物門（第62頁）的一個綱，其中包括體內寄生蟲（第110頁）條蟲。條蟲有複雜的生活史，其寄主（第110頁）一個以上，通常均為脊椎動物（第74頁）。它們的頭部長有吸盤和有力的抓鈎，以吸附到寄主的腸（第98頁）壁上。蟲體分為幾部份，具有一層堅韌的角皮（第145頁），以阻止寄主的消化作用（第98頁）。

綫蟲門（名） 多細胞（第9頁）無脊椎動物（第75頁）的一個門。它包括各種綫蟲（↓）。該門動物有陸生（第219頁）、水生和寄生（第92頁）三種類型；沒有纖毛（第12頁），但有一條包括有口和肛門（第103頁）的消化道（第98頁）。

蛔蟲（名） 綫蟲門（↑）中的任何一個成員。其特徵是具有圓柱形、不分節、兩端尖細的身體，運動時，整個身體甩動，因而呈S形。雌、雄通常異體；雌蟲可產大量的蟲卵。蛔蟲蟲體周圍可分泌（第106頁）一種保護層，從而使它們能承受不利的生活條件。

綫蟲（名） 同蛔蟲（↑）

假體腔（名） 蛔蟲（↑）消化（第98頁）道與其他器官之間充滿液體的體腔。

環節動物門（名） 多細胞（第9頁）無脊椎動物（第75頁）的一個門。該門動物主要營自由生活，一般係海生動物。其中，包括各種"真正的"分節蠕蟲（↓）。它們有一個中樞神經系統（第149頁），一層薄的角皮（第145頁），其體表上還長有鬃狀的剛毛（↓）。

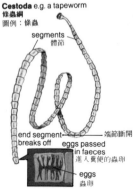

Cestoda e.g. a tapeworm
條蟲綱
圖例：條蟲

- segments 體節
- end segment breaks off 端節斷開
- eggs passed in faeces 進入糞便的蟲卵

- eggs 蟲卵

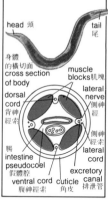

an adult roundworm 蛔蟲成蟲

- head 頭
- tail 尾
- 身體的橫切面 cross section of body
- muscle blocks 肌塊
- dorsal cord 背神經索
- lateral nerve 側神經
- 腸 intestine
- 側神經索 lateral cord
- pseudocoel 假體腔
- ventral cord 腹神經索
- cuticle 角皮
- excretory canal 排泄管

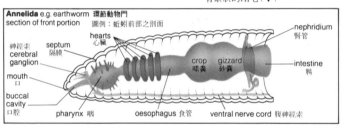

Annelida e.g. earthworm 環節動物門
section of front portion 圖例：蚯蚓前部之剖面

- 神經索 cerebral ganglion 腦神經節
- septum 隔膜
- hearts 心臟
- mouth 口
- buccal cavity 口腔
- pharynx 咽
- oesophagus 食管
- crop 嗉囊
- gizzard 砂囊
- ventral nerve cord 腹神經索
- nephridium 腎管
- intestine 腸

segmented worm any of the members of the Annelida (↑). They have a body which is divided into obvious ring-like segments. Digestion (p. 98) takes place in a simple, tube-like gut (p. 98) which runs from the mouth at the front to the anus (p. 103) at the rear. Between the body wall and the gut is a fluid-filled coelom (p. 167). They are hermaphrodite (p. 175).

nephridium (n) an organ which is used for excretion (p. 134) in some invertebrates (p. 75), e.g. Annelida (↑). It consists of a tube which opens to the exterior at one end and, at the other, to flame cells (p. 62) or to the coelom (p. 167). **nephridia** (pl.).

chaeta (n) one of a number of bristle-like structures, composed of chitin (p. 49) which are present, and arranged segmentally, along the outside of the bodies of Annelida (↑). They may assist in locomotion (p. 143) and, for example, help earthworms (p. 66) to grip the soil in which they live. **chaetae** (pl.).

分節蠕蟲 環節動物門(↑)中的任何一個成員。其蟲體分成多個明顯的環狀體節。其消化作用(第98頁)在結構單一的管狀腸(第98頁)內進行。腸由前面的口開始一直延伸到後面的肛門(第103頁)爲止。在體壁和腸之間是充滿液體的體腔(第167頁)。分節蠕蟲是雌雄同體的(第175頁)。

腎管(名) 某些無脊椎動物(第75頁)如環節動物(↑),用於排泄(第134頁)的一種器官。腎管由一端開口於體外,另一端開口於焰細胞(第62頁)或體腔(第167頁)的一根管子所構成。(複數形式爲 nephridia)

剛毛(名) 由含幾丁質(第49頁)物質組成的許多鬃狀結構物中的一類。剛毛沿着環節動物(↑)身體表面分節排列。它們可以協助行動(第143頁),例如協助蚯蚓(第66頁)抓緊其穴居的土壤。(複數形式爲 chaetae)
剛毛(↑)的另一英文名稱爲 seta。

chaeta transverse section of earthworm body wall
剛毛
蚯蚓體壁的橫切面

retractor muscle 牽縮肌
longitudinal muscles 縱肌
protractor muscle 牽引肌
circular muscles 環肌
epidermis 頂皮
cuticle 表皮
formative cell 毛原細胞
chaeta 剛毛

seta[a] (n) = chaeta (↑).

cerebral ganglion one of the pair of solid strands of nervous tissue (p. 91) which runs ventrally and forms part of the central nervous system (p. 149) in Annelida (↑) and other invertebrates (p. 75) and to which the ganglia (p. 155) are connected segmentally.

nerve cord = cerebral ganglion (↑).

Polychaeta (n) a class of marine Annelida (↑) which includes the bristleworms, ragworms, lugworms etc, that have many chaetae (↑). The sexes are usually separate.

parapodium (n) in members of the Polychaetae (↑), one of many extensions of the body wall on which the chaetae (↑) are found. **parapodia** (pl.).

腦神經節 環節動物(↑)和其他無脊椎動物(第75頁)中的一對連續索狀神經組織(第91頁)。腦神經節沿着腹部縱行,是中樞神經系統(第149頁)的組成部份。各體節中的神經節(第155頁)均與之相連。

神經索 同腦神經節(↑)。

多毛綱(名) 環節動物門(↑)中的一個綱。該綱動物均係海生的:包括毛蟲、釣餌蟲和拖拉蟲等。它們長有許多剛毛(↑)。通常雌雄異體。

疣足(名) 在多毛綱(↑)所屬的成員中,長有剛毛(↑)的體壁上的諸多延伸物。(複數形式爲 parapodia)

trochosphere larva the larva (p. 165) of Annelida (p. 64) and some other groups of invertebrates (p. 75) which may be related through evolution (p. 208). It is free-swimming, planktonic (p. 227), and covered with cilia (p. 12), especially around the mouth which leads to the digestive (p. 98) tract and anus (p. 103).

Hirudinea (*n*) a class of ectoparasitic (p. 110), freshwater Annelida (p. 64) which includes the leeches (↓). They have no chaetae (p. 65) or parapodia (p. 65) and are hermaphrodite (p. 175).

leech (*n*) any of the Hirudinea (↑) which are flattened and have a small sucker (p. 62) at the front end and a larger, more obvious one at the hind end. Some are carnivorous (p. 109) but most are parasitic (p. 92), feeding on the blood of their host (p. 110).

Oligochaeta (*n*) a class of mainly terrestrial (p. 219) and freshwater Annelida (p. 64) which includes the earthworms (↓). They have few chaetae (p. 65), no parapodia (p. 65) and are hermaphrodite (p. 175).

earthworm (*n*) any of the members of the Oligochaeta (↑) which comprise the genus (p. 40) *Lumbricus*. They live by burrowing in the soil and digesting (p. 98) any organic matter in it and are important in improving the structure of the soil. They have a few chaetae (p. 65) and secrete (p. 106) mucus (p. 99) from their skin.

clitellum (*n*) a saddle-like swelling of the epidermis (p. 131) in earthworms (↑) which binds the worms together during copulation (p. 191) and then secretes (p. 106) the cocoon (↓).

cocoon (*n*) the protective covering , e.g. for the eggs of an earthworm (↑) which is secreted (p. 106) by the clitellum (↑).

擔輪幼蟲 環節動物(第64頁)和其他一些經由進化(第208頁)可能有關聯的種羣的無脊椎動物(第75頁)的幼蟲(第165頁)。擔輪幼蟲是自由游泳的浮游生物(第227頁),周身尤其是在與消化(第98頁)道和肛門(第103頁)相通的口的四周長有纖毛(第12頁)。

蛭綱(名) 環節動物門(第64頁)的一個綱。該綱動物生活於淡水中,營外寄生(第110頁),包括各種水蛭(↓)。蛭綱動物的身體上無剛毛(第65頁)或疣足(第65頁)。它們是雌雄同體的(第175頁)。

水蛭(名) 蛭綱(↑)中的任何一個成員。其身體扁平,前端有一個小的吸盤(第62頁);後端有一個較大的、更有明顯的吸盤。某些水蛭是食肉動物(第109頁);但大多數水蛭營寄生(第92頁)生活,以寄主(第110頁)的血液為食。

寡毛綱(名) 環節動物門(第64頁)的一個綱。該綱動物主要生活在陸地(第219頁)上和淡水中,包括各種蚯蚓(↓)。寡毛綱動物身體上的剛毛(第65頁)甚少,無疣足(第65頁)。它們是雌雄同體的(第175頁)。

蚯蚓(名) 包括正蚯屬(第40頁)的寡毛綱中的任何一個成員。它們穴居在土壤裏,靠消化(第98頁)土壤中的任何有機物為生,在改進土壤結構方面起着重要作用。其身體上有少許剛毛(第65頁);皮膚可分泌(第106頁)黏液(第99頁)。

生殖帶、環帶(名) 蚯蚓表皮(第131頁)上的鞍狀隆起。在交配(第191頁)時,這種隆起將兩蚯蚓體黏合在一起,然後分泌(第106頁)卵繭(↓)。

卵繭(名) 保護性覆蓋層,例如由蚯蚓(↑)的生殖帶(↑)所分泌(第106頁)的卵的保護性覆蓋層。

leech 水蛭 sucker 吸盤

clitellum 生殖帶
copulation between earthworms
蚯蚓的交配

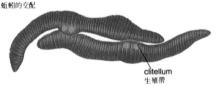

clitellum
生殖帶

parts of an insect
昆蟲身體的各組成部份

antenna
觸角

head
頭部

thorax
胸部

abdomen
腹部

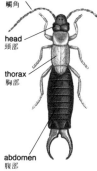

Arthropoda (*n*) a phylum of multicellular (p. 9), invertebrate (p. 75) animals that occupy aerial, terrestrial (p. 219), freshwater and marine environments (p. 218) and make up some 80 per cent of known animal life. Their bodies are highly organized: a *head*, segments at the forward end, containing the organs for feeding and sensation as well as the brain (p. 155); a *thorax*, segments between the head and the abdomen which bear the jointed appendages (↓), and when present, the wings; and an *abdomen*, segments at the rear end. They are bilaterally symmetrical (p. 62), and protected by a tough exoskeleton (p. 145) which is segmented for mobility. They often have compound eyes. Growth takes place by ecdysis (p. 165). Each segment usually bears a pair of jointed appendages. The coelom (p. 167) is small and the main body cavity is a haemocoel (p. 68) containing a tube which is able to contract and function as a heart (p. 124). There is a nerve cord (p. 65) lying below the gut (p. 98) connected to paired ganglia (p. 155) for each segment. The best-known members of this phylum are the insects (p. 69) and spiders (p. 70).

metameric segmentation the condition in which the body of an animal, especially certain invertebrates (p. 75) such as Annelida (p. 64) and Arthropoda (↑), is divided into a series of clearly definable units which are essentially similar to one another and repeat their patterns of blood vessels (p. 127), organs of excretion (p. 134) and respiration (p. 112), nerves (p. 149) etc. In the Arthropoda, the similarity between the units is reduced especially at the head end.

appendage (*n*) any relatively large protuberance or projection from the main body of an organism.

jointed appendage any one of the projections from the body of an arthropod (↑) which is divided into a number of segments, seven in insects (p. 69), and which are hinged between the segments to allow for articulation (bending) in different planes. The appendages are modified for different functions e.g. locomotion (p. 143), feeding, reproduction (p. 173), and respiration (p. 112).

節肢動物門（名）　多細胞（第9頁）無脊椎動物（第75頁）的一個門。該門動物分佈在空中、陸地（第219頁）、淡水和海洋等環境（第218頁）中，約佔已知動物的80％。它們的機體組成已達到高級的程度：頭部；即位於前端的各體節，包括攝食和感覺器官以及腦（第155頁）；胸部，即位於頭部和腹部之間的各體節，具有分節的附肢（↓），某些節肢動物中還有翅；腹部，即位於後端的各體節。節肢動物的身體呈兩側對稱（第62頁）；由堅韌的、爲便於運動而分節的外骨骼（第145頁）加以保護。它們常長有複眼。藉助脫皮（第165頁）進行生長。每一個體節通常都有一對分節的附肢。體腔（第167頁）甚小，主體腔爲血腔（第68頁），含有一根能夠收縮和起心臟（第124頁）作用的管子。在消化道（第98頁）下面，有一條神經索（第65頁），與每一體節的成對神經節（第155頁）相連。該動物門中最爲人所知的動物是昆蟲（第69頁）和蜘蛛（第70頁）。

分節現象　動物的身體尤其是諸如環節動物（第64頁）和節肢動物（↑）的某些無脊椎動物（第75頁）的身體分成一系列具有明顯界限單元的狀態。各單元相互間基本類似；其血管（第127頁）、排泄（第134頁）和呼吸（第112頁）器官、神經（第149頁）等的型式均呈重覆現象。在節肢動物中，尤其是在頭部，各單位間的相似性減少。

附肢（名）　着生在生物軀幹上的任何較大的隆起物或突出物。

分節附肢　着生在節肢動物（↑）軀幹上的任何一個突出物。這種突出物分成若干節，在昆蟲（第69頁）中，它有7節；各節之間相互鉸合，可使關節（彎曲）成不同的平面。分節附肢發生變異，以適應不同的功能，例如行動（第143頁）、攝食、生殖（第173頁）和呼吸（第112頁）功能。

antenna (*n*) one of the pair of highly mobile, thread-like, jointed appendages (p. 67) which occur on the head of an arthropod (p. 67) which are used mainly for touch and smell although, in some members of the group, they may assist in locomotion (p.143). **antennae** (*pl.*)

haemocoel (*n*) a blood-filled (p. 90) cavity which forms the main body cavity in arthropods (p. 67). The coelom (p. 167) is reduced to cavities surrounding the gonads (p. 187) etc while the haemocoel is essentially an expanded part of the blood system.

Crustacea (*n*) a class of the Arthropoda (p. 67), which includes the aquatic shrimps and crabs and the terrestrial (p. 219) woodlice. Typically, the body is divided into a head with two pairs of antennae (↑) and compound eyes, a thorax (p.115), and an abdomen (p. 116). The exoskeleton (p. 145) may be hardened by calcite ($CaCO_3$).

copepod (*n*) any of the group of small, aquatic crustaceans (↑) which form an important part of the marine plankton (p. 227). They have no carapace (↓) or compound eyes and the first pair of appendages (p. 67) on the head are modified for filter feeding (p. 108) while the six pairs on the thorax (p. 115) are used for swimming. The abdomen (p. 116) has no appendages.

isopod (*n*) any of the flattened, terrestrial (p. 219), freshwater, marine and often parasitic (p. 92) members of the Crustacea (↑) which have no carapace (↓) e.g. woodlice and shore slaters.

decapod (*n*) any of the terrestrial (p. 219), freshwater and mainly marine members of the Crustacea (↑) e.g. the highly specialized crabs, lobsters and prawns. They often have an elongated abdomen (p. 116) which ends in a tail that enables them to escape predation (p. 220) by swimming rapidly backwards. The head and thorax (p. 116) may be fused and protected by a carapace (↓). There are five pairs of jointed appendages (p. 67) on the thorax which are used in locomotion (p. 143) and three pairs used in feeding. One or two of the pairs of the legs may bear pincers which are used in courtship and for defence.

觸角（名） 節肢動物（第67頁）頭部的一對十分靈活的絲狀分節附肢（第67頁）。它主要行使觸角和嗅覺的功能，雖然在該種羣的某些成員中，還可輔助行動（第143頁）。（複數形式爲 antennae）

血腔（名） 構成節肢動物（第67頁）主體腔的充血（第90頁）腔。體腔（第167頁）縮成爲包括生殖腺（第187頁）等的一些空腔，而血腔實質成了血液系統的擴展部份。

甲殼綱（名） 節肢動物門（第67頁）的一個綱。它包括水生蝦、蟹和陸生（第219頁）地蝨。甲殼綱動物的身體一般分爲長有兩對觸角（↑）和複眼的頭部、胸部（第115頁）和腹部（第116頁）。外骨骼（第145頁）因含方解石（$CaCO_3$）而變得堅硬。

橈足類（名） 小的水生甲殼動物（↑）種羣中的任何一個成員。這些甲殼動物構成海洋浮游生物（第227頁）的一個重要組成部份。它們無頭胸甲（↓）和複眼。頭部的第一對附肢（第67頁）變異以適應濾食（第108頁）；胸部（第115頁）的六對附肢用於游泳；腹部（第116頁）無附肢。

等足類（名） 甲殼綱（↑）中，背腹扁平、陸生（第219頁）、淡水生、海生、通常是寄生的（第92頁）任何一個成員。它們無頭胸甲（↓），例如地蝨和鼠婦。

十足類（名） 甲殼綱（↑）中，陸生（第219頁）、淡水生、主要是海生的任何一個成員，例如高度特化的蟹、龍蝦和對蝦。它們常有一個延伸至尾端的腹部（第116頁），使它們能迅速向後游泳，免遭捕食（第220頁）。頭部和胸部（第116頁）愈合，並由頭胸甲（↓）加以保護。胸部着生五對用於運動（第143頁）的分節肢體（第67頁）和三對用於攝食的分節附肢，另有一或兩對足，長有螯，可用於求偶和防衛。

Crustacea
甲殼綱

main groups
主要種羣

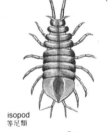

copepod
橈足類

isopod
等足類

decapod
十足類

Myriapoda 多足綱

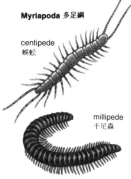

centipede
蜈蚣

millipede
千足蟲

internal structure of an insect 昆蟲的內部結構

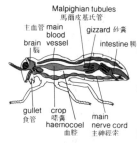

Malpighian tubules
馬爾皮基氏管
主血管 main
blood
vessel
gizzard 砂囊
brain
腦
intestine 腸

gullet
食管
crop
嗉囊
haemocoel
血腔
main
nerve cord
主神經索

front wing 前翅

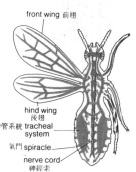

hind wing
後翅
管系統 tracheal
system
氣門 spiracle
nerve cord
神經索

carapace (n) a toughened, shield-like part of the exoskeleton (p. 145) which protects the back and sides of the head and thorax (p. 115) in some arthropods (p. 67), such as the crabs.

crayfish (n) any of the relatively small, freshwater decapods (↑), which resemble and are related to the marine lobsters. It has an elongated carapace (↑) and a flexible abdomen (p. 116). The first of the five pairs of jointed appendages (p. 67) on the thorax (p. 115) are modified to form large pincers that are used for feeding and defence. The remaining four pairs are used for locomotion (p. 143).

Myriapoda (n) a class of terrestrial (p. 219) arthropods (p. 67), including centipedes (↓) and millipedes (↓), which have elongated bodies with many segments, each bearing one or more pairs of jointed appendages (p. 67). They have a definite head bearing antennae (↑) and mouth parts.

chilopod (n) any of the flattened, carnivorous (p. 109) myriapods (↑), including the centipedes (↓), which have one pair of legs on each segment of which the first pair contains poison glands (p. 87).

centipede (n) any of the chilopods (↑), but especially members of the genus (p. 40) *Lithobius*, the members of which live beneath stones.

millipede (n) any of the rounded, herbivorous (p. 109) myriapods (↑) which have four single segments at the front end of the body and numerous double segments each with two pairs of legs.

Insecta (n) the largest, most diverse (p. 213) and important class of the Arthropoda (p. 67), among which the majority are terrestrial (p. 219) or aerial. Insects include the majority of all the known animals with more than a million species (p. 40) described and perhaps another thirty million awaiting identification. The body is characteristically divided into a head with a pair of antennae (↑), compound and simple eyes and mouth parts adapted for different feeding methods; a thorax (p. 115) with three pairs of legs and, usually, two pairs of wings for flight; and a limbless abdomen (p. 116). They have a waterproof exoskeleton (p. 145) and a very efficient means of respiration (p. 112) by means of tracheae (p. 115).

背甲、頭胸甲(名) 外骨骼(第145頁)的堅硬、盾狀部份。保護某些節肢動物(第67頁)如蟹的背部、頭部和胸部(第115頁)的兩側。

蝲咕(名) 任何一個較小的淡水生十足類(↑)的動物。這些動物與海生龍蝦相類似和相關聯。蝲咕有長形的頭胸甲(↑)和可屈曲的腹部(第116頁)。胸部(第115頁)五對分節肢(第67頁)中的第一對變換成大螯,用於攝食和防衛。其餘四對分節附肢用於運動(第143頁)。

多足綱(名) 陸生(第219頁)節肢動物(第67頁)的一個綱,其中包括各種蜈蚣和千足蟲(↓)。其身體的軀幹部呈長形,有諸多體節,每一體節生有一對或多對分節附肢(第67)。頭部明顯,其上生有觸角(↑)和口器。

唇足動物(名) 任何一個背腹扁平的多足綱(↑)食肉動物(第109頁),包括蜈蚣(↓)。這些動物的每一個體節上均生有一對足,在軀幹的第一對足上含有毒腺(第87頁)。

蜈蚣(名) 任何一個唇足動物(↑),但尤指石蜈蚣屬(第40頁)的成員。石蜈蚣均穴居在石頭下面。

千足蟲(名) 任何一個圓柱形的多足綱(↑)食草動物(第109頁)。這些動物的軀幹前端有4個單體節,另外還有諸多的雙體節,每一雙體節上有兩對足。

昆蟲綱(名) 節肢動物門(第67頁)中最大、最具多樣性(第213頁)和最為重要的一個綱。其中,大多數是陸生(第219頁)或空生的。昆蟲佔據了已為人所知的全部動物中的大多數,已作叙述的種(第40頁)數在100萬個以上,另外也許還有3,000萬種有待鑒定。昆蟲身體的特徵是,分頭、胸(第115頁)和腹(第116頁)三個組成部份。頭部長有一對觸角(↑),有複眼和單眼,以及適用於不同攝食方式的口器;胸部生有三對足,其中通常有兩對是用於飛翔的翅;腹部則無附肢。昆蟲有防水的外骨骼(第145頁)和藉氣管(第115頁)進行呼吸(第112頁)的十分有效的手段。

metamorphosis (*n*) the process by which, under the control of hormones (p. 130) some animals, e.g. insects (p. 69) change rapidly in form from the larva (p.165) into the adult with considerable destruction of the larval tissue (p. 83).

proboscis (*n*) an extension of the mouth parts of an insect (p. 69) which is used for feeding.

commissure (*n*) a nerve cord (p. 65) which connects the segmental ganglia (p. 155) in the arthropods (p. 67).

Arachnida (*n*) the class of mainly terrestrial (p.219) arthropods (p. 67), which includes the spiders (↓), scorpions and pseudoscorpions. The body is divided into two main regions, the prosoma (↓) and opisthosoma (↓), and there are four pairs of walking legs, a pair of chelicerae and a pair of pedipalps (↓). The head has simple eyes and no antennae (p. 68). Respiration (p. 112) is achieved by book lungs (↓).

prosoma (*n*) the front region of the body of an arachnid (↑) which is made up of the head and thorax (p. 115) fused together.

opisthosoma (*n*) the hind region or abdomen (p. 116) of an arachnid (↑).

spider (*n*) an arachnid (↑) in which the prosoma (↑) and opisthosoma (↑) are separated by a narrow waist and which has spinnerets (↓).

pedipalps (*n.pl.*) the second pair of jointed appendages (p. 67) on the prosoma (↑) of arachnids (↑). They may be used for seizing prey, adapted as antennae (p. 68), or used for purposes of copulation (p. 191).

book lung one of the organs of respiration (p. 112) in arachnids (↑). They are composed of many fine layers of tissue (p. 83), resembling a book, through which blood (p. 90) flows and absorbs (p. 81) oxygen.

web (*n*) the thin, silken material which is spun by spiders (↑) in a variety of forms and used to capture prey such as flying insects, as in a net.

spinneret (*n*) one of the pair of appendages (p. 67) which is situated on the opisthosoma (↑) of spiders (↑) and secretes (p. 106) a liquid that hardens into silk for making webs (↑), wrapping the cocoon (p. 66), or binding the prey.

變態作用（名） 如昆蟲（第69頁）的某些動物從幼蟲（第165頁）發育成成蟲所經歷的形態迅速變化過程。這種過程是在激素（第130頁）的控制下，伴隨着幼蟲組織（第83頁）重大破壞而進行的。

吻（名） 昆蟲（第69頁）口器的延長部，它供攝食之用。

接索（名） 節肢動物（第67頁）中連接各體節神經節（第155頁）的神經索（第65頁）。

蜘蛛綱（名） 節肢動物（第67頁）的一個綱。該綱動物主要是陸生的（第219頁），包括各種蜘蛛（↓）、蝎和擬蝎。其身體分爲兩個主要部份，即前體（↓）和後體（↓）。長有四對步足，一對螯肢和一對鬚肢（↓）。頭部長有單眼，無觸角（第68頁）。藉書肺（↓）進行呼吸（第112頁）。

前體（名） 蜘蛛綱（↑）動物身體的前部。它由愈合在一起的頭部和胸部（第115頁）組成。

後體（名） 蜘蛛綱（↑）動物身體的後部或腹部（第116頁）。

蜘蛛（名） 一種蜘蛛綱（↑）動物。這種動物的前體（↑）和後體（↑）被一個陝窄的腰部分開。它有多個吐絲器（↓）。

鬚肢（名、複） 蜘蛛綱（↑）動物前體（↑）上着生的第二對分節附肢（第67頁）。它們可用於捕食，也可用作觸角（第68頁），還可用於交配（第191頁）。

書肺 蜘蛛綱（↑）動物的呼吸（第112頁）器官，由諸多精細的組織（第83頁）層構成，猶如一本書。血液（第90頁）流過該器官並吸收（第81頁）氧氣。

網（名） 由蜘蛛（↑）織成的，具有種種形態的薄絲狀物。它可用來捕捉飛蟲之類的食物，使之落入網中。

吐絲器（名） 位於蜘蛛（↑）後體（↑）上的一對附肢（第67頁），可分泌一種能硬化成爲用於織網（↑）的絲的液體，以包裹卵繭（第66頁），或黏住捕獲物。

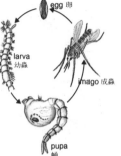

metamorphosis 變態作用
e.g. mosquito life cycle 圖例：蚊的生活史
egg 卵
larva 幼蟲
imago 成蟲
pupa 蛹

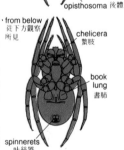

structure of a spider 蜘蛛的結構
from above 從上方觀察所見
prosoma 前體
leg 1 足1
leg 2 足2
pedipalp 鬚肢
eyes 眼
leg 3 足3
leg 4 足4
opisthosoma 後體

from below 從下方觀察所見
chelicera 螯肢
book lung 書肺
spinnerets 吐絲器

書肺

Mollusca 軟體動物門

Gastropoda e.g. a snail
腹足綱 圖例：蝸牛

Cephalopoda
e.g. a squid
頭足綱
圖例：槍烏賊

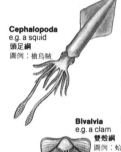

Bivalvia
e.g. a clam
雙殼綱
圖例：蛤

adult snail and larva
(trochophore)
蝸的成體和幼蟲（擔輪幼蟲）

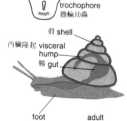

trochophore
擔輪幼蟲

殼 shell

內臟隆起 visceral
hump
腸 gut

foot
足

adult
成體

Mollusca (n) a phylum of bilaterally symmetrical (p. 62), invertebrate (p. 75), multicellular (p. 9) animals that occupy terrestrial (p. 219), freshwater, and marine environments (p. 218) and include cockles, slugs, snails etc. They have soft, unsegmented bodies which are divided into a head region, a visceral hump (↓), and a foot (↓). In some groups of molluscs, the mantle (↓) secretes (p. 106) a hard shell. The coelom (p. 167) is reduced and there is a haemocoel (p. 68).

visceral hump the soft mass of tissue (p. 83) which makes up the bulk of a mollusc (↑) and which contains the main digestive (p. 98) system.

foot[a] (n) a soft, muscular (p. 143) development of the underside of the body of a mollusc (↑) which is used for locomotion (p. 143).

mantle (n) a fold of the body wall which covers the visceral hump (↑). In some molluscs (↑) it secretes (p. 106) a shell composed of calcium carbonate while, in others, it is folded to form a cavity which encloses the organs of respiration (p. 112).

radula (n) a tongue-like strip in most molluscs (↑) which is covered with horny teeth and is used to grind away food particles. As it is worn away, it is continuously replaced.

tentacle (n) a flexible appendage (p. 67). In cephalopods (p. 72), there are normally eight or ten extending from the foot (↑) which is incorporated into the head. Each tentacle bears many suckers and they are used for sense organs, for defence, and for grasping prey.

trochophore larva the free-swimming, ciliated (p. 12) larva (p. 165) of aquatic molluscs (↑).

Gastropoda (n) a class of terrestrial (p. 219), freshwater and marine molluscs (↑), including winkles, slugs, and snails, in which the visceral hump (↑) is coiled. This torsion (↓) of the visceral hump is reflected in the coiling of the shell. There is a muscular foot (↑) which is used in locomotion (p. 143), the eyes are on tentacles (↑), and gastropods feed using a radula (↑).

torsion (n) the act or condition of being twisted.

軟體動物門（名） 身體呈兩側對稱的（第62頁）多細胞（第9頁）無脊椎動物（第75頁）的一個門。該門動物包括蛤類、蛞蝓類、螺類等，分佈在陸地（第219頁）、淡水和海洋環境（第218頁）中。其身體柔軟、不分節，分成頭部、內臟隆起（↓）和足（↓）三部。某些軟體動物的種羣中，外套膜（↓）可分泌（第106頁）一種硬殼。體腔（第167頁）退化，而存在血腔（第68頁）。

內臟隆起 構成軟體動物（↑）軀體的軟組織（第83頁）塊。它含有主要的消化（第98頁）系統。

足[a]（名） 軟體動物（↑）身體下側形成的軟肌肉（第143頁）結構。它可用於運動（第143頁）。

外套膜 覆蓋內臟隆起（↑）的體壁褶膜。在某些軟體動物（↑）中，它分泌（第106頁）出由碳酸鈣組成的外殼，而在另一些軟體動物中，它折疊成包圍呼吸（第112頁）器官的腔。

齒舌（名） 大多數軟體動物（↑）具有的舌狀條帶。其上充滿角質齒，用來磨碎食物顆粒。當它磨損後，會不斷地再生出來。

觸手（名） 一種柔軟的附肢（第67頁）。在頭足綱（第72頁）動物中，與頭結合在一起的足（↑）上通常伸出八個或十個這類附肢。每隻觸手上有諸多吸盤，用於感覺，防衞和捕食。

擔輪幼蟲 能自由游泳、長有纖毛（第12頁）的水生軟體動物（↑）的幼蟲（第165頁）。

腹足綱（名） 陸生（第219頁）、淡水和海洋軟體動物（↑）的一個綱，其中包括蛾螺類、蛞蝓類和螺類等。該綱動物體的內臟隆起（↑）是繞成盤狀的。內臟隆起的這種扭轉（↓）在外殼的螺旋中得到反映。腹足綱動物有肌肉質的足（↑），用於運動（第143頁）；其眼長在觸手（↑）上；攝食則利用齒舌（↑）。

扭轉（名） 盤繞動作或盤繞狀態。

snail (*n*) any of the terrestrial (p. 219) members of the Gastropoda (p. 71), which have no gills (p. 113) but the mantle (p. 71) cavity functions as a lung (p. 115). This includes the slugs which lack the shell found in true snails.

Bivalvia (*n*) a class of flattened freshwater and marine molluscs (p. 71) in which the mantle (p. 71) occurs in two parts and secretes (p. 106) a shell consisting of two hinged valves which may be pulled together by powerful muscles (p. 143). They have a poorly developed head and are filter feeders (p. 108). Some bivalves burrow into sand, mud, rock, or wood, some are attached to the substrate by strong threads, and others may be free swimming, propelling themselves backwards by forcibly opening and closing the valves.

mussel (*n*) any of a group of typical members of the Bivalvia (↑) which include freshwater and marine forms that have powerful muscles (p. 143) to clamp their valves tightly closed for protection. They are attached firmly to the substrate, such as rocks, by strong threads.

siphon (*n*) a tube, e.g. one of two tubes which protrude at the posterior end between the open valves of a bivalve (↑) mollusc (p. 71) and which form part of the system that circulates water through the mantle (p. 71) cavity for feeding and respiration (p. 112).

Cephalopoda (*n*) a class of marine molluscs (p. 71) with a well-developed head containing a complex brain (p. 155) and eyes. The head is surrounded by a ring of sucker-covered tentacles (p. 71) which is a modification of the foot (p. 71). There is a muscular siphon (↑) for respiration (p. 112) and the shell is much reduced and usually internal.

octopus (*n*) any of the members of the Cephalopoda (↑) with eight arm-like tentacles (p. 71) and a soft, ova¹ly-shaped body.

螺（名） 腹足綱（第71頁）中的任何一個陸生的（第219頁）成員。它們皆無鰓（第113頁），但其外套（第71頁）腔起肺（115頁）作用。蛞蝓類的情況也是如此，但它們並沒有真螺中所具有的那種外殼。

雙殼綱（名） 身體扁平的淡水和海洋頓體動物（第71頁）的一個綱。在該綱動物中，外套膜（第71頁）分成兩部份，分泌（第106頁）出由兩片鉸合的殼瓣構成的貝殼。這兩片殼瓣由強韌的肌肉（第143頁）牽引在一起。雙殼綱動物為濾食性動物（第108頁），其頭部退化。某些雙殼綱動物穴居在沙、泥、岩石或木頭中；還有一些雙殼綱動物藉強有力的細絲固著在底物上；其他一些則可自由游泳，藉殼瓣的有力開、閉，使身體向後運動。

貝類（名） 包括淡水和海洋兩類雙殼綱（↑）的一羣典型成員中的任何一個。這些成員具有強韌的肌肉（第143頁），可使其瓣殼夾緊閉合，以保護自己。它們藉強有力的細絲，牢固地附着在底物例如岩石上。

水管（名） 一種管子，例如雙殼綱（↑）頓體動物（第71頁）的兩片張開瓣殼之間的後端處突出的兩根管子。這兩根管子構成攝食和呼吸（第112頁）所需的、使外套（第71頁）腔進行循環的系統的一部份。

頭足綱（名） 海洋頓體動物（第71頁）的一個綱。該綱動物頭部發育良好，有複雜的腦（第155頁）和眼睛。頭部被一圈長有吸盤的、由足（第71頁）演變成的觸手（第71頁）所圍繞。有一根用於呼吸（第112頁）的肌肉質水管（↑）。貝殼嚴重退化，通常僅留體內部份。

章魚（名） 頭足綱（↑）中具有八隻腕狀觸手（第71頁）和一個柔頓的卵圓形身體的任何一個成員。

bivalve e.g. razor clam
雙殼綱頓體動物 圖例：剃刀蛤

siphons
水管

foot
足

burrow
穴

octopus 章魚

tentacle
觸手

sucker
吸盤

Echinodermata
棘皮動物門

a starfish
海星

a sea urchin
海膽

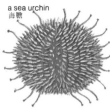

a sea lily
海百合

Echinodermata (*n*) a phylum of radially symmetrical (p. 60) and usually five-rayed (↓), multicellular (p. 9), invertebrate (p. 75) animals that occupy marine environments (p. 218) and include the starfishes (↓) and sea urchins (p. 74). They have no head and a simple nervous system (p. 149). Part of the coelom (p. 167) is adapted to become a water vascular system (↓) which is unique to the group and connects with the tube feet (↓) which are used in locomotion (p. 143) and feeding. They have an internal skeleton (p. 143) of plates composed of calcite ($CaCO_3$) and most of them have spines.

spiny-skinned animal any of the members of the Echinodermata (↑) in which the ectoderm (p. 166) is covered with sharp, moveable, calcareous ($CaCO_3$) spines which connect with the calcareous ossicles (↓).

five-rayed radial symmetry radial symmetry (p. 60) which is typical of the Echinodermata (↑) in which there are five axes of symmetry.

tube foot any of the mobile, hollow, tube-like appendages (p. 67) which connect with the water vascular system (↓) and may end in suckers. They are used for locomotion (p. 143), feeding, and, in the sedentary sea-lilies, have cilia (p. 12) and are used for collecting food particles.

water vascular system a vascular system (p. 127) which is unique to the Echinodermata (↑) and consists of a series of canals containing sea water which, under pressure, operate the tube feet (↑).

madreporite (*n*) a sieve plate on the upper surface of echinoderms (↑) which is the opening of the water vascular system (↑) to the exterior.

calcareous ossicle any of the bone-like plates, made of calcium carbonate which make up the internal skeleton (p. 143) of the Echinodermata (↑).

starfish (*n*) any of the group of flattened, star-shaped Echinodermata (↑) which, typically, have five flexible arms radiating from the central disc which contains the main organs and the mouth on the underside. The arms have tube feet (↑) on the underside which are used for locomotion (p. 143) and for gripping prey. They usually live in the littoral (p. 219) zone.

棘皮動物門（名）　多細胞（第 9 頁）無脊椎動物（第 75頁）的一個門。該門動物的成體呈輻射對稱（第60頁），通常是五放射型（↓）。它們生活在海洋環境（第218頁）中，其成員包括各種海星（↓）和海膽（第74頁）。它們沒有頭，但有一個簡單的神經系統（第149頁）。部份體腔（第167頁）變成爲該類動物獨有的、與管足（↓）相連的水管系統（↓）。管足用於運動（第143頁）和攝食。它們具有由方解石（$CaCO_3$）組成的多塊骨片構成的內骨骼（第143頁）。其大多數體上生有棘。

棘皮動物　棘皮動物門（↑）中的任何一個成員。其外胚層（第166頁）被尖銳可動的鈣質（$CaCO_3$）棘所覆蓋。棘則與鈣質小骨片（↓）相連。

五放射型輻射對稱　棘皮動物（↑）所特有的輻射對稱（第60頁）。這種輻射對稱有五個對稱軸。

管足　任何一個可移動的中空管狀肢體（第67頁）。這些管狀附肢與水管系統（↓）相連接，其末端爲吸盤。管足用於運動（第143頁）和攝食。固着的海百合的管足有長纖毛（第12頁），用於收集食物顆粒。

水管系統　棘皮動物（↑）所特有的一種導管系統（第127頁）。由一系列可容納海水的水管構成。這些水管在壓力作用下，操縱管足的動作。

篩板（名）　棘皮動物（↑）上表面的一塊篩節孔板。它是水管系統（↑）通往外部的出口。

鈣質小骨片　任何一個由碳酸鈣組成的骨狀小片。這些骨狀小片構成棘皮動物（↑）的內骨骼（第143頁）。

海星（名）　棘皮動物門（↑）中身體扁平、呈星狀的種羣中的任何一個成員。其特徵是，從身體中盤伸出五個柔軟的腕。中盤內包含主要的器官，在下面有一個口。各個腕的下面有管足（↑），可供運動（第143頁）和捕食用。海星通常生活在沿岸（第219頁）帶。

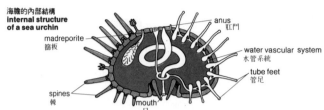

海膽的內部結構
**internal structure
of a sea urchin**

madreporite
篩板

spines
棘

mouth
口

anus
肛門

water vascular system
水管系統

tube feet
管足

sea urchin any of the group of usually, globular, heart-shaped, or disc-shaped Echinodermata (p. 73) which have no arms and in which the calcareous ossicles (p. 73) are fused together to form a rigid, shell-like skeleton (p. 143) to which are attached spines that can be moved by the water vascular system (p. 73). They usually live on or buried in the sea bed feeding on plants and other debris through the mouth on the underside.

Chordata (*n*) a phylum of bilaterally symmetrical (p. 60), invertebrate (↓) and vertebrate (↓) multicellular (p. 9) animals, that includes humans and other mammals (p. 80) and is characterized by possessing a stiff, rod-like notochord (p. 167) during some stage of their life cycle.

cranium (*n*) the part of the skeleton (p. 143), composed of bone, of a vertebrate (↓) member of the Chordata (↑) which is also referred to as the skull and which contains the brain (p. 155).

vertebral column the part of the skeleton (p. 143) of a vertebrate (↓) member of the Chordata (↑) which is situated along the dorsal length of the body, from the cranium (↑) to the tail (↓), and is made of a linked chain of small bones or cartilages (p. 90), the vertebrae. It is flexible and allows for movement and locomotion (p. 143). It replaces the notochord (p. 167) and is a hollow column containing the spinal cord (p. 154). Also known as **spine** or **backbone**.

visceral cleft one of the paired openings in the pharynx (p. 99) which occur at some stage in the life cycle of members of the Chordata (↑) and persist in the aquatic species (p. 40). They lead from the exterior to the gills (p. 113) and are involved with filter feeding (p. 108) and gas exchange (p. 112) as water is pumped through them.

vertebrate (*n*) an animal with a vertebral column (↑).

海膽　棘皮動物門（第73頁）中身體通常呈球形、心形或圓盤形的種羣中的任何一個成員。它們沒有腕；其鈣質小骨片（第73頁）併合在一起形成堅硬的殼狀骨骼（第143頁）。該骨骼上附生許多可由水管系統（第73頁）使其活動的棘。它們通常棲息或穴居於海底，經由下面的口攝取植物和其他碎屑。

脊索動物門（名）　身體呈兩側對稱（第60頁）的多細胞（第 9 頁）無脊椎動物（↓）和脊椎動物動物的一個門。該門動物包括人和其他一些哺乳動物（第80頁）。脊椎動物的特徵是，在其生活史的某一階段，具有一根堅韌的棒狀脊索（第167頁）。

顱　（名）　脊索動物門（↑）的脊椎動物（↓）成員由骨構成的骨骼（第143頁）的一部份。這一部份骨骼也稱爲頭顱，其內包含腦（第155頁）。

脊柱　脊索動物門（↑）的脊椎動物（↓）其成員骨骼（第143頁）的一部分。這部份骨骼的位置是，沿着身體的背部從顱（↑）延伸至尾（↓）。脊柱是由一條小骨或軟骨（第90頁）構成的連接鏈即脊椎組成。脊柱可彎曲，便於活動或運動（第143頁）。脊椎動物的脊柱取代了脊索（第167頁），是一根含有脊髓（第154頁）的空心柱。它也稱爲脊骨或脊椎骨。

鰓裂　位於咽（第99頁）部的成對裂隙。脊索動物門（↑）一些成員，只在其生活史的某一階段出現鰓裂，而在水生種（第40頁）則持續地存在鰓裂。鰓裂從外部通向鰓（第113頁）；當水被汲入而流經鰓裂時，它即參於濾食（第108頁）和氣體交換（第112頁）。

脊椎動物（名）　具有脊柱（↑）的動物

the three main groups of fish 魚的三個主要種羣

Agnatha e.g. lamprey
無頷類 圖例：七鰓鰻

cartilaginous fish e.g. shark
軟骨魚 圖例：鯊魚

teleost fish 眞骨魚 e.g. perch
圖例：鱸魚

invertebrate (*n*) an animal without a vertebral column (↑).

tail (*n*) an extension of the vertebral column (↑) which continues beyond the anus (p.103) in most vertebrate (↑) members of the Chordata (↑.) It may be used for locomotion (p.143), for balance and manoeuvrability (↓), or as a fifth limb.

manoeuvrability (*n*) the ability to make controlled changes of movement and direction.

Gnathostomata (*n*) a subphylum or superclass of the vertebrate (↑) Chordata (↑) which are characterized by the possession of a jaw (p.105). The notochord (p.167) is not retained throughout the life history.

Agnatha (*n*) a subphylum or superclass of the vertebrate (↑) Chordata (↑) which are characterized by having no jaw (p.105). They are aquatic and primitive (p.212).

Pisces the class of the Chordata (↑) which contains the fish. Fish are freshwater and marine animals with streamlined bodies that are usually covered with scales (p.76). They have a powerful, finned (↓) tail (↑) which is used to propel them through the water, while their pairs of pelvic (↓) and pectoral (↓) fins are used for stability and manoeuvrability (↑). Gas exchange (p.112) takes place in the gills (p.113) and fish are exothermic (p.130).

fin (*n*) a flattened, membraneous external organ on the body of a fish which usually occurs in pairs. It is used for steering, stability, and propulsion.

pectoral (*adj*) of the chest, e.g. the pectoral fins (↑) of a fish are attached to the shoulder and are used for steering up or down in the water and for counteracting pitching and rolling.

pelvic (*adj*) of the pelvic girdle (p.147), e.g. the pelvic fins (↑) of a fish are attached to the pelvic girdle and are used for steering up or down and for counteracting pitching and rolling.

dorsal (*adj*) at, near or towards the back of an animal i.e. that part which is normally directed upwards (or backwards in humans).

ventral (*adj*) at, near or towards the part of an animal that is normally directed downwards (or forwards in humans).

無脊椎動物（名） 無脊柱（↑）的動物。

尾（名） 脊索動物門（↑）的大多數脊椎動物（↑），其脊柱（↑）延伸於肛門（第103頁）以後的部份。尾可用於運動（第143頁），保持身體平衡和靈活性（↓），或作爲第五肢使用。

靈活性（名） 控制運動和方向變化的能力。

有頷類（名） 脊索動物門（↑）的脊椎動物（↑）的一個亞門或總綱。其特徵是具有頷（第105頁）。在其整個生活史中，都未保留脊索（第167頁）。

無頷類（名） 脊索動物門（↑）的脊椎動物（↑）的一個亞門或總綱。其特徵是沒有頷（第105頁）。它們是水生的和原始的（第212頁）脊椎動物。

魚綱 包含魚的脊椎動物門（↑）的一個綱。魚爲淡水動物和海洋動物，身體爲流線型，體表通常覆鱗（第76頁）。魚類有一強力的帶鰭（↓）的尾（↑），用來推動魚體在水中前進；而其成對的腹（↓）鰭和胸（↓）鰭則用來保持魚體的穩定性和靈活性（↑）。氣體交換（第112頁）在其鰓（第113頁）內進行。魚屬變溫（第130頁）動物。

鰭（名） 魚體上的一種扁平、膜狀外部器官，這種器官通常成對生長。鰭可用來操縱方向，保持魚體平穩和推動魚體前進。

胸的（形） 指胸而言，例如魚的胸鰭（↑）着生在肩上，用來操縱魚體在水中的上下和消除魚體前後左右的搖擺。

腰的、腹的（形） 指腰帶（第147頁）而言，例如魚的腹鰭（↑）着生在腰帶上，用來操縱魚體在水中的上下和消除魚體前後左右的搖擺。

背部的、背面的（形） 指在動物的背部，接近動物的背部或朝着動物的背部而言，即通常所指的朝上的部份（對人而言，則指朝後的部份）。

腹部的、腹面的（形） 指通常所說在動物的朝下部份，靠近動物的朝下部份或朝着動物的朝下部份而言（對人而言，則指朝前的部份）。

scale (*n*) one of the many bony or horny plates which are made in the skin of fish and which may be above or beneath the skin. They overlap to form a protective and streamlined covering for the fish. Under the microscope (p. 9), it can be seen that they have a ring-like structure which represents the growth rate of the fish and can be used for aging purposes.

Chondrichthyes (*n*) a subclass of the Pisces (p. 75) which are entirely marine and include the sharks and rays. They are characterized by having an internal skeleton (p. 143) made of cartilage (p. 90) and are, therefore, also referred to as the cartilaginous fish. They have no swim bladder (↓) so that they sink if they cease moving.

cartilaginous fish = Chondrichthyes (↑).

Osteichthyes (*n*) a subclass of the Pisces (p. 75) which includes both freshwater and marine forms. They are characterized by having an internal skeleton (p. 143) and scales (↑) made from bone and are, therefore, referred to as bony fish. They possess a swim bladder (↓).

bony fish = Osteichthyes (↑).

teleost fish any of the main group of Osteichthyes (↑) in which the body tends to be laterally flattened and which have a swim bladder (↓) to adjust their buoyancy (↓). Their fins (p. 75) are composed of a thin, membraneous skin supported on bony rays. Their jaws (p. 105) are shortened so that the mouth can open widely and the visceral clefts (p. 74) are protected by a covering operculum (p. 113). The scales (↑) are thin, bony, and rounded. There is a wide variety of types of teleost fish and they occupy most aquatic environments (p. 218).

鱗、鱗片（名） 魚皮膚中形成的諸多骨質或角質片。它們可以生在魚的皮膚上或皮膚下。鱗片互相重疊，構成一層對魚起保護作用的流綫型被覆物。在顯微鏡（第 9 頁）下可以看到鱗有一環狀結構，這種結構可顯示魚的生長率，用來判斷其成熟度。

軟骨魚亞綱（名） 魚綱（第75頁）的一個亞綱。該亞綱魚類完全是海生的，也包括鯊魚和魟。其特徵是，具有由軟骨（第90頁）構成的內骨骼（第143頁），因此也稱為軟骨魚。它們無鰾（↓），所以如果停止游動，就會下沉。

軟骨魚 同軟骨魚亞綱（↑）。

硬骨魚亞綱（名） 魚綱（第75頁）的一個亞綱，該亞綱魚類包括淡水生和海生兩種類型。其特徵是，具有由骨構成的內骨骼（第143頁）和鱗片（↑），因此也稱為硬骨魚。它們具有一個鰾（↓）。

硬骨魚 同硬骨魚亞綱（↑）。

眞骨魚 硬骨魚亞綱（↑）的主要種羣中的任何一個成員。在眞骨魚中，魚體往往是側面扁平的，並有一個可調節浮力（↓）的鰾。鰭（第75頁）由支承在骨質鰭刺上的薄膜狀皮膚構成。頜（第105頁）短縮，因此口可以張得很大。鰓裂（第74頁）得到鰓蓋骨（第113頁）覆被的保護。鱗片（↑）較薄，呈圓形，係由骨質所構成。眞骨魚種類繁多，分佈於大部份的水生環境（第218頁）之中。

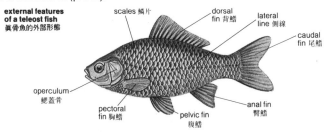

external features of a teleost fish
眞骨魚的外部形態

scales 鱗片
dorsal fin 背鰭
lateral line 側線
caudal fin 尾鰭
operculum 鰓蓋骨
pectoral fin 胸鰭
pelvic fin 腹鰭
anal fin 臀鰭

mermaid's purse
"美人魚的錢包"

homocercal
tail fin 正尾鰭

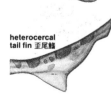

heterocercal
tail fin 歪尾鰭

Amphibia 兩棲綱
e.g. salamander 圖例：蠑螈

swim bladder a sac situated within the abdominal (p. 116) cavity of bony fish (↑). It contains a mixture of oxygen and nitrogen and oxygen can be pumped into it from the blood (p. 90) to increase the fish's buoyancy (↓) or vice versa so that the fish's depth can be controlled. It also functions as a sound detector and producer and, in lung fish, enables respiration (p. 112) out of water.

buoyancy (*n*) the ability to float in a liquid.

mermaid's purse the protective egg case which encloses the small number of eggs produced by cartilaginous fish (↑).

homocercal (*adj*) of a tail (p. 75), such as that of the teleost fish (↑), which is symmetrical (p. 60) in shape.

heterocercal (*adj*) of a tail (p. 75), such as that of the cartilaginous fish (↑), which is asymmetrically (p. 60) shaped such that the lower fin (p. 75) is larger than the upper fin to give the fish additional lift thereby compensating for the lack of a swim bladder (↑).

tetrapod (*n*) any of the vertebrate (p. 74) members of the Chordata (p. 74), such as a mammal (p. 80), which have two pairs of limbs for support, locomotion (p. 143), etc.

pentadactyl (*adj*) of the limb of a tetrapod (↑) which terminates in five digits, although the digits may be reduced or fused together as adaptations to various modes of life.

Amphibia (*n*) a class of primitive (p. 212) tetrapod (↑) chordates (p. 75), such as the frogs and toads, among which fertilization (p. 175) is external so that they must return to water to breed. Their larval (p. 165) forms are all aquatic and have gills (p. 113) but the majority of the adults are able to survive in damp conditions on land because they have a lung (p. 115) and are able to breathe air, respiring (p. 112) mainly through the thin, porous skin. Because of the thin skin, body fluids are easily lost. Like fish, they are exothermic (p. 130).

salamanders (*n.pl.*) members of the order Urodela of the Amphibia (↑) which includes tailed amphibia. The order Urodela also includes newts.

鰾　位於硬骨魚(↑)腹(第116頁)腔內的一個囊。鰾內含有氧和氮的混合物；氧可從血液(第90頁)注入其內，以增加魚體浮力(↓)，反過來也如此，這樣，就可控制魚在水中所處的深度。鰾還可起測音器和發音器的作用；在肺魚中，鰾使呼吸(第112頁)能在水中進行。

浮力(名)　在液體中浮起的能力。

"美人魚的錢包"　將軟骨魚(↑)所產生的爲數很少的卵包裹於其中的保護性卵鞘。

正形尾的(形)　指其形狀是對稱的(第60頁)魚尾(第75頁)而言，例如眞骨魚(↑)的魚尾。

歪形尾的(形)　指其形狀是不對稱的(第60頁)魚尾(第75頁)而言，例如軟骨魚(↑)的尾。由於其下尾鰭(第75頁)較上尾鰭大，使魚獲得附加浮力，從而可對鰾(↑)的欠缺作出補償。

四足動物(名)　脊索動物門(第74頁)中任何一種脊椎動物(第74頁)成員，例如哺乳動物(第80頁)。四足動物有兩對肢，供支撐身體，進行運動(第143頁)等之用。

五趾的、五指的(形)　指其端部爲五趾的四足動物(↑)的肢而言，雖然隨着適應不同的生活方式，趾的數目可能會減少，或者會併合在一起。

兩棲綱(名)　脊索動物(第75頁)中的原始(第212頁)四足動物(↑)的一個綱，例如蛙和蟾蜍。該綱中的動物都是在體外進行受精(第175頁)的，因此，它們必須要回到水中生殖。其幼體(第165頁)的形態全部是水生形態，並且都有鰓(第113頁)；但大多數成體皆能在潮濕的陸地環境條件下生存，這是因爲它們都有肺(第115頁)，可以呼吸空氣，不過，大部份的呼吸作用(第112頁)還是靠薄而有孔的皮膚進行的。由於皮膚薄，其體液極易消失。像魚類一樣，兩棲綱動物都屬變溫(第130頁)動物。

蠑螈(名、複)　包括有尾兩棲動物在內的兩棲綱(↑)有尾目的一些成員。有尾目也包括水螈。

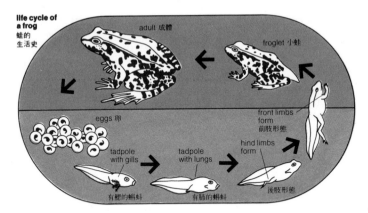

life cycle of a frog 蛙的生活史

adult 成體

froglet 小蛙

eggs 卵

front limbs form 前肢形態

hind limbs form 後肢形態

tadpole with gills 有鰓的蝌蚪

tadpole with lungs 有肺的蝌蚪

Anura (*n*) the order of the Amphibia (↑) which includes the aquatic, tree-dwelling, or damp-loving frogs and the warty skinned toads which can survive in drier conditions. Their hind limbs are elongated and powerful for jumping and have webbed feet for swimming.

Reptilia (*n*) a class of the Chordata (p. 74) which includes the most primitive (p. 212) tetrapods (p. 77), such as the snakes and lizards, that are wholly adapted to terrestrial (p. 219) environments (p. 218) although some, such as the turtles, have returned to an aquatic existence. They are air breathing and possess a true lung (p. 115). Their skin is scaly so that it is resistant to loss of body fluids. Since fertilization (p. 175) is internal, there is no need to return to the water to breed, and they lay amniote (p. 191) eggs with a leathery skin from which the young develop without passing through a larval (p. 165) stage. Like the fish and amphibians (p. 77), they are exothermic (p. 130).

cleidoic (*adj*) of an egg, such as that of a reptile (↑), which has a waterproof covering or shell that is permeable to air.

Chelonia (*n*) an order of the Reptilia (↑), which contains the turtles and tortoises, that are characterized by the plates of bone, overlaid with further horny plates, that enclose the body.

無尾目（名） 兩棲綱（↑）的一個目，包括水生、樹棲或喜歡潮濕地的蛙類和可在較乾燥環境中生存的瘤皮蟾蜍類。它們的後肢長而有力，適於跳躍，並有適於游泳的蹼足。

爬行綱（名） 脊索動物門（第74頁）的一個綱。該綱動物也包括大多數原始（第212頁）四足動物（第77頁），例如蛇和蜥蜴；這些動物完全適應陸地（第219頁）環境（第218頁），雖然某些動物如海龜已返回水中生存。它們呼吸空氣，具有真正的肺（第115頁）。其表皮爲鱗片狀，因此能阻止體液的散失。由於它們行體內受精（第175頁），所以無需回到水中進行繁殖。它們生產具有革質外皮的羊膜（第191頁）卵，從卵發育成小爬行動物無需經過幼體（第165頁）期。它們像魚和兩棲動物（第77頁）一樣，是變溫（第130頁）動物。

有殼的（形） 指卵而言，例如爬行動物（↑）的卵，其卵有一防水的、可透空氣的包覆物或外殼。

龜鱉目（名） 爬行綱（↑）的一個目。該目動物包括海龜和龜，其特徵是，身體被有骨質板，骨質板上又覆有角質板。

the four main groups of Reptilia 爬行綱的四個主要種羣

Chelonia e.g. turtles 龜鱉目 圖例：海龜

Ophidia e.g. snakes 蛇亞目 圖例：蛇

Lacertilia e.g. lizards 蜥蜴亞目 圖例：蜥蜴

Crocodylia e.g. crocodiles 鱷目 圖例：鱷

Squamata (*n*) an order of the Reptilia (↑), which contains the scaly skinned lizards and snakes.

Lacertilia (*n*) a suborder of the Squamata (↑) which contains the lizards. Most are truly tetrapod (p. 77) with a long tail (p. 75), have opening and closing eyelids, an eardrum (p. 158), and normal articulation of the jaws (p. 105).

Ophidia (*n*) a suborder of the Squamata (↑) which contains the snakes. They have elongated bodies with no limbs, no eardrum (p. 158) and no moveable eyelid. The jaws (p. 105) can be dislocated to allow a very wide gape so that large prey can be swallowed whole.

cloaca (*n*) the chamber which terminates the gut (p. 98) in all vertebrates (p. 74), other than the placental (p. 192) mammals (p. 80), and into which the contents of the alimentary canal (p. 98), the kidneys (p. 136), and the reproductive (p. 173) organs are discharged. There is a single opening leading to the exterior.

poison gland one of the modified salivary glands (p. 87) which may be present in some species (p. 40) of, for example, the Ophidia (↑), and which secrete (p. 106) toxic substances that may be, for example, injected into prey through the fangs.

Aves (*n*) a class of the Chordata (p. 74), which contains the birds. They are characterized by the possession of feathers (p. 147) for insulation and flight (p. 147) and other adaptations for flying. Although they are similar in many ways to the Reptilia (↑) from which they evolved (p. 208), they are endothermic (p. 130). There are some flightless species (p. 40). They lay amniote (p. 191) eggs with a calcareous ($CaCO_3$) shell.

鱗蜥目（名）　爬行綱（↑）的一個目。該目動物包括被有鱗皮的蜥蜴類和蛇類。

蜥蜴亞目（名）　鱗蜥目（↑）的一個亞目。該亞目包括各種蜥蜴。大多數蜥蜴爲長有長尾（第75頁）的眞正四足動物（第77頁），有可開、閉的眼瞼、鼓膜（第158頁）和正常的頜（第105頁）關節。

蛇亞目（名）　鱗蜥目（↑）的一個亞目。該亞目包括各種蛇。蛇類軀體細長，無肢，無鼓膜（第158頁），無可開閉的眼瞼，頜（第105頁）可移位，形成很寬的間距，從而能將大的捕獲物整個吞入腹內。

泄殖腔（名）　除了有胎盤（第192頁）哺乳動物（第80頁）以外的所有脊椎動物（第74頁）消化道（第98頁）末端處的空腔。消化道（第98頁）、腎臟（第136頁）和生殖（第173頁）器的排泄物都排放到該腔內。該腔有通向體外的單個開口。

毒腺　一種變性唾腺（第87頁）。這種腺體存在於如蛇亞目（↑）的某些種（第40頁）內。它能分泌（第106頁）有毒的物質通過毒牙將此毒物注入捕獲物內。

鳥綱（名）　脊索動物門（第74頁）的一個綱。該綱包括各種鳥。鳥類的特徵是有保溫和飛翔（第147頁）所需的羽毛（第147頁），以及有其他適於飛翔的適應性結構。雖然鳥類是從爬行類（↑）進化（第208頁）而來的，在許多方面類似於爬行類，但鳥類卻屬恒溫（第130頁）動物。鳥類也有一些不會飛行的種（第40頁）。鳥類產的卵爲具有鈣質（$CaCO_3$）外殼包裹的羊膜（第191頁）卵。

Internal organs 鳥的內部器官 **of a bird**

bill 咮
oesophagus 食管
crop 嗉囊
gizzard 砂囊
stomach 胃
keel 龍骨突
urinary tract 尿道
intestine 腸
cloaca 泄殖腔

bill (n) the horny structure which encloses the jaws (p. 105) of birds. It lacks teeth and may take a variety of forms adapted to different methods of feeding. Also known as **beak**.

keel (n) a bony projection of the sternum (p. 149) of birds to which the powerful pectoral (p. 75) muscles (p. 148) are attached for flight (p. 147).

air sac one of a number of thin-walled, bladder-like sacs in birds, which are connected to the lungs and which are present in the abdominal (p. 116) and thoracic (p. 115) cavities. They even penetrate into some of the bones of the skeleton (p. 143) to lighten the body of the bird without reducing its strength. The tracheae (p. 115) of some insects (p. 69) contain air sacs.

Mammalia (n) a class of the Chordata (p. 74) which contains all the mammals, e.g. dogs, cats and apes. They are endothermic (p. 130), have a glandular (p. 87) skin, and are covered with hair for insulation. They are characterized by possessing mammary glands which secrete (p.106) milk to feed the young. They possess heterodont dentition (p. 104), a secondary palate which enables them to eat and breathe at the same time and relatively large brains (p. 155).

Monotremata (n) a subclass of the Mammalia (↑) which includes the primitive (p. 212) spiny anteater and duck-billed platypus. They possess a cloaca (p. 79) and lay eggs. The young are transferred to a pouch and fed from milk which is secreted (p. 106) on to a groove in the abdomen (p. 116). They are covered with hair but have a relatively low body temperature. They have a poorly developed brain (p. 155).

Metatheria (n) a subclass of the Mammalia (↑) which includes the marsupial or pouched forms, such as the kangaroo. They are viviparous (p. 192) but the poorly developed live young are born after only a brief period of gestation (p. 192) and then transferred to a pouch where they are suckled and complete their growth.

Eutheria (n) a subclass of the Mammalia (↑) which contains the 'true' viviparous (p. 192), placental (p. 192) mammals.

鳥嘴（名） 鳥類上下顎（第105頁）的角質結構。鳥嘴無牙，但可取各種形狀，適應於不同的攝食方法。鳥嘴也稱爲喙。

龍骨突（名） 鳥胸骨（第149頁）上的骨質突出物。在該骨質突出物上，附有用於飛翔（第147頁）的强有力的胸（第75頁）肌（第148頁）。

氣囊 鳥類中的諸多個薄壁泡狀囊，這些囊與肺相通，包含於腹（第116頁）腔和胸（第115頁）腔中。它們甚至可進入到骨骼系統（第143頁）中的某些骨頭之內以減輕鳥體重量，而不致降低其强度。某些昆蟲（第69頁）的氣管（第115頁）也含有氣囊。

哺乳綱（名） 脊索動物門（第74頁）的一個綱。該綱包括所有哺乳動物，例如狗、貓和猿。哺乳動物屬恒溫（第130頁）動物，具有腺（第87頁）皮，體表被有用於保溫的毛。其特徵是有可分泌（第106頁）乳汁以哺育幼仔的乳腺。哺乳動物還具有異型齒系（第104頁），以及使它們能同時進食和呼吸的次生腭和較大的腦（第155頁）。

單孔亞綱（名） 哺乳綱（↑）的一個亞綱。該亞綱包括原始（第212頁）帶刺食蟻動物和鴨嘴獸。單孔亞綱的動物具有泄殖腔（第79頁），並可產卵。幼仔被轉移到育兒袋中，由分泌（第106頁）到腹部（第116頁）一溝中的乳汁哺養。它們的體表被有毛，但體溫較低。其腦（第155頁）不發達。

後獸亞綱（名） 哺乳綱（↑）的一個亞綱。該亞綱包括有袋類或有囊類動物，例如袋鼠。後獸亞綱動物是胎生的（第192頁），但成活的幼仔是僅經過短暫的姙娠（第192頁）期後產出的，發育較差，所以此後被轉移到育兒袋中，在那裏得到哺育後，再完成其生長。

眞獸亞綱（名） 哺乳綱（↑）的一個亞綱。該亞綱包含"眞正"胎生的（第192頁）有胎盤（第192頁）哺乳動物。

the three main subclasses of mammals
哺乳動物的三個主要亞綱
Monotremata (monotremes)
e.g. duck-billed platypus
單孔亞綱（單孔類動物）
圖例：鴨嘴獸

Metatheria (marsupials)
e.g. kangaroo 後獸亞綱
（有袋類動物）
圖例：袋鼠

Eutheria (placental mammals)
e.g. elephant
眞獸亞綱（有胎盤哺乳動物）
圖例：象

tap root
直根

fibrous roots
鬚根

adventitious roots
on corm
球莖上的不定根

contractile   adventitous
root 收縮根   root 不定根

root cap L.S. 根冠縱切面

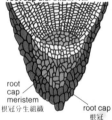

root
cap
meristem
根冠分生組織

root cap
根冠

anatomy (*n*) the study or science of the internal structure of animals and plants.

histology (*n*) the study or science of tissues (p. 83).

morphology (*n*) the study or science of the external structure and form of animals and plants without particular regard to their function and internal structure or anatomy (↑).

physiology (*n*) the study or science of the processes which take place in animals and plants.

root (*n*) the structure of a plant which anchors it firmly to the soil and which is responsible for the uptake of water containing mineral salts from the soil and passing them into the stem. A root may also function as a food store. Unlike underground stems, a root does not contain chlorophyll (p. 12) and cannot bear leaves or buds.

tap root a main, usually central root which may be clearly distinguished from the other roots in a root system.

adventitious root one of a number of roots which grow directly from the stem of the plant as in bulbs (p. 174), corms (p. 174), and rhizomes (p. 174) and which do not grow from a main root.

fibrous root one of a number of roots which grow at the same time as the germination (p. 168) of a plant, such as a grass, and from which other lateral roots grow.

root cap a layer of cells at the tip of a root which protects the growing point from abrasion and wear by soil particles etc.

root hair a fine, thin-walled, tube-shaped structure which grows out from the epidermis (p. 131) just behind the root tip and which is in intimate contact with the soil surrounding a root. It greatly increases the root's surface area for the uptake of water. Water is drawn into the root hair by osmosis (p. 118) because the root hairs and the piliferous layer (p. 82) contain fluid with a lower osmotic potential (p. 118) than the water in the soil.

absorb (*v*) to take in liquid through the surface. **absorption** (*n*).

解剖學（名） 研究動植物內部結構的科學。

組織學（名） 研究組織（第83頁）的科學。

形態學（名） 研究動植物的外部結構和形態，而不專門涉及其功能和內部結構或解剖學（↑）。

生理學（名） 研究在動植物體內發生的過程的科學。

根（名） 能將植物牢固地固着於土壤中的那部份植物結構。根的功能是從土壤中吸收含有無機鹽的水份，並將它們輸送至莖內。根還可以起食物貯藏所的作用。與地下莖不同，根不含有葉綠素（第12頁），也不能長葉或芽。

直根、主根 一種主要的、通常是中央的根。在根系中，這種根可明顯地與其他一些根區別開來。

不定根 如同在鱗莖（第174頁）、球莖（第174頁）和根狀莖（第174頁）中那樣，直接從植物的莖部長出的許多根。這些根並不從主根上長出。

鬚根 植物如禾本科植物發芽（第168頁）的同時所長出的許多根。從這種根可長出其他一些側根。

根冠 根尖處的一層細胞。這層細胞可保護生長點免於被土粒等擦破和磨損。

根毛 緊接於根尖後面的表皮（第131頁）處長出的一種纖細、薄壁、管狀的結構。這種結構與根周圍的土壤緊密接觸，大大地增加了根吸收水份的表面積。水份可藉滲透作用（第118頁）被吸入根毛，因為根毛和根毛層（第82頁）所含液體的滲透勢（第118頁）低於土壤中水份的滲透勢。

吸收（動） 通過表面吸入液體。（名詞形式為absorption）

piliferous layer a single layer of cells which surrounds the root tip and part of the root of a plant and from which the root hairs (p. 81) grow. The cells contain fluid with a lower osmotic potential (p. 118) than that of soil water so that it is the main region of absorption (p. 81) of the root.

根毛層 圍繞植物根尖和根的一部份的單層細胞。根毛(第81頁)由這層細胞長出。根毛層細胞所含液體的滲透勢(第118頁)低於土壤水份的滲透勢，因此，根毛層是根吸收(第81頁)水份的主要部位。

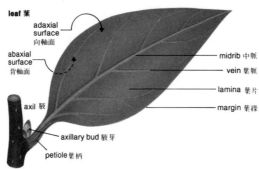

leaf 葉
adaxial surface 向軸面
abaxial surface 背軸面
midrib 中脈
vein 葉脈
lamina 葉片
margin 葉緣
axil 腋
axillary bud 腋芽
petiole 葉柄

leaf (n) a usually flattened, green structure which may or may not be joined to the stem of a plant by a stalk or petiole (↓). Its function is to make food for the plant in the form of carbohydrates (p. 17) by photosynthesis (p. 93).

vein[p] (n) one of a network of structures found in a leaf which provide support for the leaf and also transport the water, which is used during photosynthesis (p. 93), and organic (p. 15) solutes, into and out of the leaf tissue (↓).

lamina (n) the flat, thin, blade-shaped structure which comprises the major part of the foliage leaf.

petiole (n) the stalk or stem which may join the lamina (↑) to the stem of a plant.

midrib (n) the central or middle rib of a leaf which is an extension of the petiole (↑) into the leaf blade.

stem (n) that part of a plant which is usually erect and above ground and which bears the leaves, buds and flowers of the plant. Its function is to transport water and food throughout the plant, space out the leaves, and hold any flowers in a suitable position for pollination (p. 183).

葉(名) 一種通常爲扁平狀的綠色結構。這種結構可以由柄或葉柄(↓)，也可以不由柄或葉柄與植物的莖相連。其功能是，經由光合作用(第93頁)以碳水化合物(第17頁)的形式，爲植物製造養料。

葉脈[p](名) 葉上的網狀結構，這種結構可支持葉。在進行光合作用(第93頁)的過程中，它運輸水份和有機(第15頁)溶解物進出葉組織(↓)。

葉片(名) 構成營養葉主要部份的扁平、薄片狀結構。

葉柄(名) 將植物的葉(↑)與莖相連接的柄或梗。

中脈(名) 葉中央或中間主脈。它是葉柄(↑)進入葉片的延伸部份。

莖(名) 植物中通常有直立的、生長在地上的那一部份。其上長有植物的葉、芽和花。莖的功能是將水份和養料輸送至植物的全身，使各葉得以間隔排列，並把所有的花都固定在適當的位置上，以便於傳粉。(第183頁)。

a generalized flowering plant 普通有花植物
terminal bud 頂芽
internode 節間
axil 腋
leaf 葉
node 節
lateral or axillary bud 側芽或腋芽
ground level
shoot 枝
main or tap root 主根或直根
root 根
lateral roots 側根

whorl 輪

whorl (*n*) a group of three or more of the same organs arranged in a circle at the same level on a stem.

node (*n*) that part of the stem from which the leaves grow.

internode (*n*) the region of the stem between the nodes (↑).

bud (*n*) an undeveloped shoot which may develop into a flower or a new shoot. It consists of a short stem around which the immature leaves are folded and overlap. Buds may be at the tip of a shoot, when they are called terminal, or in the axils (↓) when they are called axillary.

shoot (*n*) the whole part of the plant which occurs above ground and which usually consists of a stem, leaves, buds and flowers.

axil (*n*) the angle between the upper side of a leaf and the stem on which the leaf is growing.

lenticel (*n*) a small raised gap or pore (p. 120) in the bark (p. 172) of a woody stem through which oxygen and carbon dioxide may pass.

tissue (*n*) a group of cells that perform a particular function in an organism.

vascular tissue a tissue (↑) which is specialized mainly for the transport of food and water throughout a plant, and is composed mainly of xylem (p. 84) and phloem (p. 84) together with sclerenchyma (p. 84) and parenchyma (↓).

ground tissue a tissue (↑), such as pith (p. 86) and cortex (p. 86), usually composed of parenchyma (↓), and which occupies all parts of the plant which do not contain the specialized tissue.

packing tissue = ground tissue (↑).

epidermal tissue a dermal (p. 131) tissue (↑) which forms a continuous outer skin over the surface of a plant. There are no spaces between the cells but it is penetrated by stomata (p. 120).

cuticle (*n*) the waterproof, waxy, or resinous outer surface of the epidermal tissue (↑) which occurs on the aerial (p. 219) parts of the plant.

parenchyma (*n*) a tissue (↑) which consists of rounded cells enclosed in a cellulose (p. 19) cell wall (p. 8) and containing air-filled intercellular (p. 110) spaces. Parenchyma supports the non-woody parts of a plant and also functions as storage tissue in the roots, stem and leaves.

parenchyma cells
in cross section
薄壁組織細胞的橫切面
cytoplasm 細胞質
cell wall 細胞壁　plastid 質體

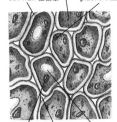

nucleus　vacuole 液泡
細胞核
intercellular space 胞間隙

輪（名）　在莖上以同一高度排列成環狀的一組三個或三個以上的同樣器官。

節（名）　從其上長出葉子的那一部份莖。

節間（名）　莖上兩個相鄰節（↑）之間的部份。

芽（名）　一種未發育的枝。這種枝可發育成花或新枝。它由一根在其周圍疊合着未成熟葉子的短莖構成。芽可長在枝的頂端，稱其爲頂芽；也可長在腋（↓）內，稱爲腋芽。

枝（名）　植物的整個地上部份。它通常由莖、葉、芽和花構成。

腋（名）　葉的正面與葉生長於其上的莖之間的夾角。

皮孔（名）　木質莖的莖皮（第172頁）內的一種小而隆起的裂縫或孔（第120頁）。氧和二氧化碳可以通常皮孔。

組織（名）　生物體內行使某一特殊功能的一羣細胞。

維管組織　主要爲整個植物運輸養料和水份的一種特化組織（↑）。這種組織主要由木質部（第84頁）和韌皮部（第84頁），以及厚壁組織（第84頁）和薄壁組織（↓）構成。

基本組織　通常由薄壁組織（↓）構成的一種組織（↑），例如髓（第86頁）和皮層（第86頁）。這種組織佔據植物中所有不包含特化組織的部份。

墊層組織　同基本組織。

表皮組織　構成覆蓋於植物表面、無間斷外皮的一種皮（第131頁）組織（↑）。這種組織的細胞之間是沒有間隙的，但有一些氣孔（第120頁）貫穿其間。

角質層（名）　植物氣生（第219頁）部份的表皮組織（↑）的外表面。這種外表面是防水的、蠟質或呈樹脂狀的。

薄壁組織（名）　由纖維素（第19頁）細胞壁（第8頁）包圍着的一些圓形細胞構成的一種組織（↑）。這種組織含有充滿空氣的胞間（第110頁）隙。薄壁組織可支持植物的非木質部份，在根、莖和葉中，也可起貯藏組織的作用。

collenchyma (*n*) a tissue (p. 83) composed of elongated cells in which the primary cell wall (p. 14) is unevenly thickened with cellulose (p. 19). Collenchyma tissue is specialized to provide support to actively growing parts of the plant which may also need to be flexible.

sclerenchyma (*n*) a tissue (p. 83) which has a secondary cell wall (p. 14) of lignin (p. 19) and which is composed of sclereids (↓) and fibres (↓). Its function is to provide support.

sclereid (*n*) one of the two types of cells which comprise the sclerenchyma (↑). It is not always easy to differentiate between a sclereid and a fibre (↓) although they are generally very little longer than they are broad and are the stone cells of the shells of nuts and the stones of fruits.

fibre[p] (*n*) one of the two types of cells which comprise the sclerenchyma (↑). They are elongated lignified (p. 19) cells with no living contents and provide great support.

xylem (*n*) a vascular tissue (p. 83) consisting of hollow cells with no living contents and additional supporting tissue (p. 83) including fibres (↑), sclereids (↑) and some parenchyma (p. 83). The cell walls (p. 8) are lignified (p. 19), the thickness varying in shape and extent. The two main cell types found in xylem are vessels (↓) and tracheids (↓).

tracheid (*n*) one of the two types of cells found in xylem (↑). A tracheid is elongated and has tapering ends and cross walls. Tracheids run parallel to the length of the organ which contains them. Each tracheid is connected to its neighbour by pairs of pits (p. 14) through which water can easily pass.

phloem (*n*) a vascular tissue (p. 83) which transports food throughout the plant (translocation (p. 122)). It contains sieve tubes (↓) and companion cells (↓), and, in some plants, may also contain other cells, such as parenchyma (p. 83) and fibres (↑).

sieve tube a column of thin-walled, elongated cells which are specialized to transport food materials through the plant.

厚角組織（名） 由細長細胞組成的一種組織（第83頁）。在這些細胞中，其初生細胞壁（第14頁）由纖維素（第19頁）不均勻地加厚。厚角組織在結構上的特化，可對植物中生長旺盛，而又需要具有柔韌性的那些部份提供支持。

厚壁組織（名） 具有木質素（第19頁）的次生細胞壁（第14頁）的一種組織（第83頁）。這種組織由石細胞（↓）和纖維（↓）組成，其功能是提供支持。

石細胞（名） 構成厚壁組織（↑）的兩類細胞之一。雖然一般來說，石細胞的長度比寬度要稍微大一些，它們都是堅果殼和果核的石細胞，但是，區分石細胞和纖維（↓）卻未必總是輕而易舉的。

纖維（名） 構成厚壁組織（↑）的兩類細胞之一。這類細胞是具有無生命內含物的細長木質（第19頁）化細胞，具有較大的支持功能。

木質部（名） 一種維管組織（第83頁）。這種組織由具有無生命內含物的中空細胞和包括纖維（↑）、石細胞（↑）和某種薄壁組織（第83頁）在內的附加支持組織（第83頁）組成。木質部細胞的細胞壁（第8頁）木質（第19頁）化，其厚度隨細胞的形狀和大小而異。木質部中存在的兩類主要細胞是導管（↓）和管胞（↓）。

管胞（名） 木質部（↑）含有的兩類細胞之一。管胞細長，兩端漸陝，並有橫壁。管胞與包含它們的器官的長度方向相平行。每一個管胞經由成對的容易導水的紋孔（第14頁）與其鄰接的管胞相通。

韌皮部（名） 一種在整個植物體內具有運輸養料功能（轉移作用（第122頁））的維管組織（第83頁）。韌皮部含有篩管（↓）和伴細胞（↓）。在某些植物中，它還含有諸如薄壁組織（第83頁）和纖維（↑）的其他一些細胞。

篩管 一種由多個薄壁長形細胞構成的管柱。其結構特化，用於運輸整個植物體內養料。

collenchyma cells
in cross section 厚角細胞的橫切面

cellulose thickening 纖維素加厚

cell types in xylem
木質部中的細胞類型

tracheids 管胞

tracheid 管胞

pits 紋孔

tracheid 管胞

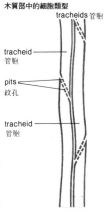

vessels 導管

no end walls 無端壁

spiral thickening of cell wall 細胞壁的螺紋加厚

phloem 朝皮部（縱切面）
(L.S.)

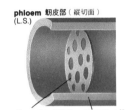

sieve plate 帶孔的   wall of sieve 篩管分
with pores 篩板   element 子的壁
position of xylem and phloem
in young and old roots
木質部和朝皮部在幼根和老根中的位置

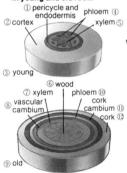

① pericycle and
endodermis   phloem ④
②cortex   xylem ⑤

③ young

⑥ wood
⑦ xylem   phloem ⑩
vascular   cork
⑧ cambium   cambium ⑪
     cork ⑫

⑨ old

position of xylem and phloem
in young and old stems
木質部和朝皮部在幼莖和老莖中的位置

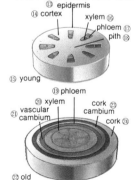

⑬ epidermis
⑭ cortex   xylem ⑯
     phloem ⑰
     pith ⑱

⑮ young

⑲ phloem
⑳ xylem   cork ㉓
vascular   cambium
㉑ cambium   cork ㉔

㉒ old

① 中柱鞘和內皮層   ⑭ 皮層
② 皮層   ⑮ 幼莖
③ 幼根   ⑯ 木質部
④ 朝皮部   ⑰ 朝皮部
⑤ 木質部   ⑱ 髓
⑥ 木材   ⑲ 朝皮部
⑦ 木質部   ⑳ 木質部
⑧ 維管形成層   ㉑ 維管形成層
⑨ 老根   ㉒ 老莖
⑩ 朝皮部   ㉓ 木栓形成層
⑪ 木栓形成層   ㉔ 木栓
⑫ 木栓
⑬ 表皮

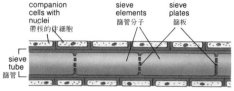

companion
cells with   sieve   sieve
nuclei   elements   plates
帶核的伴細胞   篩管分子   篩板

sieve
tube
篩管

sieve plate the perforated end wall of a sieve tube
(↑) through which strands of cytoplasm (p. 10)
pass to connect the neighbouring cells.

companion cell a small, thin-walled cell
containing dense cytoplasm (p. 10) and a well-
defined nucleus (p. 13) situated alongside the
sieve tube (↑) and which may aid the
metabolism (p. 26) of the sieve tube.

vessel[P] (*n*) one of the two types of cells found in
xylem (↑). Each vessel consists of a series of
cells arranged into a tube-like form with no
cross walls. It runs parallel to the length of the
organ containing it and is found mainly in the
angiosperms (p. 57). When mature, the vessel
has no living contents, and has thick lignified
(p. 19), walls for strength. There are several
types of thickening: *annular* which has rings of
lignin along the length of the cell; *spiral* which
has a spiral or coil of lignin round the inner
surface of the cell wall (p. 8); *scalariform* which
has a ladder-like series of bars of lignin on the
inner surface of the cell wall; *reticulate* which
has a network of lignin over the inner surface of
the cell wall; and *pitted* which has lignin over
the whole inner surface of the cell wall except
for many small pits (p. 14) or pores (p. 120).

篩板 篩管（↑）的穿孔端壁。胞質（第10頁）絲穿過
篩板，可使相鄰的細胞相連接。

伴細胞 排列於篩管（↑）旁邊的一種含濃厚細胞質
（第10頁）和顯著細胞核（第13頁）的小薄壁細
胞。這種細胞有助於篩管的新陳代謝（第26
頁）。

導管[P]（名） 木質部（↑）中含有的兩類細胞之一。
每一根導管均由一系列排列成管狀、不具橫壁
的細胞構成。導管與包含它的器官長度方向相
平行。它主要存在於被子植物（第57頁）體內。
在植物成熟時，導管並不具有生命的內含物，
而具有厚的木質化（第19頁）壁，以增加其強
度。導管的加厚情況有幾種類型：沿細胞長度
方向具有木質素環的環紋；圍繞細胞壁（第8
頁）內表面具有木質素的螺旋和卷曲的螺紋；
在細胞壁內表面上具有一系列木質素梯狀條的
梯紋；在細胞壁內表面上具有木質素網的網
紋；除諸多個小紋孔（第14頁）或小孔（第120
頁）之外，在細胞壁整個內表面上都有木質素
的孔紋。

vessels types of thickening 導管加厚類型
annular   spiral   scalariform   reticulate   pitted
環紋   螺紋   梯紋   網紋   孔紋

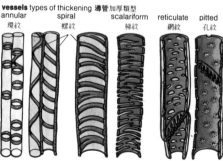

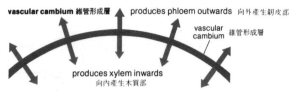

vascular cambium 維管形成層　produces phloem outwards 向外產生朝皮部

vascular cambium 維管形成層

produces xylem inwards
向內產生木質部

cambium (*n*) the layer of narrow, thin-walled cells which are situated between the xylem (p. 84) and phloem (p. 84) and give rise by division to secondary xylem (p. 172) and secondary phloem (p. 172). The cambium does not lose its ability to make new cells and is responsible for lateral growth in plants.

secondary tissue the additional tissue (p. 83) formed by the cambium (↑) leading to an increase in the lateral dimensions of the stem or root of a plant.

stele (*n*) the core or bundle of vascular tissue (p. 83) in the centre of the roots and stems of plants.

exodermis (*n*) the outer layer of thickened cells which may replace the epidermis (p. 131) in the older parts of roots.

endodermis (*n*) the layer of cells surrounding the stele (↑) on the innermost part of the cortex (↓) of a root.

cortex (*n*) the tissue (p. 83), usually of parenchyma (p. 83), which occurs in the stems and roots of plants between the stele (↑) and the epidermis (p. 131). It tends to make the stem more rigid.

pith (*n*) the central core of the stem composed of parenchyma (p. 83) tissue (p. 83) and found within the stele (↑).

medullary ray one of a number of plates of parenchyma (p. 83) cells which are arranged radially and pass from the pith (↑) to the cortex (↑) or terminate in secondary xylem (p. 84) and phloem (p. 84).

pericycle (*n*) the outermost layer of the stele (↑) with the endodermis (↑) and composed of parenchyma tissue (p. 83).

mesophyll (*n*) the tissue (p. 83) which lies between the epidermal (p. 131) layers of a leaf lamina (p. 82) and is involved in photosynthesis (p. 93).

形成層（名）　木質部（第84頁）和朝皮部（第84頁）之間的一層陝窄的薄壁細胞。這一層細胞經分裂產生次生木質部（第172頁）和次生朝皮部（第172頁）。形成層保持其產生新細胞的能力；其功能是進行植物的橫向生長。

次生組織　由形成層（↑）構成的附加組織（第83頁），可致使植物的莖或根增加橫向尺寸。

中柱（名）　位於植物根和莖的中軸部份、由維管組織（第83頁）構成的維管柱或維管束。

外皮層（名）　可以替代根較老部份表皮（第131頁）的外層加厚細胞。

內皮層（名）　根部皮膚（↓）最裏面、圍繞中柱（↑）的一層細胞。

皮層（名）　植物莖和根內位於中柱（↑）和表皮（第131頁）之間的那種通常為薄壁組織（第83頁）的組織（第83頁）。它可使莖變得更為堅硬。

髓（名）　莖的中柱（↑）內由薄壁組織（第83頁）構成的中心部份。

髓射線　由薄壁組織（第83頁）細胞構成的許多呈輻射狀排列的片狀結構。這些片狀結構從髓（↑）部起通到皮膚為止，亦即終止於次生木質部（第84頁）和朝皮部（第84頁）。

中柱鞘（名）　中柱（↑）的最外層。它與內皮層（↑）附在一起，由薄壁組織（第83頁）構成。

葉肉（名）　葉片（第82頁）表皮（第131頁）層之間的組織（第83頁）。這一組織參與光合作用（第93頁）。

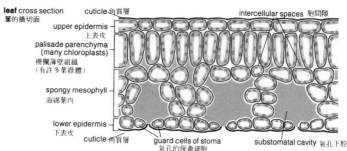

leaf cross section 葉的橫切面
cuticle 角質層
upper epidermis 上表皮
palisade parenchyma (many chloroplasts) 柵欄薄壁組織（有許多葉綠體）
spongy mesophyll 海綿葉肉
lower epidermis 下表皮
cuticle 角質層
intercellular spaces 胞間隙
guard cells of stoma 氣孔的保衛細胞
substomatal cavity 氣孔下腔

types of epithelium
上皮的類型

pavement
(squamous)
扁平上皮
（鱗狀上皮）

nucleus 核
cytoplasm
細胞質

basement membrane 基底膜

columnar
柱狀上皮

ciliated
纖毛上皮

glandular 腺上皮

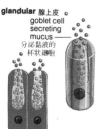

goblet cell secreting mucus
分泌黏液的杯狀細胞

palisade mesophyll the mesophyll (↑) composed of cylindrical cells at right-angles to the leaf surface and situated just below the upper epidermis (p.131). It contains numerous chloroplasts (p.12) and is concerned with photosynthesis (p. 93).

spongy mesophyll the mesophyll (↑) composed of loosely and randomly arranged cells with few chloroplasts (p. 12) and large air spaces which are connected with the atmosphere through the stomata (p. 120).

epithelium (n) an animal tissue (p. 83) composed of a sheet of cells which are densely packed and which covers a surface or lines a cavity.

endothelium (n) the epithelium (↑) which lines the heart (p. 124) and blood vessels (p. 127).

basement membrane a membrane (p. 14) composed of a thin layer of cement to which one of the cells of the epithelium (↑) is fixed.

ciliated epithelium an epithelium (↑) bearing cilia found in the trachea (p.115) and bronchi (p.116).

glandular epithelium an epithelium (↑) which is specialized to form secretory (p. 106) glands (↓).

goblet cell a wine-glass-shaped cell which secretes (p. 106) mucus (p. 99) on to the outside of columnar epithelium (↑) to protect it.

gland (n) an organ which secretes (p. 106) chemicals to the outside. **glandular** (adj).

compound epithelium an epithelium (↑) which is made up from more than one layer of cells with columnar cells attached to the basal membrane (p. 14) and squamous (flattened) cells furthest from it. It is found at areas of stress such as the epidermis (p. 131) of skin.

柵欄狀葉肉　由與葉表面成直角、緊接上表皮（第131頁）下方的圓筒形細胞組成的葉肉（↑）。這種葉肉含有大量葉綠體（第12頁），參與光合作用（第93頁）。

海綿葉肉　由鬆散而且無規則地排列着的細胞和一些較大的氣隙組成的葉肉（↑）。這些細胞內含有少量葉綠體（第12頁）。那些氣隙則通過氣孔（第120頁）與大氣相通。

上皮（名）　由一層緊密排列，覆蓋於表面或空腔上的細胞組成的一種動物組織（第83頁）。

內皮（名）　覆蓋在心臟（第124頁）和血管（第127頁）內表面的上皮（↑）。

基底膜　由一薄層膠結物所組成的一種膜（第14頁）。上皮（↑）的細胞之一固定於其上。

纖毛上皮　氣管（第115頁）和支氣管（第116頁）有纖毛的一種上皮（↑）。

腺上皮　特化形成分泌（第106頁）腺（↓）的一種上皮（↑）。

杯狀細胞　一種酒杯狀細胞。這種細胞向柱狀上皮（↑）外面分泌（第106頁）黏液（第99頁），以對其進行保護。

腺（名）　將化學物質分泌（第106頁）於其外的一種器官。（形容詞形式為 glandular）

複層上皮　由一層以上的細胞所組成的一種上皮（↑）。這些細胞有固定在基底膜（第14頁）上的柱狀細胞和離基底膜最遠的鱗狀（扁平）上皮細胞。複層上皮存在於像皮膚表皮（第131頁）那樣的應力區。

stratified epithelium = compound epithelium (p.87).

transitional epithelium a stratified epithelium (↑) that is also capable of stretching, and found in areas such as the bladder (p. 135).

connective tissue the tissue (p. 83) that functions for support or packing purposes in animals. It has a few quite small cells with greater amounts of intercellular (p. 110) or matrix (↓) material.

matrix (*n*) the intercellular (p. 110) ground substance in which cells are contained.

areolar tissue a connective tissue (↑) which surrounds and connects organs. It is composed of collagen (↓) and elastic fibres (↓) in an amorphous matrix (↑).

fibroblast (*n*) an irregularly shaped but often elongated and flattened cell which functions in the production of collagen (↓).

mast cell a cell present in the matrix (↑) of areolar tissue (↑) which produces anticoagulant (p. 128) substances and is also found in the endothelium (p. 87) of blood vessels (p. 127).

macrophage (*n*) a large cell found widely in animals but particularly in the connective tissue (↑). Macrophages wander freely through the tissue and in the lymph nodes (p. 128) by amoeboid movement (p. 44) and destroy harmful bacteria (p. 42) by engulfing them as well as helping to repair any damage to tissue (p. 83).

collagen fibre a non-elastic fibre with high tensile strength found in connective tissue (↑), particularly in tendons (p. 146), skin and skeletal (p. 145) material. Also known as a **white fibre**.

elastic fibre a highly elastic fibre (p. 143) found in connective tissue (↑), particularly in ligaments (p. 146) and organs, such as lungs (p. 115). Also known as a **yellow fibre**.

adipose tissue a connective tissue (↑) similar to areolar tissue (↑) but containing closely packed fat cells and found under the skin and associated with certain organs to provide insulation, protection, and to store energy.

bone (*n*) a hard connective tissue (↑) composed of osteoblasts (p. 90) in a matrix (↑) made up of collagen fibres (p. 88) and calcium phosphate. It makes up the majority of the skeleton (p. 145).

疊層上皮　亦稱複層上皮(第87頁)。

移行上皮　一種也能牽張的複層上皮(↑)。這種上皮存在於如膀胱(第135頁)之類部位。

結締組織　在動物中起支持或填充作用的組織(第83頁)。這種組織具有幾個很小的細胞，這些細胞帶有大量的胞間(第110頁)物質和基質(↓)。

基質(名)　其內含有細胞的胞間(第110頁)基礎物質。

蜂窩組織　圍繞和連接器官的一種結締組織(↑)。這種結締組織是由在無定形基質(↑)中的膠原纖維(↓)和彈性纖維(↓)組成的。

成纖維細胞、纖維母細胞(名)　一種形狀不規則但通常爲細長和扁平狀的細胞。這種細胞的功能是產生膠原(↓)。

肥大細胞　存在於蜂窩組織(↑)基質(↑)中的一種細胞。這種細胞可產生抗凝血(第128頁)物質；它也存在於血管(第127頁)的內皮(第87頁)中。

巨噬細胞(名)　廣泛存在於動物體內，特別是存在於結締組織(↑)中的一種大型細胞。巨噬細胞可藉變形運動(第44頁)，自由地穿過結締組織，在淋巴結(第128頁)中游動；並藉吞噬作用消滅有害的細菌(第42頁)以及幫助修復組織(第83頁)的任何損傷。

膠原纖維　結締組織(↑)中，特別是腱(第146頁)、皮膚和骨骼(第145頁)物質中含有高抗張強度的一種非彈性纖維。它也稱爲白纖維。

彈性纖維　存在於結締組織(↑)中，特別是存在於韌帶(第146頁)和像肺(第115頁)那樣的一些器官中的一種彈力大的纖維(第143頁)。它也稱爲黃纖維。

脂肪組織　類似於蜂窩組織(↑)，但含有緊密排列的脂肪細胞的一種結締組織(↑)。這種組織存在於皮下，並與某些器官相聯合，以起到保溫、保護和貯存能量的作用。

骨(名)　由成骨細胞(第90頁)組成的一種堅硬的結締組織(↑)。成骨細胞處於以膠原纖維(第88頁)和磷酸鈣組成的基質(↑)之中。骨構成骨骼(第145頁)的大部份。

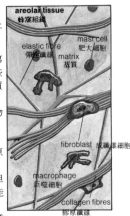

areolar tissue
蜂窩組織

mast cell 肥大細胞
elastic fibre 彈性纖維
matrix 基質
fibroblast 成纖維細胞
macrophage 巨噬細胞
collagen fibres 膠原纖維

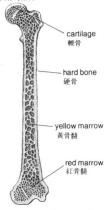

structure of a long bone
長骨的結構

cartilage 軟骨
hard bone 硬骨
yellow marrow 黃骨髓
red marrow 紅骨髓

compact bone bone in which the Haversian canals (↓) are densely packed.

periosteum (*n*) connective tissue (↑) surrounding the bone and containing osteoblasts (p. 90) as well as many collagen fibres (↑) making it tough. The muscles (p. 143) and ligaments (p. 146) are attached to the periosteum.

Haversian canal a canal running along the length of bone and containing the nerves (p. 149) and blood vessels (p. 127) as well as the lymph vessels (p. 128) which secrete (p. 106) the osteocytes (p. 90).

Haversian system the system of Haversian canals (↑) surrounded by rings of bone and which connect with the surface of the bone and with its marrow (p. 90).

canaliculus (*n*) one of the fine canals linking the lacunae (↓) and containing the branches of the osteocytes (p. 90).

endosteum (*n*) a thin layer of connective tissue (↑) within a bone next to the cavity containing the marrow (p. 90).

lacuna (*n*) one of the spaces between the bone lamellae (↓) in which the osteoblasts (p. 90) are found. **lacunae** (*pl.*).

bone lamellae ring-like layers of calcified matrix (↑) in bone and surrounding the Haversian canals (↑).

密質骨　其內緊密地排列着哈氏管（↓）的骨。

骨膜（名）　圍裹着骨的結締組織（↑），其中含有可使之變得堅韌的成骨細胞（第90頁）和許多膠原纖維（↑）。肌肉（第143頁）和韌帶（第146頁）附着在骨膜上。

哈氏管　一種沿着骨縱行的管道。其內含有神經（第149頁）、血管（第127頁）和分泌（第106頁）骨細胞（第90頁）的淋巴管（第128頁）。

哈氏系統　由骨環圍繞的哈氏管（↑）的系統。哈氏管與骨的表面及其骨髓（第90頁）相通。

骨小管（名）　連接腔隙（↓）並包含骨細胞（第90頁）分支的細管。

骨內膜（名）　骨內鄰近含骨髓（第90頁）的腔的一薄層結締組織（↑）。

骨腔隙、骨陷窩（名）　內含成骨細胞（第90頁）的骨板（↓）之間的間隙。（複數形式爲 lacunae）

骨板　骨肉圍繞哈氏管（↑）的、成環狀的各層鈣化基質（↑）。

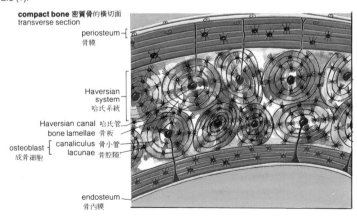

compact bone 密質骨的橫切面
transverse section

periosteum
骨膜

Haversian system
哈氏系統

Haversian canal　哈氏管
bone lamellae　骨板
osteoblast　canaliculus　骨小管
成骨細胞　lacunae　骨腔隙

endosteum
骨內膜

chondroblast (*n*) a cell which occurs in cartilage (p. 90) and secretes (p. 106) the matrix (p. 88) of clear chondrin (↓).

cartilage (*n*) a skeletal (p. 145) tissue (p. 83) composed of chondroblasts (↑) in a matrix (p. 88) of clear chondrin (↓). There are also many collagen fibres (p. 88) contained within it.

chondrin (*n*) a bluish-white clear gelatinous material which forms the ground substance of cartilage (↑). Chondrin is elastic.

hyaline cartilage cartilage (↑) which contains collagen fibres (p. 88) and which forms the embryonic (p. 166) skeleton (p. 145).

osteoblast (*n*) a cell present in the hyaline cartilage (↑) which is responsible for the laying down of the calcified matrix (p. 88) of bone.

osteocyte (*n*) an osteoblast (↑) which has become incorporated in the bone during its formation and has stopped dividing.

spongy bone bone which contains a network of bone lamellae (p. 89) surrounding irregularly placed lacunae (p. 89) containing red marrow (↓).

epiphysis (*n*) the end of the limb (p. 147) bone in mammals (p. 80) which enters and takes part in the joint (p. 146).

marrow (*n*) the soft, fatty tissue (p. 83) which is present in some bones and which produces the white blood cells (↓).

blood (*n*) the specialized fluid in animals which is found in vessels (p. 127) contained within endothelial (p. 87) walls and which may contain a pigment (p. 126) used in the transport of respiratory (p. 112) gases as well as transporting food and other materials throughout the body.

plasma (*n*) the clear, almost colourless fluid part of the blood (↑) which carries the white blood cells (↓), the red blood cells (↓) and the platelets (p. 128). It consists of 90 per cent water and 10 per cent other organic (p. 15) and inorganic (p. 15) compounds.

serum (*n*) the clear, pale-yellow fluid which remains after blood (↑) has clotted (p. 129) and consists essentially of plasma (↑) without the clotting agents.

成軟骨細胞（名） 軟骨（第90頁）中的一種細胞。這種細胞可分泌（第106頁）透明軟骨膠（↓）的基質（第88頁）。

軟骨（名） 由透明軟骨膠（↓）基質（第88頁）中的成軟骨細胞（↑）構成的一種骨骼（第145頁）組織（第83頁）。基質內也含有許多膠原纖維（第88頁）。

軟骨膠 構成軟骨（↑）基質的一種藍白色透明膠狀物質。

透明軟骨 含有膠原纖維（第88頁）的軟骨（↑）。這種軟骨構成胚胎（第166頁）骨骼（第145頁）。

成骨細胞（名） 透明軟骨（↑）中的一種細胞。這種細胞的功能是形成骨的鈣化基質（第88頁）。

骨細胞（名） 在其形成間就已結合於骨中並停止分裂的一種成骨細胞（↑）。

鬆質骨 內含網狀骨板（第89頁）的骨。這些骨板包圍着不規則排列，含有紅骨髓（↓）的骨腔隙（第89頁）。

骺（名） 哺乳動物（第80頁）肢（第147頁）骨的端部。這種骨端插入並構成關節（第146頁）的一部份。

骨髓（名） 某些骨中含有柔軟脂肪組織（第83頁）。這種組織能製造白血細胞（↓）。

血液（名） 動物體內的特化液體。這種液體存在於內皮（第87頁）壁所包圍的管（第127頁）中，並含有一種色素（第126頁）。血液的功能是為動物體全身輸送呼吸（第112頁）氧體，以及養料和其他物質。

血漿（名） 血液（↑）中透明的，幾乎是無色的液體部份。這種液體部份含有白血細胞（↓）、紅血細胞（↓）和血小板（第128頁）。它由90％的水和10％的其他有機化合物（第15頁）和無機化合物（第15頁）組成。

血清（名） 血液（↑）凝固（第129頁）後剩餘的透明、淡黃色液體。這種液體基本上由不含凝固物的血漿（↑）組成。

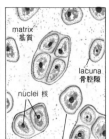

cartilage
軟骨

chondroblast (cartilage cell)
成軟骨細胞（軟骨細胞）

blood cells 血細胞

red blood cell or erythrocyte
紅血細胞或紅血球

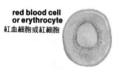

white blood cells 白血細胞

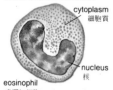

cytoplasm
細胞質

nucleus
核

eosinophil
嗜曙紅細胞

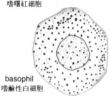

basophil
嗜鹼性白細胞

lymphocyte
淋巴球

cytoplasm
細胞質

nucleus
核

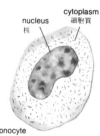

monocyte
單核白血胞

red blood cell a blood cell which contains the respiratory (p. 112) pigment (p. 126), such as haemoglobin (p. 126).

erythrocyte (*n*) = red blood cell (↑).

white blood cell a blood cell which contains no respiratory (p. 112) pigment (p. 126). White blood cells are important in defending the body against disease because they are able to engulf bacteria (p. 42) as well as producing antibodies (p. 233).

leucocyte (*n*) = white blood cell (↑).

polymorphonuclear leucocyte a white blood cell (↑) with a dark staining, lobed nucleus (p. 13) and granular cytoplasm (p. 10). They are produced in the bone marrow (p. 10).

granulocyte (*n*) = polymorphonuclear leucocyte (↑).

eosinophil (*n*) a polymorphonuclear leucocyte (↑) which can be stained with acid (p. 15) dyes such as eosin. Their numbers are normally quite low in the blood (↑) but increase in number if the body becomes infected with parasitic (p. 92) or allergic (p. 234) disease.

basophil (*n*) a polymorphonuclear leucocyte (↑) which can be stained with basic (p. 15) dyes. Their numbers are normally very low in the blood (↑) but they are able to engulf bacteria (p. 42).

neutrophil (*n*) the commonest type of leucocytes (↑) which are able to migrate out of the blood (↑) stream into the tissues (p. 83) of the body to engulf bacteria (p. 42) wherever they invade. On their death, they give rise to pus.

lymphocyte (*n*) a white blood cell (↑) which is produced in the lymphatic system (p. 128) and is important in defending the body against disease. It has a large nucleus (p. 13) and clear cytoplasm (p. 10).

monocyte (*n*) the largest type of white blood cell (↑) and is produced in the lymphatic system (p. 128). It has a spherical nucleus (p. 13) and clear cytoplasm (p. 10). It actively engulfs and devours any invading foreign bodies such as bacteria (p. 42).

nervous tissue tissue (p. 83) containing the nerve cells (p. 149), which are specialized for the transmission of nervous impulses (p. 150), together with the supporting connective tissue (p. 88).

紅血細胞、紅血球　含有血紅蛋白(第126頁)之類呼吸(第112頁)色素(第126頁)的一種血細胞。

紅細胞、紅血球(名)　同紅血細胞(↑)。

白血細胞、白血球　不含呼吸(第112頁)色素(第126頁)的一種血細胞。白血細胞在身體防禦疾病方面起重要作用，因為它們能吞噬細菌(第42頁)和產生抗體(第233頁)。

白細胞、白血球(名)　同白血細胞(↑)。

多形核白細胞　一種帶暗斑、具淺裂狀(第13頁)和顆粒狀細胞質(第10頁)的白血細胞(↑)。多形核白細胞產生於骨髓(↑)之中。

顆粒白(血)細胞、顆粒白血球(名)　同多形核白細胞(↑)。

嗜曙紅細胞、嗜酸性白細胞(名)　可用酸性(第15頁)染料例如曙紅着色的一種多形核白細胞(↑)。血液(↑)中嗜曙紅細胞的數量通常很少，但是，如果身體感染了寄生蟲(第92頁)病或變應性(第234頁)疾病，則其數量便會增加。

嗜鹼性白血細胞(名)　可用鹼性(第15頁)染料着色的一種多形核白細胞(↑)。血液(↑)中嗜鹼細胞的數量通常很少，但是，它們能夠吞噬細菌(第42頁)。

嗜中性白細胞(名)　最普通類型的白細胞(↑)。無論細菌從何處入侵，這種白細胞均能從血(↑)流中遷移出來，進入身體的組織(第83頁)內，吞噬細菌(第42頁)。這種細胞死亡時，會產生膿。

淋巴細胞、淋巴球(名)　淋巴系統(第128頁)中所產生的一種白血細胞(↑)。這種白血細胞在防禦疾病方面起重要作用。它有一個較大的核(第13頁)和透明的細胞質(第10頁)。

單核白細胞、單核細胞(名)　淋巴系統(第128頁)中所產生的最大類型的白血細胞(↑)。這種白血細胞有一個球形核(第13頁)和透明的細胞質(第10頁)。它能積極地吞噬和毀滅任何入侵的外來物體，例如細菌(第42頁)。

神經組織　含有神經細胞(第149頁)的組織(第83頁)。這些神經細胞與起支持作用的結締組織(第88頁)一起，專用於傳導神經衝動(第150頁)。

nutrition (*n*) the means by which an organism provides its energy by using nutrients (↓).

nutrient (*n*) any material which is taken in by a living organism and which enables it to grow and remain healthy, replace lost or damaged tissue (p. 83), and provide energy for these and other functions.

holophytic (*adj*) of nutrition (↑), such as that of plants, in which simple inorganic compounds (p. 15) can be taken in and built up into complex organic compounds (p. 15) using the energy of light, either to provide energy for metabolism (p. 26) or growth or to make living protoplasm (p. 10).

chemosynthetic (*adj*) of nutrition (↑) in which energy is obtained by a simple inorganic (p. 15) chemical reaction such as the oxidation (p. 32) of ammonia to a nitrite by a bacterium (p. 42).

autotrophic (*adj*) of nutrition (↑) in which simple inorganic compounds (p. 15) are taken in and built up into complex organic compounds (p. 15).

heterotrophic (*adj*) of nutrition (↑), such as that in animals and some fungi (p. 46), in which the organic compounds (p. 15) can only be made from other complex organic compounds which have to be first taken into the body.

saprozoic (*adj*) of nutrition (↑) in which the organism takes in organic compounds (p. 15) only in solution (p. 118) rather than in solid form.

holozoic (*adj*) of nutrition (↑), as found in animals, in which complex organic compounds (p. 15) are broken down into simpler substances which are then used to make body structures or oxidized (p. 32) to supply the organism's energy needs.

saprophytic (*adj*) of nutrition (↑) in which the organism obtains complex organic compounds (p. 15) in solution (p. 118) from dead and/or decaying plant or animal material.

parasitic (*adj*) of nutrition (↑) in which the organism derives its food directly from another living organism at the expense of the host (p. 111) but without necessarily killing it.

營養（名） 生物體利用營養素（↓）為自身提供能量的方式。

營養素、養分（名） 生物體吸收的任何物質。這些物質使生物體生長發育有保持健康，替換失去的或損壞了的組織（第83頁），並為行使這些功能和其他一些功能提供能量。

植物式營養的（形） 指營養（↑）如植物的營養而言，這種營養方式攝入簡單的無機化合物（第15頁），利用光能將之合成為複雜的有機化合物（第15頁），從而為新陳代謝（第26頁）和生長發育提供能量，或者用於建造有生命的原生質（第10頁）。

化能合成的（形） 指營養（↑）而言，利用這種營養方式時，能量的獲得，可以依靠簡單的無機（第15頁）化學反應，例如由細菌（第42頁）將氨變成亞硝酸的氧化（第32頁）反應。

自養的（形） 指營養（↑）而言，以這種營種方式攝入簡單的無機化合物（第15頁），將之合成為複雜的有機化合物（第15頁）。

異養的（形） 指營養（↑）如動物和某些真菌（第46頁）的營養而言，利用這種營養方式時，有機化合物（第15頁）的製造，只能從首先攝入生物體內的其他複雜的有機化合物來着手。

腐生動物式營養的（形） 指營養（↑）而言，生物體利用這種營養方式時，僅以溶液（第118頁）的形式，而不是以固體的形式，攝入有機化合物（第15頁）。

動物式營養的（形） 指營養（↑）如動物中的營養而言，利用這種營養方式時，複雜的有機化合物（第15頁）首先被分解成簡單的物質，然後再被用來製造身體結構物質，或者被氧化（第32頁），提供生物體所需要的能量。

腐生植物式營養的（形） 指營養（↑）而言，生物體利用這種營養方式時，是從死亡的及/或腐爛的植物或動物體中，以溶液（第118頁）的形式來獲取複雜的有機化合物（第15頁）。

寄生的（形） 指營養（↑）而言，生物體利用這種營養方式時，靠消耗寄主（第111頁），但未必殺死寄主的途徑，直接從另一生物體獲取食物。

types of nutrition 營養的類型

holophytic/autotrophic 植物式營養的/自養的
green plant 綠色植物

heterotrophic/saprophytic 異養的/腐生植物式
ink cap fungus 營養的鬼傘真菌

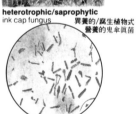

heterotrophic/parasitic 異養的/寄生的
pathogenic bacterium 病原菌

heterotrophic/holozoic 異養的/動物式
bird 鳥

葉綠體內的希爾反應（水的光解）

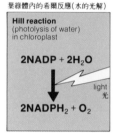

Hill reaction
(photolysis of water)
in chloroplast

$$2NADP + 2H_2O$$

light
光

$$2NADPH_2 + O_2$$

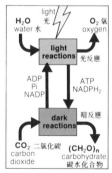

the links between the light
reactions and dark reactions
of photosynthesis
光合作用的光反應和暗反應
之間的偶聯

macronutrient (*n*) any nutrient (↑) which is required by an organism in substantial amounts. *See* p. 240.

major elements = macronutrients (↑).

micronutrient (*n*) a nutrient (↑) which is required in only minute or trace amounts. *See* p. 241.

chlorosis (*n*) the yellowing of the leaves of green plants caused by the loss of chlorophyll (p. 12).

active mineral uptake the uptake and transport through a plant, across a cell membrane (p. 14), of mineral ions from regions of low concentration to regions of high concentration. This process requires energy both to take up minerals and to retain them.

passive mineral uptake the uptake and transport of mineral ions through a plant, usually across a cell membrane (p. 14), from regions of high concentration to regions of low concentration by diffusion (p. 119) without using energy.

photosynthesis (*n*) the process that takes place in green plants in which organic compounds (p. 15) are made from inorganic compounds (p. 15) using the energy of light. It takes place in two main stages: in the light-dependent or photochemical stage, light is absorbed by chlorophyll (p. 12) in the chloroplasts (p. 12), located mainly on the leaves of plants, and used to produce ATP (p. 33) and to supply hydrogen atoms by oxidizing (p. 32) water. These are then used in the reduction (p. 32) of carbon dioxide. In the dark or chemical stage, carbon dioxide is reduced and carbohydrates (p. 17) are made. Photosynthesis will only take place at suitable temperatures and in the presence of chlorophyll, carbon dioxide, water, and light.

limiting factor one of a number of factors which controls the rate at which a chemical reaction, such as photosynthesis (↑), takes place. The rate is limited by that factor which is closest to its minimum or smallest value.

photosynthetic pigment one of the pigments (p. 126) which make up chlorophyll (p. 12) and absorb (p. 81) light. The following substances are photosynthetic pigments: chlorophyll a; chlorophyll b; carotene; and xanthophyll.

大量營養素、常量元素（名） 生物體需要相當數量的任何營養素（↑）。見第240頁。

主要元素、大量元素 同大量營養素（↑）。

微量營養素、微量元素（名） 生物體僅需要少量或微量的營養素（↑）。見第241頁。

黃化綠葉病（名） 綠色植物由於失去葉綠素（第12頁）而引起的葉子黃化現象。

主動礦質吸收 植物體吸收和運輸礦質離子的一種過程。此過程中礦物質離子從低濃度區遷移到高濃度區，穿過細胞膜（第14頁）進入植物體。在此過程中，無論是對礦質的吸收，還是使礦質滯留，都需要能量。

被動礦質吸收 植物體吸收和運輸礦質離子的一種過程。此過程通常是不使用能量的情況下，以礦質離子從高濃度區擴散（第119頁）至低濃度區，穿過細胞膜（第14頁），進入植物體的方式來進行的。

光合作用（名） 綠色植物中所發生的一種過程。在這種過程中，植物利用光能由無機化合物（第15頁）製造有機化合物（第15頁）。該過程按兩個主要階段進行：在依賴光的階段或光化學的階段，光被主要位於植物葉子上的葉綠體（第12頁）中的葉綠素（第12頁）吸收，產生ATP（第33頁），與此同時，通過對水的氧化（第32頁），提供氫離子。這些產物然後用於二氧化碳還原（第32頁）。在暗的階段或化學階段，二氧化碳被還原，製成碳水化合物（第17頁）。光合作用只有在合適的溫度下和有葉綠素、二氧化碳、水和光存在時方可進行。

限制因素 控制化學反應如光合作用（↑）速率的若干因素之一。該速率受到最接近於其最低值或最小值的因素限制。

光合色素 種類繁多的色素（第126頁）中的一類。葉綠素（第12頁）即為光合色素類中的一種，可吸收（第81頁）光。下列物質均屬於光合色素類：葉綠素a，葉綠素b，胡蘿蔔素和葉黃素。

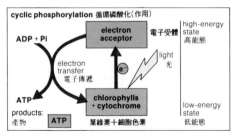

cyclic photophosphorylation a step in the light dependent stage of photosynthesis (p. 93) in which light is involved in the formation of ATP (p. 33) from ADP (p. 33) by the addition of phosphate.

non-cyclic photophosphorylation a step in the light-dependent stage of photosynthesis (p. 93) in which light is involved in the formation of ATP (p. 33) from ADP (p. 33) by the addition of phosphate and in which the water is split to provide hydrogen ions.

absorption spectrum a diagrammatic representation of the way in which a substance, such as chlorophyll (p. 12), absorbs (p. 81) radiation of different wavelengths by different amounts. Chlorophyll absorbs blue and red light readily so that it appears green.

循環光合磷酸化(作用)　光合作用(第93頁)中依賴光的階段的一個反應步驟。在這個步驟的中，光參與由 ADP(第33頁)加磷酸生成 ATP(第33頁)的反應。

非循環光合磷酸化(作用)　光合作用(第93頁)中依賴光的階段的一個反應步驟。在這個步驟中，光參與由 ADP(第33頁)加磷酸生成 ATP(第33頁)的反應，同時，使水分解，提供氫離子。

吸收光譜　一種圖譜。這種圖譜表達一種物質如葉綠素(第12頁)以不同的量吸收(第81頁)不同波長光的方式。葉綠素容易吸收藍光和紅光，因此，它呈現綠色。

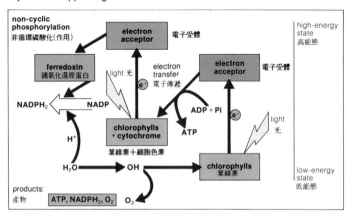

光合作用的光譜和吸收光譜
action spectra and absorption spectra in photosynthesis

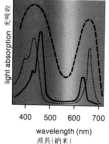

light absorption 光吸收

wavelength (nm)
波長（納米）

- - - - - action spectrum of photosynthesis
光合作用的作用光譜
·············· absorption spectrum chlorophyll a
葉綠素 a 的吸收光譜
——— absorption spectrum chlorophyll b
葉綠素 b 的吸收光譜

action spectrum a diagrammatic representation of the way in which radiation of different wavelengths affects a process, such as photosynthesis (p. 93). In this case, it shows that red and blue light are the most effective in the action of photosynthesis.

photosystem I one of the two systems of pigments (p. 126) which each contains chlorophyll a (p. 12), accessory pigments, and electron carriers (p. 31) and which are involved in electron transfer reactions coupled with phosphorylation (↑). Also known as **pigment system I** or **PSI**.

photosystem II see photosystem I (↑). Also known as **pigment system II** or **PSII**.

ferredoxin (n) any of a number of red-brown (iron-containing) pigments (p. 126) which function as electron carriers (p. 31) in photosynthesis (p. 93).

plastoquinone (n) an electron carrier (p. 31) used in photosynthesis (p. 93).

C3 plant a plant in which PGA (p. 97) containing three carbon atoms is produced in the early stage of photosynthesis (p. 93). Photosynthesis in these plants is less efficient than in C4 plants (↓).

C4 plant a plant in which a dicarboxylic acid containing four carbon atoms is produced in the early stage of photosynthesis (p. 93). The method of fixing carbon dioxide has evolved from that of C3 plants (↑) and operates more efficiently.

作用光譜　一種圖示。這種圖示表達不同波長的光對像光合作用(第93頁)這種過程產生影響的方式。在這個例子中，作用光譜顯示，紅光和藍光對光合作用的效應最為顯著。

光(合)系統 I　兩種色素(第126頁)系統之一。每種色素系統均包含葉綠素 a(第12頁)、輔助色素和電子載體(第31頁)。兩種色素系統皆參與磷酸化作用(↑)偶聯的電子傳遞反應。它也稱為色素系統 I 或 PSI。

光(合)系統 II　見光系統 I(↑)。它也稱為色素系統 II 或 PSII。

鐵氧化還原蛋白(名)　在光合作用(第93頁)中起電子載體(第31頁)作用的若干紅褐色(含鐵)色素(第126頁)中的任何一個。

質體醌(名)　用於光合作用(第93頁)的一種電子載體(第31頁)。

三碳植物　一類植物。這類植物於光合作用(93頁)的早期階段產生含有三個碳原子的 PGA(第97頁)。這些植物的光合作用效率低於四碳植物(↓)。

四碳植物　一類植物。這類植物於光合作用(93頁)的早期階段產生含有四個碳原子的二羧酸。這種固定二氧化碳的方法，是從三碳植物(↑)所用的方法演變而來的，其效率更高。

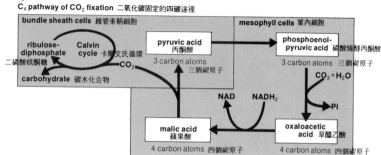

C_4 pathway of CO_2 fixation 二氧化碳固定的四碳途徑

Calvin cycle the steps in the dark stage of
photosynthesis (p. 93) in which carbon dioxide
is reduced (p. 32) using the hydrogen produced
in the light-dependent stage and synthesized
into carbohydrates (p. 17) using the energy of
ATP (p. 33) also formed during the
light-dependent stage.

卡爾文氏循環 光合作用（第93頁）中暗階段的一組
反應步驟。在這組反應步驟中，利用依賴光階
段所產生的氫，使二氧化碳還原（第32頁），並
利用也是在依賴光階段所生成 ATP（第33頁）
中的能量，將二氧化碳合成爲碳水化合物（第
17頁）。

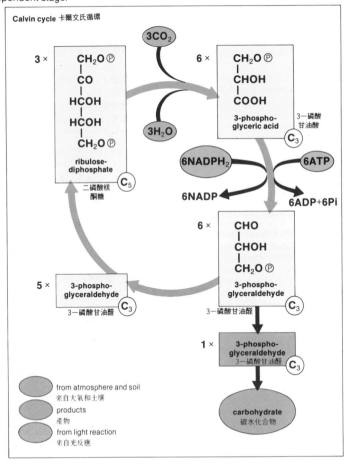

ribulose diphosphate RUDP. A pentose (p. 17) with which carbon dioxide is combined at the beginning of the Calvin cycle (↑).

phosphoglyceric acid PGA. A complex organic acid (p. 15) which is formed as the result of the combination of carbon dioxide with RUDP (↑) in fixing of carbon dioxide at the beginning of the Calvin cycle (↑).

phosphoglyceraldehyde (*n*) a compound formed as the result of the reduction (p. 32) of PGA (↑) during the Calvin cycle (↑). This is then synthesized into starch (p. 18) which is the most important product of photosynthesis (p. 93). Also known as **triose phosphate**.

phosphoenol pyruvic acid PEP. An organic compound (p. 15) which is used by C4 plants (p. 95) in the fixation of carbon dioxide instead of RUDP (↑). Using this compound, carbon dioxide can be stored in chemical form and used later. This is very useful in areas, e.g. the tropics, where carbon dioxide may be in short supply.

compensation point the point at which the intensity of light is such that the amount of carbon dioxide produced by respiration (p. 112) and photorespiration (↓) exactly balances the amount consumed by photosynthesis (p. 93).

photorespiration (*n*) a light-dependent process in which carbon dioxide is produced and oxygen used up, wasting carbon and energy.

animal nutrition heterotrophic nutrition (p. 92) in which carbohydrates (p. 17) and fats are needed for structural materials and for energy, amino acids (p. 21) are needed to supply nitrogen, and to stimulate growth etc, minerals are required to ensure that the body functions healthily, and vitamins (p. 25) are required to promote and maintain growth.

joule (*n*) the work done when the point of application of a force of 1 newton is displaced through a distance of 1 metre in the direction of the force. 1 calorie (↓) is equivalent to 4.18 joules. The joule can be used as a measure of the energy value of nutrients (p. 92).

kilojoule (*n*) = 1000 joules (↑).

calorie (*n*) *see* joule (↑).

compensation point 補償點

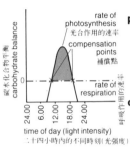

rate of photosynthesis
光合作用的速率
compensation points
補償點
碳水化合物平衡 carbohydrate balance
rate of respiration
呼吸作用的速率
time of day (light intensity)
二十四小時內的不同時刻（光強度）

photorespiration
光呼吸作用

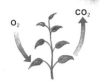

high O_2 concentration in plant tissue
植物組織中的 O_2 濃度高

low CO_2 concentration in plant tissue
植物組織中的 CO_2 濃度低

二磷酸核酮糖　即 RUDP。在卡爾文氏循環（↑）起始時，與二氧化碳相結合的一種戊糖（第17頁）。

磷酸甘油酸　即 PGA。卡爾文氏循環（↑）起始時，在固定二氧化碳過程中，二氧化碳與 RUDP（↑）相結合生成的一種複雜的有機酸（第15頁）。

磷酸甘油醛（名）　在卡爾文氏循環（↑）過程中，由 PGA（↑）還原（第32頁）生成的一種化合物。此後，這種化合物再被合成爲光合作用（第93頁）最重要產物的澱粉（第18頁）。它也稱爲 磷酸丙糖。

磷酸烯醇丙酮酸　即 PEP。四碳植物（第95頁）在二氧化碳固定過程中，用以代替 RUDP（↑）的一種有機化合物（第15頁）。利用這種化合物，二氧化碳能以化合物形式貯存起來，待以後使用。這在二氧化碳供給可能發生短缺的地區如熱帶地區是十分有用的。

補償點　呼吸作用（第112頁）和光呼吸作用（↓）產生的二氧化碳量正好與光合作用（第93頁）消耗的二氧化碳量相等時的光強度。

光呼吸（作用）（名）　一種依賴光的過程。在這種過程中，會產生二氧化碳和消耗氧氣，從而浪費碳和能量。

動物營養　一種異養（第92頁）的營養方式。利用這種營養方式時，需要碳水化合物（第17頁）和脂肪，用以建造身體的結構物質和提供能量；需要氨基酸（第21頁），用以提供氮素和促進生長等；需要礦物質，以確保身體功能健康地發揮；需要維生素（第25頁），以促進和維持生長。

焦耳（名）　1牛頓的力作用於一點時，此點在力的方向發生1米距離的位移所做的功。1卡（↓）等於4.18焦耳。焦耳可作爲營養素（第92頁）的能量值的一種計量單位。

千焦耳（名）　同1000焦耳（↑）。

卡（名）　見焦耳（↑）。

gut or alimentary canal 胃腸道或消化道

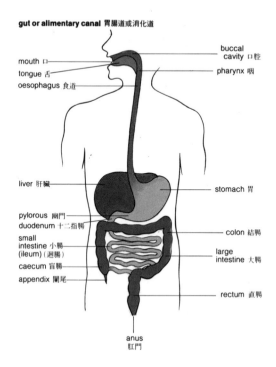

mouth 口

tongue 舌

oesophagus 食道

buccal cavity 口腔

pharynx 咽

liver 肝臟

stomach 胃

pylorous 幽門

duodenum 十二指腸

small intestine 小腸 (ileum) (迴腸)

caecum 盲腸

appendix 闌尾

colon 結腸

large intestine 大腸

rectum 直腸

anus 肛門

gut (n) a tube, the gastro-intestinal tract, usually leading from the mouth to the anus (p. 103), in animals, which in humans may be as much as 9 metres long and in which food is conveyed, digested (↓) and absorbed (p. 81).

alimentary canal = gut (↑).

ingestion (n) the process of taking nutrients (p. 92) into the body for digestion (↓).

digestion (n) the breakdown of complex organic compounds (p. 15) or nutrients (p. 92) into simpler, soluble materials which can then be used in the metabolism (p. 26) of the animal.

egestion (n) the process of eliminating or discharging food or waste products from the body.

胃腸道(名) 動物體內一種通常由口通至肛門(第103頁)的胃腸管道。人的胃腸道可長達9米。在其內進行食物的輸送、消化(↓)和吸收(第81頁)。

消化道 同胃腸道(↑)。

攝食(名) 將營養素(第92頁)攝入體內供消化(↓)的過程。

消化(作用)(名) 將複雜有機化合物(第15頁)或營養素(第92頁)分解成較簡單的可溶性物質的作用。這些物質此後可用於動物的新陳代射(第26頁)。

排遺作用(名) 從體內排除或放出食物殘渣的過程。

faeces (*n.pl.*) the substances that remain after digestion (↑) and absorption (p. 81) of food in the alimentary canal (↑).

defaecation (*n*) the process of egesting (↑) unwanted food from the body. Material which is defaecated has not taken part in the metabolism (p. 26) of the organism and is therefore not an example of excretion (p. 134).

assimilation (*n*) after digestion (↑), the process of taking into the cells the simple, soluble organic compounds (p. 15) which can then be converted into the complex organic compounds from which the organism is made. **assimilate** (*v*).

buccal cavity the mouth cavity. In mammals (p. 80), that part of the alimentary canal (↑) into which the mouth opens and in which food particles are masticated (p. 104) before they are swallowed.

mucus (*n*) any slimy fluid produced by the mucous membranes (p. 14) of animals and used for protection and lubrication.

saliva (*n*) a fluid secreted into the buccal cavity (↑) by the salivary gland (p. 87) in response to the presence of food. It consists mainly of mucus (↑) and lubricates the food before swallowing. It contains an enzyme (p. 28) in some animals to aid the digestion (↑) of starch (p. 18).

pharynx (*n*) the part of the alimentary canal (↑) between the buccal cavity (↑) and the oesophagus (↓) into which food that has been masticated (p. 104) is pushed by the tongue. The pharynx then contracts by muscular (p. 143) action to force the food into the oesophagus. The gill slits (p. 113) open into the pharynx in fish.

oesophagus (*n*) the part of the alimentary canal (↑) between the pharynx (↑) and the stomach (p.100). It is lined with a folded mucous membrane (p.14) and has layers of smooth muscle fibres (p.144) which contract to force food into the stomach.

epiglottis (*n*) a flap which closes the trachea (p. 175) during swallowing so that food passes into the oesophagus (↑) and not into the trachea.

bolus (*n*) a rounded mass consisting of masticated (p. 104) food particles and saliva (↑) into which food is formed in the buccal cavity (↑) before swallowing.

糞便（名、複）　食物在消化道（↑）中經消化（↑）和吸收（第81頁）以後殘留的物質。

排糞（名）　將不需要的食物從體內排泄（↑）出去的過程。作爲糞便排出的物質並未參與生物體的新陳代謝（第26頁），因此，它不是一種排泄物（第134頁）。

同化（作用）（名）　經消化作用（↑），將簡單的可溶性有機化合物（第15頁）吸收到細胞內的過程。這種簡單的可溶性有機化合物此後可以轉化爲組成生物體的複雜有機化合物。（動詞形式爲assimilate）

口腔　哺乳動物（第80頁）的口通向消化道（↑）的那一部份。食物顆粒在被下嚥之前，先在其內受到咀嚼（第104頁）。

黏液（名）　由動物的黏膜（第14頁）產生的任何黏性液體。這種液體可起保護和潤滑作用。

唾液、涎（名）　因食物的存在而由唾腺（第87頁）分泌到口腔（↑）內的一種液體。這種液體主要由黏液（↑）組成，在食物下嚥之前，可對其起潤滑作用。某些動物的唾液中含有一種有助於澱粉（第18頁）消化（↑）的酶（第28頁）。

咽（名）　口腔（↑）和食管（↓）之間的消化道（↑）部份。經過咀嚼（第104頁）的食物由舌推入其內。然後由於肌肉（第143頁）的作用，引起咽收縮，又迫使食物進入食管。魚的鰓裂（第113頁）通向咽部。

食道（名）　咽（↑）和胃（第100頁）之間的消化道（↑）部份。其內襯有一層折疊的黏膜（第14頁），並具有多層平滑肌纖維（第144頁），這些肌層可收縮迫使食物進入胃中。

會厭（名）　在吞嚥時關閉氣管（第175頁）的一個活瓣，使食物在吞嚥時進入食管（↑），而不進入氣管。

團（名）　咀嚼（第104頁）後的食物顆粒和唾液（↑）組成的一種圓形團塊。這種團塊是食物在吞嚥之前於口腔（↑）內形成的。

stomach (*n*) the part of the alimentary canal (p. 98) between the oesophagus (p. 99) and the duodenum (↓) into which food is passed and can be stored in quite large amounts for long periods so that it is not necessary for the animal to be eating continuously. Food is mixed with gastric juices and, although little of it is absorbed (p. 81), materials, such as minerals or vitamins (p. 25), may be taken into the blood (p. 90) stream. The stomach is muscular (p. 143) and is lined with a mucous membrane (p. 14).

peristalsis (*n*) the waves of rhythmical contractions which take place in the alimentary canal (p. 98) by muscular (p. 143) action and which force food through the canal.

chief cell one of a number of cells found in the gastric glands (p. 87) which secrete (p. 106) the enzymes (p. 28) pepsin (p. 107) and rennin (p. 106) which digest (p. 98) proteins (p. 21) and milk protein (in young mammals (p. 80)) respectively. Also known as **peptic cell**.

fundis gland a gland (p. 87) in the stomach (↑) which secretes (p. 106) mucus (p. 99) to protect and lubricate the wall of the stomach.

胃（名）食道（第99頁）和十二指腸（↓）之間的消化道（第98頁）部份。食物可在其內大量貯存較長時間，使動物無需不停進食。食物與胃液混合後，雖然被吸收（第81頁）的量極少，但其中一些物質諸如礦物質和維生素（第25頁）則可被吸收而進入血（第90頁）流。胃是肌肉質的（第143頁），其內壁襯有一層黏膜（第14頁）。

蠕動（名）消化道（第98頁）中由於肌肉（第143頁）的節律性及收縮而形成的一些波動。這些波動可迫使食物通過消化道。

主細胞 胃腺（第87頁）中含的若干類細胞中的一種。這類細胞可分泌（第106頁）出分別消化（第98頁）蛋白質（第21頁）和乳蛋白（在幼小的哺乳動物（第80頁）中）的一些酶（第28頁）——胃蛋白酶（第107頁）和凝乳酶（第106頁）。也稱為胃酶細胞。

胃底腺 胃（↑）內的一種腺（第87頁）。這種腺可分泌出（第106頁）黏液（第99頁），以保護和潤滑胃壁。

peristalsis 蠕動

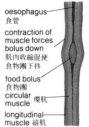

oesophagus
食管
contraction of muscle forces bolus down
肌肉收縮促使食物團下移
food bolus
食物團
circular muscle 環肌
longitudinal muscle 縱肌

wave of contraction passing down oesophagus
沿食管而下的收縮波

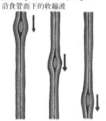

section of stomach wall 胃壁的切面

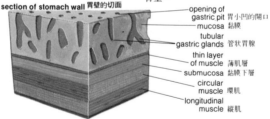

opening of gastric pit 胃小凹的開口
mucosa 黏膜
tubular gastric glands 管狀胃腺
thin layer of muscle 薄肌層
submucosa 黏膜下層
circular muscle 環肌
longitudinal muscle 縱肌

oxyntic cell one of a number of cells found in the gastric glands (p. 87) which secrete (p. 106) hydrochloric acid (HCl) which in turn kills harmful bacteria (p. 42), makes available calcium and iron salts and provides a suitably low pH (p. 15) for the formation of pepsin (p.107)

chyme (*n*) a partially broken-down, semi-fluid mixture of food particles and gastric juices which is then released in small quantities into the duodenum (↓).

泌酸細胞 胃腺（第87頁）中含的若干類細胞中的一類。這種細胞可分泌（第106頁）鹽酸（HCl），鹽酸能殺死有害細菌（第42頁），並獲得鈣鹽和鐵鹽，為胃蛋白酶（第107頁）的生成提供適當低的 pH（第15頁）值。

食糜（名）（在胃中）由食物顆粒和胃液組成的一種已部份分解的半流體狀混合物。這種混合物隨後即不斷地以少量釋入十二指腸（↓）。

detail of gastric gland 胃腺結構詳圖

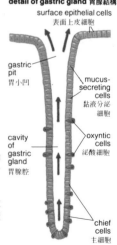

surface epithelial cells
表面上皮細胞
gastric pit
胃小凹
mucus-secreting cells
黏液分泌細胞
cavity of gastric gland
胃腺腔
oxyntic cells
泌酸細胞
chief cells
主細胞

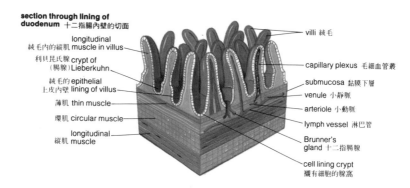

section through lining of duodenum 十二指腸內壁的切面

絨毛內的縱肌 longitudinal muscle in villus

利貝昆氏腺 crypt of (腸腺) Lieberkuhn

絨毛的 epithelial 上皮內壁 lining of villus

薄肌 thin muscle

環肌 circular muscle

longitudinal 縱肌 muscle

villi 絨毛

capillary plexus 毛細血管叢

submucosa 黏膜下層

venule 小靜脈

arteriole 小動脈

lymph vessel 淋巴管

Brunner's gland 十二指腸腺

cell lining crypt 襯有細胞的腺窩

duodenum (n) the part of the alimentary canal (p. 98) between the stomach (↑) and the ileum (p. 102) which forms the first part of the small intestine (p. 102) into which the chyme (↑) passes from the stomach. Digestion (p. 98) continues in the duodenum with the aid of intestinal juices (p. 102) and, in addition, it receives secretions (p. 106) from the pancreas (p. 102) and the liver (p. 103). Its lining is covered with villi (p. 103) and the glands (p. 87) that secrete the intestinal juices.

chyle (n) lymph (p. 128) containing the results of digestion (p. 98). The liquid looks milky because it contains emulsified (p. 26) fats and oils.

bile (n) a secretion (p. 106) from the liver (p. 103) which contains some waste material from the liver and bile salts which emulsify fats, increase the activity of certain enzymes (p. 28), aid in the absorption (p. 81) of some vitamins (p. 25) and is rich in sodium bicarbonate which neutralizes stomach (↑) acid.

bile duct the tube through which bile (↑) is passed from the liver (p. 103) to the duodenum (↑).

gall bladder a sac-like bladder extending from the bile duct (↑) and situated within or near the liver (p. 103). It functions as a store for bile (↑) when it is not required for digestive (p. 98) purposes and then, by muscular (p. 143) contractions, empties into the duodenum (↑) through the bile duct.

十二指腸（名） 胃（↑）和迴腸（第102頁）之間的消化道（第98頁）部份。這一部份構成小腸（第102頁）的第一段，來自胃的食糜（↑）即釋入其中。在十二指腸內，藉腸液（第102頁）的幫助繼續進行消化（第98頁）。此外，十二指腸還接受來自胰腺（第102頁）和肝臟（第103頁）的分泌物（第106頁）。其內壁覆有絨毛（第103頁）和能分泌腸液的多種腺體（第87頁）。

乳糜（名） 含有消化（第98頁）產物的淋巴（第128頁）。由於其含有乳化（第26頁）脂肪和油脂，所以這種液體呈乳白色。

膽汁（名） 肝臟（第103頁）的一種分泌物（第106.頁）。這種分泌物含有來自肝臟的某些廢物和膽汁鹽。膽汁鹽能乳化脂肪，增加某些酶（第28頁）的活性，幫助吸收（第81頁）某些維生素（第25頁），並富含可中和胃（↑）酸的碳酸氫鈉。

膽管（名） 膽汁（↑）從肝臟（第103頁）輸送至十二指腸（↑）的管道。

膽囊 由膽管（↑）擴大而成的囊狀物。該囊狀物位於肝臟（第103頁）的內部或處於靠近肝臟的地方。當不需要其參與消化（第98頁）時，它就起貯存膽汁（↑）的作用；到時候它就藉肌肉（第143頁）的收縮，使膽汁經膽管流入十二指腸（↑）。

pancreatic juice a solution (p. 118) in water of alkaline salts, which neutralize the acid (p. 15) from the stomach (p. 100), and enzymes (p. 28) to aid in digestion (p. 98).

pancreas (*n*) a gland (p. 87) which is connected to the duodenum (p. 101) by a duct and which produces pancreatic juice (↑) and insulin (↓).

islets of Langerhans cells contained within the pancreas (↑) which produce insulin (↓).

insulin a hormone (p. 130) which controls the sugar level in the blood (p. 90) and, if it is deficient, the sugar level rises while, if it is in excess, the level falls leading to a coma.

jejunum (*n*) the part of the small intestine (↓) between the duodenum (p. 101) and the ileum (↓).

small intestine the narrow tube which forms part of the alimentary canal (p. 98) between the stomach (p. 100) and the colon (↓). It is the main region of digestion (p. 98) and absorption (p. 81) and includes the duodenum (p. 101).

ileum (*n*) the longest, usually coiled part of the small intestine (↑) between the jejunum (↑) and the colon (↓). It is muscular (p. 143) and causes food particles to pass along it by peristalsis (p. 100). Its lining is folded and covered with large numbers of villi (↓) which increase the surface area for absorption (p. 81).

intestinal juice a secretion (p. 106) produced in the glands (p. 87) of the intestine (↑) which contains a mixture of enzymes (p. 28) of digestion (p. 98), such as amylase (p. 106) and sucrase (p. 107). Also known as **succus entericus**

Brunner's glands the deep-lying glands (p. 87) in the walls of the duodenum (p. 101) which secrete (p. 106) intestinal juices.

crypts of Lieberkuhn the glands (p. 87) found within the walls of the small intestine (↑) which secrete (p. 106) the intestinal juices.

appendix (*n*) in humans, a blind-ended tube at the end of the caecum (↓).

caecum (*n*) a blind-ended branch of the gut (p. 98) at the junction of the small and large intestines (↑). It is very large and important in the digestions (p. 98) of some mammals (p. 80), not humans.

胰液 一種鹼鹽的水溶液(第118頁)。它能中和胃(第100頁)酸(第15頁),並含有多種有助於消化(第98頁)作用的酶(第28頁)。

胰腺(名) 有一導管與十二指腸(第101頁)相通的一種腺(第87頁)。這種腺體分泌胰液(↑)和胰島素(↓)。

胰島 胰島(↑)內含產生胰島素(↓)的細胞羣。

胰島素 能控制血(第90頁)糖含量的一種激素(第130頁)。胰島素分泌不足,血糖含量會昇高;胰島素分泌過剩,則血糖含量會下降,導致昏迷。

空腸(名) 十二指腸(第101頁)和迴腸(↓)之間的小腸(↓)部份。

小腸 胃(第100頁)和結腸(↑)之間構成消化道(第98頁)一部份的窄管。小腸是進行消化(第98頁)和吸收(第81頁)的主要部位。它也包括十二指腸(第101頁)。

迴腸(名) 小腸(↑)中最長的、通常呈盤曲狀的那一部份。它位於空腸(↑)和結腸(↓)之間。迴腸是肌肉質的(第143頁),藉蠕動(第100頁)作用可使食物顆粒得以通過。迴腸內壁有皺褶,並覆有大量可增加其吸收(第81頁)表面積的絨毛(↓)。

腸液 小腸(↑)中的腺體(第87頁)所產生的一種分泌液(第106頁)。這種分泌液含有諸如澱粉酶(第106頁)和蔗糖酶(第107頁)這樣一些消化(第98頁)酶(第28頁)組成的一種混合物。腸液的拉丁文名稱爲 succus entericus

十二指腸腺(名) 深埋於十二指腸(第101頁)腸壁內,可分泌(第106頁)腸液的腺(第87頁)。

利貝昆氏腺窩、腸腺(名) 在小腸(↑)腸壁內,可分泌(第106頁)腸液的腺(第87頁)。

闌尾(名) 人體盲腸(↓)末端的一根盲管。

盲腸(名) 位於大、小腸(↑)連接處的、消化道(第98頁)的一個盲端分支。某些哺乳動物(第80頁)的盲腸很大,在消化(第98頁)中起重要作用;但是,人的盲腸則已退化。

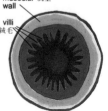

section across 小腸(迴腸)的橫切面
small intestine (ileum)

muscular 肌壁
wall

villi 絨毛

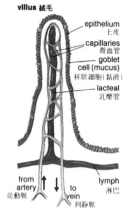

villus 絨毛

epithelium
上皮
capillaries
微血管
goblet
cell (mucus)
杯狀細胞(黏液)
lacteal
乳糜管

from
artery
從動脈
to
vein
到靜脈
lymph
淋巴

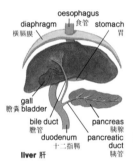

oesophagus
食管
diaphragm
橫膈膜
stomach
胃
gall
bladder
膽囊
pancreas
胰腺
bile duct
膽管
pancreatic
duct
胰管
duodenum
十二指腸
liver 肝

colon (n) the first part of the large intestine. It secretes (p. 106) mucus (p. 99) and contains the remains of food materials which cannot be digested (p. 98) as well as the digestive juices. From this material, water and vitamins (p. 25) are reabsorbed into the blood (p. 90) leaving the faeces (p. 99) which are moved on to the rectum (↓).

rectum (n) the part of the large intestine in which the faeces (p. 99) are stored before release through the anus (↓).

anus (n) the posterior opening to the alimentary canal (p. 98) through which the faeces (p. 99) may pass at intervals and which is closed by a ring of muscles (p. 143) called the anal sphincter (p. 127).

villi (n.pl.) the rod-like projections which cover the lining of the small intestine (↑) to increase the surface area for absorption (p. 81). **villus** (sing.).

liver (n) a gland (p. 87) which lies close to the stomach (p. 100) and is connected with the small intestine (↑) by the bile duct (p. 101) through which it secretes (p. 106) bile (p. 101) for digestive (p. 98) purposes. The liver also removes damaged red cells (p. 91) from the blood (p. 90), stores iron, synthesizes vitamin (p. 25) A, and stores vitamins A, D, and B, synthesizes blood proteins (p. 21), removes blood poisons, and synthesizes agents which help blood clot (p. 129), breaks down alcohol, stores excess carbohydrates (p. 17) and metabolizes (p. 26) fats.

hepatic portal vein one of a system of veins (p. 127) which carry blood (p. 90) rich in absorbed (p. 81) food materials, such as glucose (p. 17), direct from the intestine (↑) to the liver (↑).

liver cell one of the cells making up the liver (↑). Each cell is in direct contact with the blood (p. 90) so that material diffuses between blood and liver very rapidly. Liver cells are cube shaped with granular cytoplasm (p. 10).

reticulo-endothelial system a system of macrophage (p. 88) cells which is present in the liver (↑) and other parts of the body and which is in contact with the blood (p. 90) and other fluids. These macrophage cells are able to engulf foreign bodies and thus protect the body from infection, damage and disease.

結腸(名) 大腸的第一段。結腸可分泌(第106頁)黏液(第99頁)，內含不能消化(第98頁)的食物殘留物以及消化液。結腸還可從此殘留物中再吸收水分和維生素(第25頁)，將之送入血液(第90頁)留下待送直腸(↓)的糞便(第99頁)。

直腸(名) 大腸的一段。糞便(第99頁)經肛門(↓)排出之前，積聚於這一段大腸中。

肛門(名) 消化道(第98頁)的後部開口。糞便(第99頁)每隔一段時間可經此開口排出。它由肛門括約肌(第127頁)的環形肌(第143頁)控制關閉。

絨毛(名、複) 覆蓋小腸(↑)內壁，以增加其吸收(第81頁)表面積的許多棒狀突出物。(單數形式爲 villus)

肝(臟)名 靠近胃(第100頁)的一種腺(第87頁)。這種腺經由膽管(第101頁)與小腸(↑)相通，它可分泌(第106頁)膽汁(第101頁)，通過膽管的輸送，起消化(第98頁)作用。肝臟能除去血液(第90頁)送來的已損壞的紅細胞(第91頁)，貯存鐵，合成維生素(第25頁)A，貯存維生素A、D和B，合成血液中的蛋白質(第21頁)，清除血液毒物，合成有助於血液凝固(第129頁)的各種因子，分解醇，貯存過剩的碳水化合物(第17頁)，以及使脂肪進行新陳代謝(第26頁)。

肝門靜脈 靜脈(第127頁)系中的一條靜脈。這條靜脈從腸(↑)通往肝臟(↑)，其中所輸送的血液(第90頁)含有豐富的、已吸收的(第81頁)多種營養物質，例如葡萄糖(第17頁)。

肝細胞 組成肝(↑)的細胞。每一個肝細胞都直接與血液(第90頁)接觸，使物質在血液和肝之間的擴散就非常迅速。肝細胞呈立方體形，具有顆粒狀的細胞質(第10頁)。

網狀皮內系統 由巨噬細胞(第88頁)組成的系統。這種系統存在於身體的肝臟(↑)和其他部份，與血液(第90頁)及其他一些液體保持接觸。這些巨噬細胞能吞噬外來物體，從而保護身體免受感染、傷害和疾病。

tooth (*n*) one of a number of hard, resistant structures found growing on the jaws (↓) of vertebrate (p. 74) animals and which are used to break down food materials mechanically. They may be specialized for different functions among different animals and even within the same animal. **teeth** (*pl.*).

dentition (*n*) the kind, arrangement and number of the teeth (↑) of an animal.

heterodont dentition the condition in which the teeth (↑) of an animal, typically mammals (p. 80), are differentiated into different forms, such as molars (↓) and canines (↓), to perform different functions, such as grinding up the food or killing prey.

homodont dentition the condition in which the teeth (↑) of an animal are all identical.

mastication (*n*) the process which takes place in the buccal cavity (p. 99) whereby food is mechanically broken down by the action of teeth (↑), tongue, and cheeks into a bolus (p. 99) for swallowing.

dental formula a formula which indicates by letters and numbers the types and numbers of teeth (↑) in the upper and lower jaws (↓) of a mammal (p. 80). For example, the dental formula of a human would be
i2/2, c1/1, p2/2, m3/3
which indicates that both the upper and lower jaws have two incisors (↓), one canine (↓), two premolars (↓), and three molars (↓) on each side of the jaw.

incisor (*n*) one of the chisel-shaped teeth (↑), very prominent in rodents, that occur at the very front of a mammal's (p. 80) jaw (↓) and which have a single root and a sharp edge with which to sever portions of the food.

canine (*n*) one of the 'dog teeth' or pointed conical teeth (↑) which occur between the incisors (↑) and the premolars (↓) and which are used to kill prey in carnivorous (p. 109) animals such as dogs and cats.

carnassial (*n*) one of the large flesh-cutting teeth (↑) found in terrestrial (p. 219) carnivores (p. 109).

牙、齒(名) 脊椎動物(第74頁)頜骨(↓)上生長的若干個堅硬而耐久的結構物。這些結構物可用於機械地破碎食物。不同類動物間，甚至在同一類動物中，牙的功能也不盡相同。(複數形式爲 teeth)

齒系、牙列(名) 動物牙(↑)的種類、排列和數目。

異齒系、異形牙牙列 動物，特別是一些哺乳動物(第80頁)的牙(↑)長成不同類型分化的狀態。例如，分化爲臼齒(↓)和犬齒(↓)，行使諸如磨碎食物或咬死捕獲物的不同功能。

同齒系、同形牙牙列 動物的牙(↑)長成完全同一的狀態。

咀嚼(作用)(名) 發生於口腔(第99頁)中的過程。借助於這一過程，依靠牙(↑)、舌和頰的動作，食物被機械地破碎，形成一團(第99頁)，以便吞嚥。

齒式、牙式 一種用字母和數字表示哺乳動物(第80頁)上、下頜骨(↓)上牙(↑)的種類及數目的公式。人的牙式是 i2/2，c1/1，p2/2，m3/3，它表示每一側頜的上、下頜骨上各有門齒(↓)二枚，犬齒(↓)一枚，前臼齒(↓)二枚和臼齒(↓)三枚。

門齒、切牙(名) 鑿形牙(↑)。齧齒動物的這種牙非常凸出；在哺乳動物(第80頁)中，長於頜骨(↓)的正前方。它有單個牙根和銳利的切緣，可藉助切緣將食物切斷成幾個部份。

犬齒、犬牙、尖牙(名) 長於切牙(↑)和前磨牙(↓)之間的"狗牙"或尖的圓錐形牙(↑)。食肉動物(第109頁)諸如狗和貓的犬牙用來咬死捕獲物。

裂齒、裂牙(名) 陸生(第219頁)食肉動物(第109頁)的大型切肉牙(↑)。

heterodont dentition 異齒系
e.g. carnivore (a dog) 圖例：食肉動物(

incisors 門齒
upper jaw (maxilla) 上頜 (上頜骨)
molars 臼齒
canines 犬齒
premolars 前臼齒
carnassials 裂齒
lower jaw (mandible) 下頜(下頜骨)

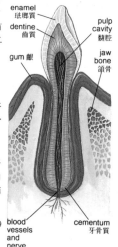

incisor 門齒

enamel 琺瑯質
dentine 齒質
gum 齦
pulp cavity 髓腔
jaw bone 頜骨
blood vessels and nerve fibres 血管和神經纖維
cementum 牙骨質

molar臼齒

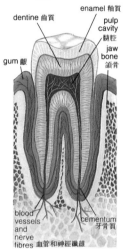

enamel 釉質
dentine 齒質
pulp cavity 髓腔
jaw bone 頜骨
gum 齦
blood vessels and nerve fibres 血管和神經纖維
cementum 牙骨質

premolar (n) one of the crushing and grinding cheek teeth (↑) that occur between the canines (↑) and molars (↓) in mammals (p. 80). They are usually ridged and furrowed with more than one root. They are represented in the first 'milk' or deciduous (p. 59) dentition (↑).

molar (n) one of the large crushing and grinding cheek teeth (↑) that occur at the back of the mouth of a mammal (p. 80). They are ridged and furrowed with more than one root. They are not represented in the first 'milk' or deciduous (p. 59) dentition (↑) and, in humans, there are four which do not erupt until later in life and are referred to as 'wisdom teeth'.

enamel (n) the hard outer layer of the tooth (↑) of a vertebrate (p. 74). It is composed mainly of crystals of the salts, carbonates, phosphates, and fluorides of calcium held together by small amounts of an organic compound (p. 15).

dentine (n) the hard substance that makes up the bulk of the tooth (↑) in mammals (p. 80). It is similar to bone but has a higher mineral content and no cells.

cementum (n) the hard substance that covers the root of the tooth (↑) in mammals (p. 80). It is similar to bone but has a higher mineral content and lacks Haversian canals (p. 89).

pulp cavity the substance within the centre of a tooth (↑) which contains the blood vessels (p. 127) and nerves (p. 149) which supply the tooth, together with connective tissue (p. 88). It connects to the tissue (p. 83) into which the tooth is fixed.

gum (n) the tissue (p. 83) that surrounds and supports the roots of the teeth (↑) and covers the jaw (↓) bones. It contains nerves (p. 149) and many blood capillaries (p. 127) giving it the characteristic pink colour when it is healthy. Also known as **gingiva**.

jaw (n) one of the bones in which the teeth (↑) are set. The jaw movements as well as the dentition (↑) of different animals are specialized for different actions, for example, tearing, snatching and chewing movements, such as crushing and grinding.

前臼齒、前磨牙（名）　哺乳動物（第80頁）中起壓碎和磨碎食物作用的頰牙（↑）。這些頰牙位於犬齒（↑）和臼齒（↓）之間；通常呈脊狀，有溝槽，牙根不止一個。初生的"奶齒"系（↑）或乳齒（第59頁）系中，就已長有這種牙。

臼齒、磨牙（名）　哺乳動物（第80頁）中，起壓碎和磨碎食物作用的、較大的頰牙（↑）。這些頰牙位於口的後面；呈脊狀，有溝槽，牙根不止一個。在初生的"奶齒"系（↑）或乳齒（第59頁）系中，並未長有這種牙。這種牙在人體中有四個，於生命的晚些時候才萌出，它們被稱爲"智牙"。

琺瑯質、釉質（名）　脊椎動物（第74頁）的牙（↑）的堅硬外層。它主要由鹽——碳酸鹽、磷酸鹽和鈣的氟化物——的結晶與一種少量的有機化合物（第15頁）黏合而成。

齒質（名）　組成哺乳動物（第80頁）牙（↑）的基本部份的堅硬物質。它與骨相類似；但是礦物質含量較高，且不含細胞。

牙骨質（名）　覆蓋哺乳動物（第80頁）牙（↑）根的堅硬物質。它與骨相類似；但是，礦物質含量較高，且沒有哈氏管（第89頁）。

髓腔　牙（↑）中心部份的物質，其中含有血管（第127頁）和神經（第149頁）。這些血管和神經與結締組織（第88頁）一起，爲牙提供營養，並爲牙植根於其內的組織（第83頁）相連。

（牙）齦（名）　包圍、支持牙（↑）根和覆蓋頜骨（↓）的組織（第83頁）。這種組織含有神經（第149頁）和許多毛細血管（第127頁）；在健康情況下，它呈特有的粉紅色。齦的拉丁文名詞爲gingiva。

頜、頜骨（名）　其內固定牙（↑）骨。不同動物的頜運動和齒系（↑），皆爲不同的動作——例如，撕裂、攫取和像壓碎和磨碎食物之類的咀嚼運動——所特化。

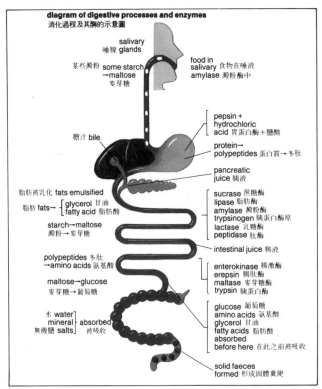

diagram of digestive processes and enzymes
消化過程及其酶的示意圖

salivary 唾腺 glands

某些澱粉 some starch →maltose 麥芽糖

food in salivary amylase 食物在唾液澱粉酶中

pepsin + hydrochloric acid 胃蛋白酶 + 鹽酸

膽汁 bile

protein→ polypeptides 蛋白質 → 多肽

pancreatic juice 胰液

脂肪被乳化 fats emulsified

脂肪 fats→ { glycerol 甘油 / fatty acid 脂肪酸 }

starch→maltose 澱粉 → 麥芽糖

sucrase 蔗糖酶
lipase 脂肪酶
amylase 澱粉酶
trypsinogen 胰蛋白酶原
lactase 乳糖酶
peptidase 肽酶

intestinal juice 腸液

polypeptides 多肽 →amino acids 氨基酸

maltose→glucose 麥芽糖 → 葡萄糖

enterokinase 腸激酶
erepsin 腸肽酶
maltase 麥芽糖酶
trypsin 胰蛋白酶

水 water mineral 無機鹽 salts } absorbed 被吸收

glucose 葡萄糖
amino acids 氨基酸
glycerol 甘油
fatty acids 脂肪酸
absorbed before here 在此之前被吸收

solid faeces formed 形成固體糞便

secretion (*n*) a material with a special function in an organism which is made within a cell and passed out of the cell to perform its function. **secrete** (*v*).

amylase (*n*) any of a number of enzymes (p. 28) which catalyze (p. 28) the hydrolysis (p. 16) of carbohydrates (p. 17), such as starch (p. 18), into simple sugars. It is secreted (↑) as saliva (p. 99), and in the pancreas (p. 102) and the small intestine (p. 102).

rennin (*n*) an enzyme (p. 28) which coagulates (p. 128) milk. It is secreted (↑) by the gastric glands (p. 87) of the stomach (p. 100).

分泌物（名） 生物體中，具有特殊功能的一種物質。這種物質在某種細胞內製造，然後從該種細胞排出，以行使其功能。（動詞形式為 secrete）

澱粉酶（名） 能催化（第28頁）如澱粉（第18頁）之類的碳水化合物（第17頁）水解（第16頁）成為單糖的若干種酶（第28頁）中的任何一種。澱粉酶以唾液（第99頁）的形式分泌（↑）出來，胰腺（第102頁）和小腸（第102頁）中也有分泌澱粉酶。

凝乳酶（名） 使乳凝固（第128頁）的一種酶（第28頁）。胃（第100頁）內的胃腺（第87頁）分泌（↑）凝乳酶。

maltase (*n*) an enzyme (p. 28) which catalyzes (p. 28) the hydrolysis (p. 16) of maltose (p. 18) into two molecules of glucose (p. 17). It is secreted (↑) by the small intestine (p. 102).

lactase (*n*) an enzyme (p. 28) which catalyzes (p. 28) the hydrolysis (p. 16) of the disaccharide (p. 18) lactose (p. 18) into glucose (p. 17) and galactose (p.18). It is secreted (↑) by the small intestine (p.102).

sucrase (*n*) an enzyme (p. 28) which catalyzes (p. 28) the hydrolysis (p. 16) of sucrose (p. 18) into glucose (p. 17) and fructose (p. 17). It is secreted (↑) by the small intestine (p. 102). Also known as **invertase**.

erepsin (*n*) a mixture of enzymes (p. 28) which catalyzes (p. 28) the breakdown of proteins (p. 21) into amino acids (p. 21). It is secreted (↑) by the small intestine (p. 102).

lipase (*n*) an enzyme (p. 28) which catalyzes (p. 28) the hydrolysis (p. 16) of fats into fatty acids (p. 20) and glycerol (p. 20). It is secreted (↑) by the pancreas (p. 102).

enterokinase (*n*) an enzyme (p. 28) which catalyzes (p. 28) the conversion of trypsinogen (p. 108) into trypsin (↓). It is secreted (↑) by the small intestine (p. 102).

chymotrypsin (*n*) an enzyme (p. 28) which catalyzes (p. 28) the conversion of proteins (p. 21) into amino acids (p. 21). It is secreted (↑) by the pancreas (p. 102).

pepsin (*n*) an enzyme (p. 28) which catalyzes (p. 28) the hydrolysis (p. 16) of proteins (p. 21) into polypeptides (p. 21) in acid solution (p.118). It is secreted (↑) by the stomach (p. 100) as pepsinogen (↓).

pepsinogen (*n*) the inactive form of pepsin (↑) which is secreted (↑) by the stomach (p. 100) and activated by hydrochloric acid (HCl).

gastrin (*n*) a hormone (p. 130) which stimulates the secretion (↑) of hydrochloric acid (HCl) and pepsin (↑) in the stomach (p. 100). It is activated by the presence of food materials.

trypsin (*n*) an enzyme (p. 28) which catalyzes (p. 28) the hydrolysis (p. 16) of proteins (p. 21) into polypeptides (p.21) and amino acids (p.21). It is secreted (↑) by the pancreas (p. 102) as trypsinogen (p. 108).

麥芽糖酶(名) 能催化(第28頁)麥芽糖(第18頁)水解(第16頁)成爲兩個分子的葡萄糖(第17頁)的一種酶(第28頁)。這種酶是由小腸(第102頁)分泌(↑)的。

乳糖酶(名) 能催化(第28頁)貳糖(第18頁)(乳糖(第18頁))水解(第16頁)成爲葡萄糖(第17頁)和半乳糖(第18頁)的一種酶(第28頁)。這種酶是由小腸(第102頁)分泌(↑)的。

蔗糖酶(名) 能催化(第28頁)蔗糖(第18頁)水解(第16頁)成爲葡萄糖(第17頁)和果糖(第17頁)的一種酶(第28頁)。這種酶是由小腸(第102頁)分泌(↑)的。它也稱爲轉化酶。

腸肽酶(名) 能催化(第28頁)蛋白質(第21頁)分解成爲氨基酸(第21頁)的一種多酶(第28頁)混合物。這種多酶混合物是由小腸(第102頁)分泌(↑)的。

脂肪酶(名) 能催化(第28頁)脂肪水解(第16頁)成脂肪酸(第20頁)和甘油(第20頁)的一種酶(第28頁)。這種酶是由胰腺(第102頁)分泌(↑)的。

腸激酶(名) 能催化(第28頁)胰朊酶原(第108頁)轉化成胰朊酶(↓)的一種酶(第28頁)。這種酶是由小腸(第102頁)分泌(↑)的。

胰凝乳蛋白酶(名) 能催化(第28頁)蛋白質(第21頁)轉變成爲氨基酸(第21頁)的一種酶(第28頁)。這種酶是由胰腺(第102頁)分泌(↑)的。

胃朊酶(名) 能在酸性溶液(第118頁)中催化(第28頁)蛋白質(第21頁)水解(第16頁)成爲多肽(第21頁)的一種酶(第28頁)。這種酶是以胃朊酶原(↓)形式由胃(第100頁)分泌(↑)的。

胃朊酶原(名) 胃朊酶(↑)的無活性形式。它是由胃(第100頁)分泌(↑)的,可被鹽酸(HCl)激活。

胃泌素(名) 能刺激胃(第100頁)內鹽酸(HCl)和胃朊酶(↑)分泌(↑)的一種激素(第130頁)。這種激素可因食物的存在而激活。

胰朊酶(名) 能催化(第28頁)蛋白質(第21頁)水解(第16頁)成爲多肽(第21頁)和氨基酸(第21頁)的一種酶(第28頁)。這種酶是以胰朊酶原(第108頁)形式,由胰腺(第102頁)分泌(↑)的。

trypsinogen (*n*) the inactive form of trypsin (p.107) which is secreted (p.106) by the pancreas (p.102) and converted into trypsin by enterokinase (p.107).

peptidase (*n*) an enzyme (p.28) which catalyzes (p.28) the hydrolysis (p.16) of polypeptides (p.21) into amino acids (p.21) by breaking down the peptide bonds (p.21). It is secreted (p.106) by the small intestine (p.102).

nucleotidase (*n*) an enzyme (p.28) which catalyzes (p.28) the hydrolysis (p.16) of a nucleotide (p.22) into its component nitrogen bases (p.22), pentose (p.17) and phosphoric acid (p.22). It is secreted (p.106) by the small intestine (p.102).

secretin (*n*) a hormone (p.130) which stimulates the secretion (p.106) of bile (p.101) from the liver (p.103) and digestive (p.98) juices from the pancreas (p.102). It is secreted by the duodenum (p.101).

pancreozymin (*n*) a hormone (p.130) which stimulates the release of digestive (p.98) juices from the pancreas (p.102). It is secreted (p.106) by the duodenum (p.101).

microphagous (*adj*) of an animal which feeds on food particles that are tiny compared with the size of the animal so that it must feed continuously to receive enough nutrients (p.92)

filter feeder a microphagous (↑) feeder which lives in water and filters suspended food particles from the water.

deposit feeder a microphagous (↑) feeder which feeds on particles that have been deposited on and perhaps mixed with the base layer of the environment (p.218) in which the animal lives.

fluid feeder a microphagous (↑) feeder which feeds by ingesting (p.98) fluids containing nutrients (p.92) from living or recently dead animals or plants.

pseudopodial feeder a microphagous (↑) feeder in which cells develop temporary, finger-like projections, pseudopodia (p.44), to engulf food particles.

mucous feeder a microphagous (↑) feeder in which food particles are trapped in mucus (p.99) secreted (p.106) by the organism and moved by ciliary (p.12) action to the mouth.

胰朊酶原（名） 胰腺（第102頁）分泌（第106頁）的胰朊酶（第107頁）的鈍性體。它可由腸激酶（107頁）轉化成胰朊酶。

肽酶（名） 可藉肽鍵（第21頁）的斷裂，催化（第28頁）多肽（第21頁）水解（第16頁）成爲氨基酸（第21頁）的一種酶（第28頁）。這種酶是由小腸（第102頁）分泌（第106頁）的。

核苷酸酶（名） 能催化（第28頁）核苷酸（第22頁）水解（第16頁）成爲其組成成份（含氮鹼（第22頁）、戊糖（第17頁）和磷酸（第22頁））的一種酶（第28頁）。這種酶是由小腸（第102頁）分泌（第106頁）的。

胰泌素（名） 能刺激肝（第103頁）分泌（第106頁）膽汁（第101頁）和促進胰腺（第102頁）分泌消化（第98頁）液的一種激素（第130頁）。這種激素是由十二指腸（第101頁）分泌的。

胰酶分泌素（名） 能刺激胰腺（第102頁）釋放消化（第98頁）液的一種激素（第130頁）。這種激素是由十二指腸（第101頁）分泌（第106頁）的。

微噬的、食微粒的（形） 指動物而言，與這種動物的身體相比，這種動物是以微小的食物顆粒為食的，這樣，該種動物就必須不斷攝食，以獲取足夠的營養素（第92頁）。

濾食性動物 一種微噬（↑）動物。這種微噬動物生活在水中，從水中濾取懸浮的食物顆粒。

食碎屑動物 一種微噬（↑）動物。這種微噬動物以已經沉積在或相混於其生活環境（第218頁）的底層的微粒為食。

食流體動物 一種微噬（↑）動物。這種微噬動物以攝食（第98頁）含營養素（第92頁）的流體為生。營養素來自活的或新近死亡的動植物。

偽足攝食動物 一種微噬（↑）動物。在這種微噬動物中，一些細胞形成暫時性的指狀突出物——偽足（第44頁），用以吞噬食物顆粒。

黏液攝食動物 一種微噬（↑）動物。在這種微噬動物中，食物顆粒被截留在機體所分泌（第106頁）的黏液（第99頁）中，然後藉纖毛（第12頁）的動作被送至口部。

types of feeding 攝食的類型
microphagous feeder 微噬動物

e.g. Right whale 圖例：露脊鯨

fluid feeder e.g. mosquito
食流體動物 圖例：蚊

pseudopodial feeder
偽足攝食動物 e.g. Amoeba
圖例：變形蟲

mucous feeder
e.g. lancelet
黏液攝食
圖例：文昌魚

mouth with tentacles
長有觸的口

types of feeding 攝食類型

setous feeder
e.g. *Daphnia*
剛毛攝食動物 圖例：水蚤

macrophagous feeders 巨噬動物

雜食動物
omnivore
e.g. American oppossum
圖例：美洲貙

食肉動物
carnivore
e.g. tiger
圖例：虎

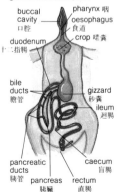

反芻食草動物
ruminant herbivore
e.g. gazelle 圖例：瞪羚

gut of a bird 鳥的消化道

buccal cavity 口腔
pharynx 咽
oesophagus 食道
crop 嗉囊
duodenum 十二指腸
bile ducts 膽管
gizzard 砂囊
ileum 迴腸
pancreatic ducts 胰管
pancreas 胰臟
caecum 盲腸
rectum 直腸

setous feeder a microphagous (↑) feeder in which the food particles are trapped by setae (p. 65) and then moved towards the mouth by beating cilia (p. 12).

macrophagous (*adj*) of an animal which feeds on relatively large food particles and usually, therefore, does not need to feed continuously.

coprophagous (*adj*) of an animal, such as some rodents, which feed on faeces (p. 99) thus improving the digestion (p. 98) of cellulose (p. 19) on second passage.

omnivore (*n*) an animal which feeds by eating a mixed diet of animal and plant food material.

carnivore (*n*) an animal which feeds by eating a diet that consists mainly of animal material. Carnivores may have powerful claws and dentition (p. 104) adapted to tearing flesh.

herbivore (*n*) an animal which feeds by eating a diet that consists mainly of plant material. Herbivores may have dentition (p. 104) and digestion (p. 98) specially adapted to deal with tough plant materials.

ruminant (*n*) one of the group of herbivores (↑) which belong to the order Artiodactyla and in which the stomach (p. 100) is complex and includes a rumen. Food is eaten but not chewed initially, and it is passed to the rumen where it is partly digested (p. 98) and then regurgitated for further chewing before swallowing and passing into the reticulum.

gizzard (*n*) part of the alimentary canal (p. 98) of certain animals. It has a very tough lining surrounded by powerful muscles (p. 143) in which food particles are broken down by grinding action against grit or stones in the gizzard lining or against spines or 'teeth' in the gizzard itself.

crop (*n*) the part of the alimentary canal (p. 98) in animals, such as birds, which either forms part of the gut (p. 98) or the oesophagus (p. 99), and in which food is stored temporarily and partly digested (p. 98).

mandible (*n*) (1) the lower jaw (p. 105) of a vertebrate (p. 74); (2) either of the pair of feeding mouth parts of certain invertebrates (p. 75).

剛毛攝食動物 一種微噬（↑）動物。在這種微噬動物中，食物顆粒被剛毛（第65頁）截獲，然後藉纖毛（第12頁）的擺動送到口部。

巨噬的、食大粒的(形) 指動物而言，這種動物以較大的食物顆粒爲食，因此，通常無需不斷攝食。

食糞的(形) 指動物諸如某些齧齒動物而言，這些動物以糞便（第99頁）爲食，使糞便中的纖維素（第19頁）在第二次通過消化道的過程中得到進一步消化（第98頁）。

雜食動物(名) 以動物性與植物性食物摻雜爲食的動物。

食肉動物(名) 以動物性物質爲主要食物的動物。食肉動物有適於撕裂肉類的强有力的爪和齒系（第104頁）。

食草動物(名) 以植物性物質爲主要食物的動物。食草動物具有特別適於攝取堅韌的植物性物質的齒系（第104頁）和消化（第98頁）系統。

反芻動物(名) 屬於偶蹄目的食草動物（↑）種羣之一。這種動物的胃（第100頁）是複雜的，它還包括一個瘤胃，所吃的食物一開始並不經過咀嚼，而是先送至瘤胃，在那裡先進行部份消化（第98頁），然後反芻進行咀嚼，再嚥入蜂巢胃。

砂囊(名) 某些動物的消化道（第98頁）的一部份。它有一層由强有力的肌肉（第143頁）所包裹的、非常堅韌的內壁。食物顆粒在砂囊內藉其內壁中的砂和石的研磨作用，或者藉砂囊本身的刺或"牙"的研磨作用而被粉碎。

嗉囊(名) 例如鳥類的某些動物其消化道（第98頁）的一部份。它或構成消化道（第98頁）的一部份，或構成食道（第99頁）的一部份。食物暫時貯存在嗉囊內並進行部份消化（第98頁）。

下頜骨、上頜(名) (1)脊椎動物（第74頁）的下頜（第105頁）；(2)某些無脊椎動物（第75頁）的一對攝食口器中的任何一個。

carnivorous plant any plant which supplements its supply of nitrates by capturing, using a variety of means, small animals, such as insects (p. 69) and digesting (p. 98) them with enzymes (p. 28) secreted (p. 106) externally.

parasitism (*n*) an association (p. 227) in which the individuals of one species (p. 40), the parasites, live permanently or temporarily on individuals of another species, the host (↓), deriving benefit and/or nutrients (p. 92) and causing harm or even death to the host.

食肉植物　能利用種種手段捕獲昆蟲(第69頁)之類小動物，並用分泌(第106頁)於外部的酶(第28頁)將之消化(第98頁)，以補充其對硝酸鹽需要的任何植物。

寄生(現象)(名)　一種結合體(第227頁)。在這種結合體中，一個生物種(第40頁)內的個體(寄生物)永久地或暫時地依靠另一個種內的個體(寄主(↓))生活，寄生物便因此能從寄主處獲取利益或營養素(第92頁)或兼得，寄主則受害，甚至死亡。

carnivorous plant 食肉植物
e.g. *Drosera* (sundew) 圖例：茅膏菜屬
（茅膏菜）

leaves clothed with sticky tentacles on which insects are trapped

parasitism e.g. infestation by hookworms
寄生現象　圖例：鈎蟲的侵染

larvae at back of mouth
幼蟲在口之後部

larvae travel down oesophagus and stomach
幼蟲沿食管和胃下行

hookworms in intestines produce eggs
鈎蟲在腸內產卵

eggs in faeces
在糞便中的卵

warm, wet soil 溫暖潮濕的土壤

larvae hatch
幼蟲孵化

larvae travel up trachea
幼蟲沿氣管上行

larvae travel up veins to lungs
幼蟲沿靜脈上行至肺

larvae enter foot
幼蟲進入足內

larvae move over ground
幼蟲在地上移動

① 葉上覆有可捕獲昆蟲的黏性觸毛

② 外寄生物圖例：寄生於豆上的筆絲子(筆絲子屬)。筆絲子圍繞寄主的莖攀繞而上，並藉吸器插入其維管系統。

③ 筆絲子的維管束

④ 寄主的維管束

⑤ 筆絲子的莖

⑥ 吸器

⑦ 寄主的莖

⑧ 內寄生物圖例：傾菌病原體。傾菌菌絲在細胞之間擺動，並用吸器插入寄主細胞，吸取其營養素

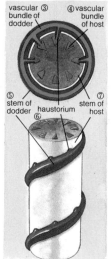

ectoparasite e.g. dodder (*Cuscuta*) on a bean. The dodder climbs around the host stem and taps into its vascular system via haustoria

vascular bundle of dodder ③
④ vascular bundle of host
⑤ stem of dodder
⑥ haustorium
⑦ stem of host

parasite (*n*) see parasitism (↑).

endoparasite (*n*) a parasite (↑) which lives within the body of the host (↓) itself, for example, the tapeworm, living within the gut (p. 98) of a vertebrate (p. 74).

ectoparasite (*n*) a parasite (↑) which lives on the surface of the host (↓) and is usually adapted for clinging on to the host on which it often feeds by fluid feeding (p. 108). Ectoparasites usually have special organs for attachment to the host.

intercellular parasite an endoparasite (↑) which lives within the material between the cells of the host (↓).

intercellular (*adj*) between cells.

寄生物(名)　見寄生(現象)(↑)。

內寄生物(名)　生活在寄主(↓)體內的一種寄生物(↑)。例如，條蟲生活在脊椎動物(第74頁)的胃腸道(第98頁)內。

外寄生物(名)　生活在寄主(↓)體表的一種寄生物(↑)。這種寄生物通常適宜附著在寄主身上，往往藉食流體(第108頁)的方式為生。外寄生物一般具有附着於寄主的特殊器官。

胞間寄生物　生活在寄主(↓)細胞之間的物質內的一種內寄生物(↑)。

胞間的(形)　在細胞之間的。

endoparasite e.g. fungal ⑧ pathogen. Fungal hyphae weave between cells and tap cells for nutrients with haustoria

host cells
寄主細胞
fungal hyphae
傾菌菌絲
haustorium
吸器

intracellular parasite an endoparasite (↑)which lives within the cells of the host (↓).

host (*n*) the species (p. 40) of organism in an association (p. 227) within which or on which a parasite (↑) lives and reaches sexual maturity, and which suffers harm or even death as a result.

secondary host a host (↑) on which or within which the young or resting stage of a parasite (↑) may live temporarily. The parasite does not reach sexual maturity on the secondary host. Also known as **intermediate host**.

transmission (*n*) the process by which a substance or an organism is transported from one place to another, e.g. a parasite (↑) is transmitted from one host (↑) to another sometimes via a secondary host (↑) and it may involve considerable risk to the parasite. **transmit** (*v*).

vector (*n*) a secondary host (↑) which is actively involved in the transmission (↑) of a parasite (↑) from one host (↑) to another or an organism which passes infectious disease from one individual to another without necessarily being affected by the disease itself. For example, the blood-sucking mosquito which transmits a malaria-causing blood parasite from one individual on which it feeds to another is a vector.

胞內寄生物　生活在寄主(↓)細胞內的一種內寄生物(↑)。

寄主、宿主(名)　在一個結合體(第227頁)中，寄生物(↑)可在其體內或體表生活並達到性成熟的生物種(第40頁)。結果，該種生物受害，甚至死亡。

次要寄主　幼體期或靜止期寄生物(↑)可暫時寄生於其體表或體內的一種寄主(↑)。在中間寄主中，寄生物不會達到性成熟。又稱為中間寄主。

傳播、傳遞(名)　將一種物質或一種生物從一處運送至另一處的過程。例如，一種寄生物(↑)有時是要從一個寄主(↑)經由一個中間寄主(↑)才能傳播到另一個寄主的，這就有可能使寄生物遭受到極大危險。(動詞形式為 transmit)

媒介　一種中間寄主(↑)或一種生物。這種中間寄主能主動參與一種寄生物(↑)從一個寄主(↑)到一個寄主的傳播過程；這種生物能將傳染病從一個個體傳播給另一個個體，而其自身則未必受該病的危害。例如，將引起瘧疾的血內寄生物從它藉以為食的一個個體傳到另一個個體的吸血蚊，便是一種媒介。

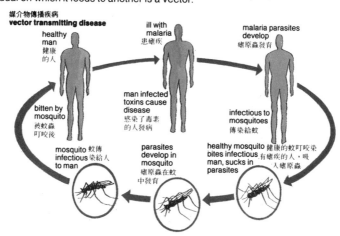

媒介物傳播疾病
vector transmitting disease

healthy man 健康的人

ill with malaria 患瘧疾

malaria parasites develop 瘧原蟲發育

bitten by mosquito 被蚊蟲叮咬後

man infected toxins cause disease 感染了毒素的人發病

infectious to mosquitoes 傳染給蚊

mosquito 蚊傳 infectious 染給人 to man

parasites develop in mosquito 瘧原蟲在蚊中發育

healthy mosquito 健康的蚊叮咬染 bites infectious 有瘧疾的人，吸 man, sucks in 入瘧原蟲 parasites

respiration (n) the process whereby energy is produced in a plant or animal by chemical reactions. In most organisms, this is achieved by taking in oxygen from the environment (p. 218) and, after transportation to the cells, its reaction with food molecules releases carbon dioxide, water and energy that is trapped in ATP (p. 33) – cellular respiration (p. 30).

respiratory quotient RQ. The ratio of the volume of carbon dioxide produced by an organism to the volume of oxygen used up during the same period of respiration.

$$RQ = \frac{\text{carbon dioxide produced}}{\text{oxygen used up}}$$

breathing (n) the process of actively drawing air or any other gases into the respiratory (↑) organs for gas exchange (↓). **breathe** (v).

gas exchange the process which takes place at a respiratory surface (↓) in which a gas, such as oxygen, from the environment (p. 218) diffuses **(p.119)** into the organism because the concentration of that gas in the organism is lower than in the environment, and another gas, such as carbon dioxide, is released from the organism into the environment. In plants, respiratory gas exchange is complicated by the gas exchange that takes place as a result of photosynthesis (p. 93).

respiratory surface the surface of an organ, such as a lung (p. 115), across which gas exchange (↑) takes place. It is usually highly folded to increase the surface area, thin, and, in organisms that live on land, damp.

inspiration (n) the process of drawing air into or across the respiratory surface (↑) by muscular (p. 143) action. The pressure within the respiratory organ is reduced to below that of the environment (p. 218) so that air flows in. Also known as **inhalation**. **inspire** (v).

expiration (n) the process of forcing air and waste gases out of the respiratory (↑) organ by muscular (p. 143) action. The pressure within the respiratory organ is increased so that air flows out. Also known as **exhalation**. **expire** (v).

呼吸(作用)（名） 植物和動物體內，藉化學反應產生能量的過程。在大多數生物中，這一過程是靠從環境(第218頁)吸入氧氣實現的，氧氣被輸送至細胞內後，即與養料分子起反應，釋放出二氧化碳、水和以 ATP（第33頁）形式貯存的能量——細胞呼吸（第30頁）。

呼吸商RQ 生物體同一呼吸時間內產生的二氧化碳體積與所消耗的氧氣體積之比。

$$RQ = \frac{\text{所產生的二氧化碳}}{\text{所消耗的氧氣}}$$

呼吸（名） 自動將空氣或其他任何氣體吸入到呼吸（↑）器官內、進行氣體交換（↓）的過程。（動詞形式爲 breathe）

氣體交換 在呼吸表面（↓）發生的過程。在這種過程中，來自環境（第218頁）的氣體如氧氣擴散（第119頁）進入生物體內，因爲生物體內該種氣體的濃度低於環境中該種氣體的濃度；而另一種氣體如二氧化酶則從生物體內釋放到環境中。在植物中，由於光合作用（第93頁）的結果也產生氣體交換，從而使呼吸的氣體交換變得複雜化。

呼吸表面 例如肺（第115頁）的器官表面。氣體交換（↑）通過該表面進行。呼吸表面通常較薄，同時呈高度皺褶狀，從而增加其表面積。而生活在陸地上的多種生物體呼吸表面，則較潮濕。

吸氣（名） 藉肌肉（第143頁）的動作，吸收空氣，使之進入或通過呼吸表面（↑）的過程。由於呼吸器官內的空氣壓力降低到環境(第218頁)中的空氣壓力以下，致使空氣進入呼吸器官。吸氣也稱爲吸入。（動詞形式爲 inspire）

呼氣（名） 藉肌肉（第143頁）的動作，迫使空氣和廢氣從呼吸（↑）器官中呼出的過程。由於呼吸器官內的氣體壓力增加，致使空氣排出呼吸器官。呼氣也稱爲呼出。（動詞形式爲 expire）

air (*n*) the mixture of gases which forms the atmosphere surrounding the Earth. It is composed of approximately 78 per cent nitrogen, 21 per cent oxygen, 0.03 per cent carbon dioxide, and little under 1 per cent of the so-called noble gases including argon, neon, etc. It also includes water vapour.

gill (*n*) one part of the respiratory surface (↑) found in most aquatic animals, such as fish. Gills are projections of the body wall or of the inside of the gut (p. 98) and may be very large and complex in relation to the animal because they are supported by water. They are very thin and well supplied with blood vessels (p. 127) so that gas exchange (↑) between the water and the blood (p. 90) of the animal is usually very efficient.

空氣（名） 形成包圍地球大氣的各種氣體的混合物。這種混合物由大約78％的氮、21％的氧、0.03％的二氧化碳和略低於１％的、包括氬、氖等在內的所謂惰性氣體組成。該混合物內還包含水蒸氣。

鰓（名） 大多數水生動物如魚的呼吸表面（↑）的一部份。鰓是體壁或胃腸道（第98頁）內側的突出物；由於鰓是靠水來維持呼吸的，所以，對於該動物來說，它可能很大、很複雜。鰓很薄，其上密佈血管（第127頁），因此，這類動物在水中其血液（第90頁）和水之間能很有效進行氣體交換（↑）。

two main gill types showing water movements
兩種主要鰓型的水流運動狀態

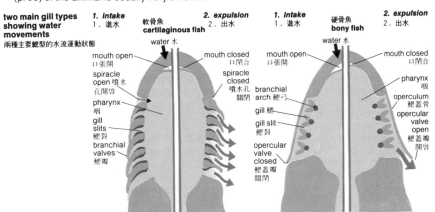

gill filament one of the numerous flattened lobes which make up a gill and increase its surface area.

gill slit one of a series of openings through the pharynx (p. 99) in fish and some amphibians (p. 77) leading to the gills (↑).

operculum[a] (*n*) a bony plate that covers the visceral clefts (p. 74) and the gill slits (↑) in bony fish (p. 76). It assists in pumping water over the gills (↑) for gas exchange (↑) by inward and outward movements.

鰓絲 構成鰓並可增加其表面積的眾多扁平葉。

鰓裂 在魚和某些兩棲動物（第77頁）中，穿過咽（第99頁）通向鰓（↑）的一系列裂隙。

鰓蓋骨[a] 硬骨魚（第76頁）中遮蓋鰓裂（（第74頁）和（↑））的一塊骨板。這塊骨板可藉向內和向外的運動，幫助將水汲送至鰓（↑）上進行氣體交換（↑）。

counter current exchange system the system found in the gills (p. 113) of bony fish (p. 76) in which the water that is pumped past the gill filaments (p. 113) flows in the opposite direction to the flow of blood (p. 90) within the gill. Gas exchange (p. 112) takes place continuously along the whole length of the gill because the levels of gases never reach equilibrium.

parallel current exchange system the system of gas exchange (p. 112) found in the gills (p. 113) of cartilaginous fish (p. 76) in which the flow of water and the flow of blood (p. 90) are in the same direction. This system is less efficient than a counter current exchange system (↑) because equilibrium is soon reached.

buccal pump one part of the double pump action which causes oxygen-containing water to flow over the gills (p. 113). Muscles (p. 143) in the floor of the buccal cavity (p. 99) cause it to rise and fall as the mouth shuts and opens forcing water over the gills and then drawing in more water through the mouth respectively.

opercular pump one part of the double pump action which causes oxygen-containing water to flow over the gills (p. 113). Muscles cause the operculum (p. 113) to open outwards as water is drawn in through the mouth.

tracheal system in insects (p. 96), a system of gas exchange (p. 112) and transport which is separate from the blood (p. 90) system. Oxygen is carried by air tubes called tracheae (↓) and some of the oxygen in the air diffuses into the body tissues (p. 83). Oxygen is also dissolved in the fluid in the tracheoles (↓).

spiracle (n) the opening to the atmosphere of an insect's (p. 69) trachea(↓).

逆流交換系統　硬骨魚(第76頁)鰓(第113頁)中具有的氣體交換系統。在這種氣體交換系統中，被汲入的水經過鰓絲(第113頁)，朝着與鰓內血(第90頁)流流向相反的方向流動。氣體交換(第112頁)沿着鰓的全長不斷進行，因爲各種氣體的含量永遠不可能達到平衡。

並流交換系統　軟骨魚(第76頁)鰓(第113頁)中具有的氣體交換(第112頁)系統。在這種氣體交換系統中，水流和血(第90頁)流處於同一方向。這種系統比逆流交換系統(↑)效率低，因爲各種氣體的含量很快就達到平衡。

口抽吸機能　雙聯泵式機能的一部份。這部份機能可使含氧的水流過鰓部(第113頁)。當口閉合和張開時，口腔(第99頁)底部的肌肉(第143頁)使口腔得以昇降，從而分別迫使水流過鰓部，然後又通過口吸進更多的水。

鰓蓋抽吸機能　雙聯泵式機能的一部份。這部份機能可使含氧的水流過鰓部(第113頁)。當通過口吸入水時，肌肉使鰓蓋骨(第113頁)向外開啓。

氣管系統　昆蟲(第96頁)體內的氣體交換(第112頁)和運輸系統。這種系統與血液(第90頁)系統分開。氧氣由被稱爲氣管(↓)的一些空氣管道進行輸送；空氣中的一些氧氣經彌散作用進入身體組織(第83頁)內。氧也溶解於微氣管(↓)中的液體內。

氣門(名)　昆蟲(第69頁)的氣管(↓)通向大氣的孔。

逆流交換系統

counter current exchange system

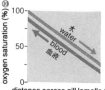

並流交換系統
parallel current exchange system

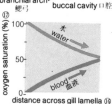

tracheal system 氣管系統

spiracle 氣門

main tracheal tube 主氣管

air sac 氣囊

tracheoles 微氣管

exoskeleton (cuticle) 外骨骼(表皮)

muscle fibre 肌纖維

① 水
② 鰓蓋腔
③ 鰓絲
④ 佈有微血管的鰓小瓣
⑤ 鰓弓
⑥ 口腔
⑦ 出鰓動脈
⑧ 入鰓動脈
⑨ 穿過鰓小瓣的距離

⑩ 氧飽和(%)
⑪ 垂直隔
⑫ 鰓絲
⑬ 佈有微血管的鰓小瓣
⑭ 入鰓動脈
⑮ 出鰓動脈
⑯ 穿過鰓小瓣的距離
⑰ 氧飽和(%)

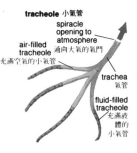

tracheole 小氣管

spiracle
opening to
atmosphere
通向大氣的氣門

air-filled
tracheole
充滿空氣的小氣管

trachea
氣管

fluid-filled
tracheole
充滿液
體的
小氣管

trachea (*n*) (1) in insects (p. 69), one of a number of air tubes which lead from the spiracles (↑) into the body tissues (p. 83); (2) in land-living vertebrates (p. 74), a tube which leads from the throat into the bronchi (p. 116). **tracheae** (*pl.*).

tracheole (*n*) one of the very fine tubes into which the tracheae (↑) of insects (p. 69) branch. They pass into the muscles (p. 143) and organs of an insect's body to allow gas exchange (p. 112) to take place.

氣管（名）　(1)昆蟲(第69頁)中從氣門(↑)通到身體組織(第83頁)內的若干空氣管道；(2)陸生脊椎動物(第74頁)中從咽喉通到支氣管(第116頁)內的一根管子。(複數形式爲 tracheae)

小氣管、微氣管(名)　由昆蟲(第69頁)的氣管(↑)分支成的一些非常微細的管子。這些管子通入昆蟲體內的肌肉(第143頁)和器官中，使氣體交換(第112頁)得以進行。

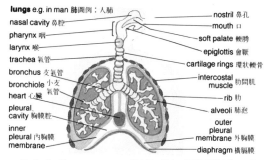

lungs e.g. in man 肺圖例：人肺

nasal cavity 鼻腔
pharynx 咽
larynx 喉
trachea 氣管
bronchus 支氣管
bronchiole 小支
氣管
heart 心臟
pleural
cavity 胸膜腔
inner
pleural 內胸膜
membrane

nostril 鼻孔
mouth 口
soft palate 軟腭
epiglottis 會厭
cartilage rings 環狀軟骨
intercostal
muscle 肋間肌
rib 肋
alveoli 肺泡
outer
pleural
membrane 外胸膜
diaphragm 橫膈膜

lungs (*n.pl.*) a pair of thin-walled, elastic sacs present in the thorax (↓) of amphibians (p. 77), reptiles (p. 78), birds and mammals (p. 80) and containing the respiratory surfaces (p. 112).

ventilation (*n*) the process whereby the air contained within the lungs (↑) is exchanged with air from the atmosphere by regular breathing (p. 112) in which muscular (p. 143) movements of the thorax (↓) varies its volume and thus the volume of the lungs. During inspiration (p. 112) the volume of the lungs increases and atmospheric pressure forces air into the lungs. During expiration (p. 112), the muscles relax and the volume of the lungs decreases by virtue of their elasticity so that air is forced out.

thorax (*n*) (1) in arthropods (p. 67), the segments between the head and the abdomen (p. 116); (2) in vertebrates (p. 74), the part of the body which contains the heart (p. 124) and lungs (↑). In mamma (p. 80) it is separated from the abdomen by the diaphragm (p. 116) and protected by the rib cage.

thoracic cavity = thorax (↑), in vertebrates (p. 74).

肺(名、複)　兩棲動物(第77頁)、爬行動物(第78頁)、鳥類和哺乳動物(第80頁)胸部(↓)內的一對薄壁彈性囊。肺內包含着許多呼吸表面(第112頁)。

換氣、通氣(名)　藉有規則的呼吸(第112頁)，使肺(↑)內空氣與環境空氣進行交換的作用過程。在這種過程中，胸部(↓)肌肉(第143頁)運動，使胸腔的體積，亦即肺內空氣的容量發生變化。吸氣(第112頁)時，肺內的空氣容量增加，大氣壓力迫使空氣進入肺內；呼氣(第112頁)時，胸部肌肉鬆弛，肺內的空氣容量由於肺彈性作用而降低，致使空氣被壓出肺外。

胸(部)、胸廓(名)　節肢動物(第67頁)的頭部和腹部(第116頁)之間的數個體節；(2)脊椎動物(第74頁)包含心臟(第124頁)和肺(↑)的身體部份。在哺乳動物(第80頁)中，胸腔被橫膈膜(第116頁)與腹腔隔開，並受肋骨籠保護。

胸腔　在脊椎動物(第74頁)中同胸(部)、胸廓(↑)。

intercostal muscle a muscle (p. 143) which connects adjacent ribs. When the external intercostal muscles contract, the ribs are moved upwards and outwards, increasing the volume of the thoracic cavity (p. 115) and thus the lungs (p. 115) so that air is forced into the lungs for inspiration (p. 112). When the internal intercostal muscles contract the volume of the thoracic cavity decreases and expiration (p. 112) takes place.

diaphragm (*n*) a sheet of muscular (p. 143) tissue (p. 83) which separates the thoracic cavity (p.115) from the abdomen (↓) in mammals (p. 80).

abdomen (*n*) (1) in arthropods (p. 67), the segments at the back of the body; (2) in vertebrates (p. 74), the part of the body containing the intestines, liver, kidney etc.

pleural cavity the narrow space, filled with fluid, between the two layers of the pleural membrane (↓)

pleural membrane the double membrane which surrounds the lungs (p. 115) and lines the thoracic cavity (p. 115). It secretes (p. 106) fluids to lubricate the two layers as the lungs expand and contract during breathing (p. 112).

larynx (*n*) a structure found at the junction of the trachea (p. 115) and the pharynx (p. 99) which contains the vocal cords (↓). During swallowing it is closed off by the epiglottis (p. 99).

vocal cord (*n*) one of the folds of the lining of the larynx (↑) which produce sound as a current of air passes over them.

bronchus (*n*) one of the two large air tubes into which the trachea (p. 115) divides and which enter the lungs (p. 115).

bronchiole (*n*) one of a number of smaller air tubes into which the bronchi (↑) divide after entering the lungs (p. 115). The bronchioles make up the 'bronchial tree' which ends in air tubes called the *respiratory bronchioles*. These divide into *alveolar ducts* (or terminal bronchioles) which give rise to the alveoli (↓).

alveolus (*n*) a pouch-like air sac which occurs in clusters at the ends of the bronchioles (↑) and which contains the respiratory surfaces (p. 112). A network of capillaries (p. 127) covers the thin, elastic epithelium (p.87). **alveoli** (*pl.*)

肋間肌　連接相鄰肋骨的肌肉(第143頁)。當肋間外肌收縮時,肋骨便向上和向外運動,從而增加胸腔(第115頁)容量,亦即增加肺(第115頁)容量,促使空氣進入肺內,進行吸氣(第112頁)。當肋間內肌收縮時,胸腔容量減少,進而產生呼氣(第112頁)。

橫膈膜(名)　哺乳動物(第80頁)體內將胸腔(第115頁)與腹腔(↓)隔開的一層肌肉(第143頁)組織(第83頁)。

腹(部)(名)　(1)節肢動物(第67頁)身體後面的數個體節;(2)脊椎動物(第74頁)體中包含腸、肝臟、腎臟等身體部份。

胸膜腔　兩層胸膜(↓)之間充滿液體的陜窄空間。

胸膜　包裹肺(第115頁)和貼於胸腔(第115頁)內壁的雙層膜。在呼吸(第112頁)過程中,當肺擴張和收縮時,它分泌(第106頁)液體,以潤滑這兩層膜。

喉(名)　氣管(第115頁)和咽(第99頁)接合處的一種結構。這種結構內包含聲帶(↓)。吞嚥食物時,由會厭(第99頁)將喉的入口關閉。

聲帶(名)　喉(↑)內壁上的一對皺襞。當氣流經過這些皺襞時,便產生聲音。

支氣管(名)　由氣管(第115頁)分出、並進入肺(第115頁)的兩大分支氣管。

小支氣管(名)　支氣管(↑)進入肺(第115頁)後分出的若干根較小的氣管。這些小支氣管形成"支氣管樹"。支氣管樹的端部是一些稱爲"呼吸支氣管"的氣管。這些呼吸小支氣管又分成爲產生肺泡(↓)的"肺泡管"(或稱末端小支氣管)。

肺泡(名)　一種袋狀氣囊。它成串地存在於小支氣管(↑)末端,也包含許多呼吸表面(第112頁)。微血管(第127頁)網覆蓋著其薄而有彈性的上皮(87頁)。(複數形式爲 alveoli)

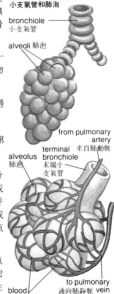

bronchiole and alveoli
小支氣管和肺泡

bronchiole 小支氣管

alveoli 肺泡

from pulmonary artery 來自肺動脈

terminal bronchiole 末端小支氣管

alveolus 肺泡

to pulmonary vein 通向肺靜脈

blood capillaries 微血管

tidal flow of the system in which inspiration (p. 112) and expiration (p. 112) take place through the same air passages so that the air passes twice over each part of the respiratory surface (p. 112). This is less efficient than a system in which there is a constant throughflow such as that which takes place over the gills (p. 113) of fish.

tidal volume the volume of air which is inspired (p. 112) or expired (p. 112) during normal regular breathing (p. 112). It is considerably less than the lung capacity (↓).

ventilation rate the rate per minute at which the total volume of air is expired (p. 112) or inspired (p. 112).

residual volume the volume of air that always remains within the alveoli (↑) because the thorax (p. 115) is unable to collapse completely. It exchanges oxygen and carbon dioxide with the tidal air.

vital capacity the total amount of air which can be inspired (p. 112) and expired (p. 112) during vigorous activity.

reserve volume the difference in volume between the total lung capacity (↓) and the vital capacity (↑).

lung capacity the total volume of air that can be contained by the lungs when fully inflated.

acclimatization (*n*) the period of time it takes for the respiration (p. 112) of an organism to get used to the reduced partial pressure of oxygen that may occur at high altitudes, for example, where the atmospheric pressure is reduced.

respiratory centre the part of the medulla oblongata (p. 156) which controls the rate of breathing (p. 112) in response to the levels of carbon dioxide dissolved in the bloodstream (p. 90).

oxygen debt the deficit in the amount of oxygen that is available for respiration (p. 112) during vigorous activity so that even when the activity ceases, breathing (p. 112) continues at a high rate until the oxygen debt is made up. Lactic acid builds up in the muscles (p. 143) from lactic acid fermentation (p. 34).

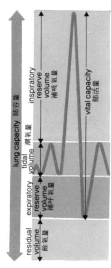

comparison of different lung volumes in man
人體不同肺容量之間的比較

潮流式（呼吸系統） 指一種呼吸系統而言，在這種呼吸系統中，吸氣（第112頁）和呼氣（第112頁）通過同一呼吸道進行，致使空氣兩次經過呼吸表面（第112頁）的每一部份。這種呼吸系統的效率比恒定貫流式呼吸系統，如在魚鰓（第113頁）運行的呼吸系統低。

潮氣量 在正常自然呼吸（第112頁）時，每次吸入（第112頁）的空氣量。該空氣量大大低於肺容量（↓）。

換氣率 每分鐘呼出（第112頁）或吸入（第112頁）的空氣總量。

餘氣量 由於胸腔（第115頁）不能完全萎陷，而始終餘留在肺泡（↑）內的空氣量。它可與潮流氣交換氧和二氧化碳。

肺活量 在一次盡力吸氣（第112頁）後，再盡力呼出（第112頁）的空氣總量。

貯氣量 肺容量（↓）和肺活量（↑）之間的空氣量之差。

肺容量 肺在完全充氣時能容納的空氣總量。

（風土）馴化（作用）、適應、習服（名） 生物體的呼吸（第112頁）習慣於較低氧分壓所需要的一段時間。例如，在海拔甚高的地方，大氧壓力降低，就會發生上述情況。

呼吸中樞 延髓（第156頁）的一個特定部位。該部位可對血流（第90頁）中溶解的二氧化碳含量作出反應，控制呼吸（第112頁）速率。

氧債 在機體劇烈活動過程中，可供呼吸作用（第112頁）耗用的氧在數量上的差額。虧欠了氧債後，即使在劇烈活動停止時，機體也仍將繼續高速率地進行呼吸（第112頁），直到償還氧債為止。乳酸發酵（第34頁）產生的乳酸，則累積在肌肉（第143頁）中。

osmosis (*n*) the process by which water passes through a semipermeable membrane (↓) from a solution (↓) of low concentration of salts to one of high concentration thereby diluting it. Osmosis will continue until the concentrations of the two solutions are equalized. In living things osmosis can take place through membranes (p. 14), e.g. tonoplast (p. 11) or plasmalemma (p. 14), in either direction. In plants, the cell walls (p. 8) are elastic so that they can contain solutions of higher concentration when osmosis ceases. **osmotic** (*adj*).

滲透(作用) 低濃度鹽水溶液(↓)中的水，通過半透性膜(↓)進入高濃度鹽水溶液使其稀釋的過程。滲透作用繼續至兩種溶液的濃度相同時為止。生物體中的滲透作用可通過液泡膜(第11頁)或質膜(第14頁)之類膜(第14頁)朝任一方向進行。植物的細胞壁(第 8 頁)富有彈性，因此，當滲透作用停止時，細胞內含有較高濃度的溶液。(形容詞形式為 osmotic)

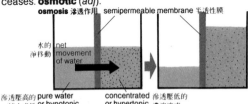

osmosis 滲透作用　semipermeable membrane 半透性膜

水的淨移動　net movement of water

滲透壓高的 純水或低 滲溶液　pure water or hypotonic solution with a high osmotic pressure

concentrated or hypertonic solution with a low osmotic pressure 滲透壓低的 濃溶液或 高滲溶液

osmotic potential the tendency of water molecules to diffuse (↓) through a semipermeable membrane (↓) from a solution (↓) of low solute concentration to a solution of high solute concentration until equilibrium is reached.

滲透勢、滲透潛能 在達到平衡之前，水分子從低濃度溶液(↓)通過半透性膜(↓)向高濃度溶液擴散(↓)的趨勢。

semipermeable membrane a membrane (p. 14), such as a tonoplast (p. 11) or plasmalemma (p. 14), with microscopic (p. 9) pores (p. 120) through which small molecules e.g. water will pass but larger molecules e.g. sucrose (p. 18), or salts will not.

半透性膜 具有極細微(第 9 頁)孔(第120頁)的膜(第14頁)，例如液泡膜(第11頁)或質膜(第14頁)。小分子，例如水分子，可通過半透性膜；而較大的分子，例如蔗糖(第18頁)或鹽分子，則不能通過該膜。

solution (*n*) a liquid (*the solvent*) with substances (*the solute*) dissolved in it. Substances that will dissolve are said to be *soluble* and those that will not, *insoluble*.

溶液(名) 含有溶解於其中的物質(溶質)一種液體(溶劑)。可溶解的物質被為是可溶的；不溶解的物質則被認為是不可溶的。

isotonic solution (*n*) a solution (↑) in which the osmotic potential (↑) is the same as that of another solution so that neither solution either gains or loses water by osmosis (↑) across a semipermeable membrane (↑).

等滲溶液(名) 其滲透勢(↑)是與另一種溶液的滲透勢相同的一種溶液(↑)。在這種情況下，兩種溶液都不能借助於通過半透膜(↑)的滲透作用(↑)來獲得水或失去水。

hypotonic (*adj*) of one solution (↑) in an osmotic (↑) system which is more dilute than another.

低滲的(形) 指滲透(↑)系統中的一種溶液(↑)而言，這種溶液比另一種溶液更稀。

hypertonic (*adj*) of one solution (↑) in an osmotic (↑) system which is more concentrated than another.

高滲的(形) 指滲透(↑)系統中的一種溶液(↑)而言，這種溶液比另一種溶液更濃。

diffusion (*n*) the process in which molecules move from an area of high concentration to an area of low concentration. Osmosis (↑) is a special type of diffusion restricted to the movement of water molecules.

diffusion pressure deficit the situation which exists between two solutions (↑) on either side of a semipermeable membrane (↑) in which a substance has been added to one of the solutions which cannot pass through the membrane and impedes the passage of water from that solution. The greater the concentration of the solution, the higher the diffusion pressure deficit and the lower the osmotic potential (↑) of that solution.

turgor (*n*) the condition in a plant cell in which water has diffused (↑) into the cell vacuole (p. 11) by osmosis (↑), causing the cell to swell, because the cell fluid was at a lower osmotic potential (↑) than that of its surroundings.

turgid (*adj*) of a plant cell in which the turgor (↑), which is resisted by the elasticity of the cell wall (p. 8), has brought the cell close to bursting and no more water can enter the cell. Turgidity provides the plant with support.

擴散作用、瀰散（名） 分子從高濃度區域向低濃度區域移動過程。滲透作用（↑）是一種只限於水分子移動的特殊類型擴散。

擴散壓虧 半透膜（↑）兩側的兩種溶液（↑）間存在的一種狀況。這種狀況是由於兩種溶液中的其中一種溶液內添加了一種不能通過半透性膜但又阻止此溶液中的水通過半透性膜的物質而致的。該溶液的濃度越高，其擴散壓虧越大，滲透勢（↑）越低。

膨壓(現象)（名） 植物細胞中出現的一種狀態。出現這種狀態時，水藉滲透作用（↑）擴散（↑）到細胞的液泡(第11頁)中，致使細胞膨脹。這是由於細胞液先前的滲透勢（↑）低於細胞液周圍的滲透勢所致。

緊脹的(形) 指植物細胞而言，在這種細胞內的膨壓現象（↑）已經使細胞接近脹破，水不能再進入細胞，該現象也受到細胞壁(第8頁)彈性的阻礙。緊脹度對植物體具有支持作用。

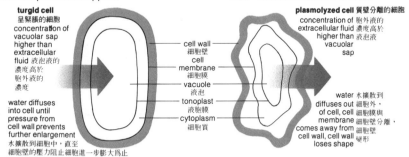

turgid cell
呈緊脹的細胞
concentration of vacuolar sap higher than extracellular fluid 液泡液的濃度高於胞外液的濃度

water diffuses into cell until pressure from cell wall prevents further enlargement 水擴散到細胞中，直至細胞壁的壓力阻止細胞進一步膨大為止

cell wall 細胞壁
cell membrane 細胞膜
vacuole 液泡
tonoplast 液胞膜
cytoplasm 細胞質

plasmolyzed cell 質壁分離的細胞
concentration of extracellular fluid 胞外液的濃度高於液泡液的 higher than vacuolar sap

water 水擴散到 diffuses out 細胞外， of cell, cell 細胞膜與 membrane 細胞壁分離， comes away from 細胞壁 cell wall, cell wall 變形 loses shape

turgor pressure the pressure exerted by the bulging cell wall (p. 8) during osmosis (↑) into the vacuole (p. 11) of a plant cell.

plasmolysis (*n*) a loss of water, and hence turgidity (↑) from a plant cell when it is surrounded by a more concentrated solution (↑). The cytoplasm (p. 10) loses volume and contracts away from the cell wall (p. 8) causing wilting.

緊脹壓 水向植物細胞的液泡(第11頁)內滲透（↑）時，緊脹的細胞壁(第8頁)所產生的壓力。

質壁分離(名) 植物細胞被濃溶液（↑）包圍時引起的失水，並隨之消失緊脹（↑）。此時，細胞質(第10頁)容積減小，並引起收縮，與細胞壁(第8頁)發生分離，從而導致細胞萎蔫。

flaccid (*adj*) of cell tissue (p. 83), weak or soft.
wilt (*v*) *of leaves and green stems* to droop.
stoma (*n*) one of the many small holes or pores
(↓) in the leaves (mainly) and stems of plants
through which gas and water vapour exchange
take place. They are able to open and close by
means of their neighbouring guard cells (↓).
stomata (*pl.*).
pore (*n*) a small opening in a surface.
guard cell one of the pair of special, crescent-
shaped cells which surround each stoma (↑)
and which enable the stoma to open or close in
response to light intensity by osmosis (p. 118).
When the guard cells are turgid (p. 119) the
stoma is open.

軟垂、萎縮的（形）　指柔弱細胞組織（第83頁）。
(使)枯萎(動)　使葉子和綠色的莖雕萎。
氣孔(名)　植物的葉(主要是葉)和莖上的許多小洞
或小孔(↓)。通過這些小洞或小孔，可進行氣
體和水蒸氣的交換。它們能借助於相都的保衛
細胞(↓)張開或關閉。(複數形式爲 stomata)

孔(名)　表面上的小開口
保衛細胞　圍繞每一氣孔(↑)的一對特化的新月形
細胞。這些細胞能根據光強度變化作出反應，
借助於滲透作用(第118頁)使氣孔張開或關
閉。當保衛細胞緊脹(第119頁)時，氣孔便張
開。

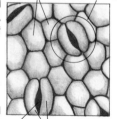

stomata 氣孔
surface view of leaf 葉表面的上視圖
epidermal cells 表皮細胞　stoma 氣孔

pore 孔　guard cells 保衛細胞

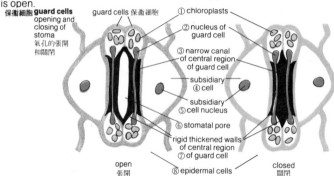

保衛細胞 **guard cells**
opening and
closing of
stoma
氣孔的張開
和關閉

guard cells 保衛細胞

① chloroplasts
② nucleus of
guard cell
③ narrow canal
of central region
of guard cell
subsidiary
④ cell
subsidiary
⑤ cell nucleus
⑥ stomatal pore
rigid thickened walls
of central region
⑦ of guard cell
⑧ epidermal cells

① 葉綠體
② 保衛細胞的核
③ 保衛細胞中部的陝窄通道
④ 副衛細胞
⑤ 副衛細胞的核
⑥ 氣孔的孔
⑦ 保衛細胞中部的硬質加厚壁
⑧ 表皮細胞

open 張開　　closed 關閉

substomatal chamber the space below the
stoma (↑).
transpiration (*n*) the loss of water from a plant
through the stomata (↑). It is controlled by the
action of the stomata. It provides a flow
(transpiration stream (p. 122)) of water through
the plant and also has a cooling effect as the
water evaporates from the plant's surface. It is
affected by temperature, relative humidity (↓),
wind speed. As air and leaf temperatures
increase so does the rate of transpiration. The
lower the humidity of the atmosphere, the
faster is the rate of transpiration. Increasing
wind speed normally increases the transpiration
rate provided the cooling effect is not greater.

氣孔下室　氣孔(↑)下的空間。

蒸騰作用(名)　植物體的水份經由氣孔(↑)逸出引
起的散失。這種水份散失受氣孔的動作控制。
蒸騰作用可形成一支貫穿植物的水流(蒸騰流
(第122頁))；當水從植物表面蒸發時，它還起
冷卻作用。蒸騰作用也受溫度、相對濕度(↓)
和風速影響。當氣溫和葉溫增高時，蒸騰速度
也增加。空氣濕度越低，蒸騰速度越快。假如
冷卻作用不太大，增加風速通常可以增加蒸騰
速度。

① 水從葉上蒸發
② 水在維管系統中向上輸送
③ 水通過根部進入植物體內

transpiration 蒸騰作用
water ① evaporates from leaves
② water transported upward in the vascular system
③ water enters plant through roots

guttation 吐水作用

high humidity
在高濕度情況下

droplets of
water exuded
from hydathodes
(ends of veins at leaf margin)
從排水器(在葉緣處的各葉脈的端部)
滲泌出的小水滴

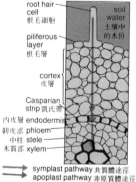

movement of water from 水份從土壤到
soil to centre of root 根部中央的移動

root hair
cell
根毛細胞

soil
water
土壤中
的水份

piliferous
layer
根毛層

cortex
皮層

Casparian
strip 凱氏帶
內皮層 endodermis
韌皮部 phloem
中柱 stele
木質部 xylem

symplast pathway 共質體途徑
apoplast pathway 非原質體途徑

**symplast and apoplast
pathways** 共質體和非原質體途徑

plasmodesmata 胞間連絲

apoplast 非原質體
substances
translocated
through cell
walls and inter-
cellular spaces
物質穿過細胞壁和
胞間隙進行運輸

symplast 共質體
substances
translocated
through living
cells and
plasmodesmata
物質穿過活細胞和
胞間連絲進行運輸

relative humidity the percentage of water vapour contained in the air. When the relative humidity is 100 per cent the air is saturated.

guttation (*n*) the process which takes place in some plants in conditions of high relative humidity (↑) in which water is actively secreted (p. 106) in liquid form by special structures called hydathodes (found at the end of the veins on the leaves) rather than being lost as water vapour. This takes place because of osmotic (p. 118) absorption (p. 81) of water by the roots.

atmospheric pressure the pressure which is exerted on the surface of the Earth by the weight of the air in the atmosphere.

vacuolar pathway a pathway for the passage of water by osmosis (p. 118). vacuoles (p. 11) contain a fluid with a lower osmotic potential (p. 118) than water so that the vacuole will take in water until it becomes turgid (p. 119).

symplast pathway a pathway for the transport of water through a plant by diffusion (p. 119) from one cell to the next through the cytoplasm (p. 10) along the threads called plasmodesmata (p. 15) which link adjacent cells.

apoplast pathway a pathway for the transport of water in a plant, particularly across the root cortex (p. 86), by diffusion (p. 119) along adjacent cell walls (p. 8).

mass flow a hypothesis (p. 235) developed by Munch in 1930 to explain the transport of substances in the phloem (p. 84). Mass flow takes place in the sieve tube (p. 84) lumina as water is taken up by osmosis (p. 118) in actively photosynthesizing (p. 93) regions where concentration is high and flows to areas where water is lost as the products of photosynthesis are being used up or stored and, therefore, concentration is low. Water is carried in the opposite direction in the xylem (p. 84) by the transpiration stream (p. 122).

root pressure the pressure in a plant which causes water to be transported from the root into the xylem (p. 84) by the plant's osmotic (p. 118) gradient.

相對濕度 空氣中所含水蒸氣的百分率。相對濕度爲100％時，空氣爲水蒸氣所飽和。

吐水(作用)、泌溢(作用)(名) 某些植物在相對濕度(↑)較高的環境中所發生的過程。在這種過程中，水以液態形式由稱之爲排水器(存在於葉上的葉脈端部)的一些特殊結構，能動地分泌(第106頁)出來，而不是作爲水蒸氣形式散失掉。產生這種過程的原因是根部對水的滲透(第118頁)吸收作用(第81頁)。

大氣壓力 大氣中的空氣重量對地球表面所產生的壓力。

液泡途徑 藉滲透作用(第118頁)運輸水份的一條途徑。液泡(第11頁)內所含液體的滲透勢(第118頁)低於水的滲透勢，因此，液泡會吸收水分，直至它緊脹(第119頁)爲止。

共質體途徑 植物體內藉擴散作用(第119頁)運輸水份的一條途徑。這條途徑是指沿着連接相鄰細胞的、稱之爲胞間連絲(第15頁)的絲狀物，通過細胞質(第10頁)從一個細胞擴散至相鄰細胞運輸水份的途徑。

非原質體途徑 植物體內輸水份的一條途徑。這條途徑特指沿着相鄰細胞的細胞壁(第8頁)，藉擴散作用(第119頁)，穿過根皮層(第86頁)運輸水份的途徑。

集體流動學說 一九三〇年由明希(Munch)所提出的、用以解釋物質在韌皮部(第84頁)中運輸機製的一種假說(第235頁)。當植物體內一些活躍地進行光合作用(第93頁)的部位，藉滲透作用(第118頁)吸收水份時，在篩管(第84頁)網眼內便會產生溶液集體流動。在這些進行光合作用的部位，溶液濃度較高，因而會流向正在消耗或貯藏光合作用產物、失去水份、而溶液濃度較低的一些區域。在木質部(第84頁)內，水份的運輸是藉蒸騰流(第122頁)朝相反的方向進行的。

根壓 植物體內藉滲透(第118頁)梯度，使水份從根部運輸至木質部(第84頁)的壓力。

Casparian strip a thickened waterproof layer which covers the radial and the transverse cell walls (p. 8) of the endodermis (p. 86) so that all water which is transported from the root cortex (p. 86) to the xylem (p. 84) must pass through the cytoplasm (p. 10) of the endodermis cells.

transpiration stream the continuous flow of water which takes place in a plant through the xylem (p. 84) as water is lost to the atmosphere by transpiration (p. 120) and taken up from the soil by the root hairs (p. 81).

cohesion theory the theory (p. 235) which explains that a column of water may be held together by molecular forces of attraction permitting the ascent of sap up a tall stem without falling back or breaking. There is stress or tension in the column of water as water is lost from the xylem (p. 84) vessels by osmosis (p. 118). Similarly, molecular forces of adhesion will cause water to cling to other substances and thereby rise up a stem by capillarity.

translocation[1] (*n*) the transport of organic material through the phloem (p. 84) of a plant. The material includes carbohydrates (p. 17) such as glucose (p. 17), amino acids (p. 21) and plant growth substances (p. 138).

active transport the method by which, with the use of energy, molecules are transported across a cell membrane (p. 14) against a concentration gradient. It probably involves the use of molecular carriers.

transcellular strand hypothesis a hypothesis (p. 235) developed by Thaine to explain why transport rates in plants appear to be greater than would be possible by diffusion (p. 119). It suggests that active transport (↑) takes place along fibrils (p. 11) of protein (p. 21) that pass through the sieve tube (p. 84).

electro-osmotic hypothesis a hypothesis (p. 235) developed by Spanner to explain why transport rates in plants appear to be greater than would be possible by diffusion (p. 119). It suggests that electro-osmotic forces exist across the sieve plates (p. 85).

凱氏帶　一種加厚的防水層。這種防水層覆蓋着內皮層（第86頁）的徑向和橫向細胞壁（第8頁），這樣，從根皮層（第86頁）輸往木質部（第84頁）的全部水份，必須通過內皮層細胞的細胞質（第10頁）。

蒸騰流　植物體內所發生的、貫穿於木質部（第84頁）的連續水流。當水份由蒸騰作用（第120頁）蒸散到大氣中，而同時根毛（第81頁）又吸收土壤中的水份，便會產生這種連續水流。

內聚力學說　這種學說（第235頁）認爲，分子之間的吸引力可以使水柱聚合在一起，使汁液有可能上昇到一根很高的莖幹上，而不會回落或中斷。當水份藉滲透作用（第118頁）從木質部（第84頁）導管流失時，水柱內存在着應力或張力。同樣，分子之間的黏附力也會使水緊貼其他物質，從而藉毛細管作用沿着莖幹上昇。

轉移（名）　有機物質經由植物韌皮部（第84頁）所進行的運輸。這類有機物質包括碳水化合物（第17頁）諸如葡萄糖（第17頁）、氨基酸（第21頁）和植物生長物質（第138頁）。

主動運輸　穿過細胞膜（第14頁）輸送各種分子的一種方式。這種輸送分子的方式，逆着濃度梯度進行，需要耗用能量，也可能涉及使用分子載體。

細胞間連絲假說　塞恩（Thaine）提出的一種假說（第235頁）。該假說對植物體內的運輸速率似乎大於擴散作用（第119頁）可能達到的運輸速率的原因作了解釋。該假說認爲，主動運輸（↑）是沿着穿過篩管（第84頁）的蛋白質（第21頁）纖絲（第11頁）進行的。

電滲假說　斯潘納（Spanner）提出的一種假說（第235頁）。這種假說對植物體內的運輸速率似乎大於由擴散作用（第119頁）可能達到的運輸速率的原因作了解釋。該假說認爲，電滲力存在於篩板（第85頁）之間。

representations of two theories of phloem transport
韌皮部運輸的兩種假說之圖示

flow of sap 汁液的流動

transcellular strand hypothesis 細胞間連絲假說
fibrils 纖絲

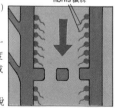

electro-osmotic 電滲假說 **hypothesis**
electro-magnetic force 電滲力

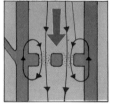

double circulatory system of a mammal 哺乳動物的雙循環系統

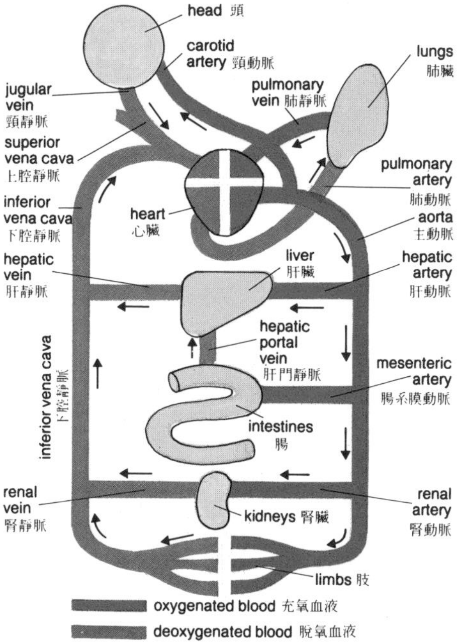

head 頭

carotid artery 頸動脈

lungs 肺臟

pulmonary vein 肺靜脈

jugular vein 頸靜脈

superior vena cava 上腔靜脈

pulmonary artery 肺動脈

inferior vena cava 下腔靜脈

heart 心臟

aorta 主動脈

hepatic vein 肝靜脈

liver 肝臟

hepatic artery 肝動脈

inferior vena cava 下腔靜脈

hepatic portal vein 肝門靜脈

mesenteric artery 腸系膜動脈

intestines 腸

renal vein 腎靜脈

renal artery 腎動脈

kidneys 腎臟

limbs 肢

▬▬ oxygenated blood 充氧血液
▬▬ deoxygenated blood 脫氧血液

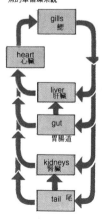

single circulation of a fish
魚的單循環系統

gills 鰓

heart 心臟

liver 肝臟

gut 胃腸道

kidneys 腎臟

tail 尾

circulatory system a system in which materials can be transported around the body of an animal, needed because the volume of the animal is usually too great for transport to be effective by diffusion (p. 119).

single circulation a circulatory system (↑), such as that which occurs in fish, in which the blood (p. 90) passes through the heart (p. 124) once on each complete circuit.

double circulation a circulatory system (↑), such as that which occurs in birds and mammals (p. 80), in which the blood (p. 90) passes through the heart (p. 124) twice on each complete circuit and so maintains the system's blood pressure. In this system, the heart is divided into left and right sides.

循環系統 可環繞動物身體運輸各種物質的一種系統。之所以需要循環系統，是因為動物體的體積通常太大，以致無法有效地藉擴散作用（第119頁）進行物質運輸。

單循環系統 一種循環系統（↑），例如魚類所具的循環系統。在這種循環系統中，血液（第90頁）每完成一次循環，只經過心臟（第124頁）一次。

雙循環系統 一種循環系統（↑），例如鳥類和哺乳類動物（第80頁）所具的循環系統。在這種循環系統中，血液（第90頁）每完成一次循環，需經過心臟（第124頁）兩次，因此，可以保持該系統的血壓。在這種循環系統中，心臟分成左、右兩邊。

open circulatory system a circulatory system (p. 123), e.g. in arthropods (p. 67), in which the blood (p. 90) is free in the body spaces for most of its circulation. The organs lie in the haemocoel (p. 68) and blood from the arteries (p. 127) bathes the major tissues (p. 83) before diffusing (p. 119) back to the open ends of the veins (p. 127). There are no capillaries (p. 127).

closed circulatory system a circulatory system (p. 123), e.g. in vertebrates (p. 74), in which the blood (p. 90) is contained within vessels (p. 127) for the greater part of its circulation.

heart (n) a muscular (p. 143) organ or specialized blood (p. 90) vessel (p. 127) which pumps around the circulatory system (p. 123).

開式循環系統 一種循環系統(第123頁)，如節肢動物(第67頁)的循環系統。此系統內的血液(第90頁)在游離狀態下在體腔內完成大部份循環。各個器官位於血腔(第68頁)內；來自各條動脈(第127頁)的血液先浸泡入一些主要組織(第83頁)，再藉擴散(第119頁)回到靜脈(第127頁)開口端。此系統無微血管(第127頁)。

閉式循環系統 一種循環系統(第123頁)，如脊椎動物(第74頁)的循環系統。此系統內的血液(第90頁)在血管(第127頁)內完成其大部份循環。

心(臟)(名) 一種由肌肉(第143頁)構成的器官或特化血(第90頁)管(第127頁)。它在整個循環系統(第123頁中起泵送作用。

open circulatory system
e.g. an insect
開式循環系統 圖例：昆蟲

pericardial cavity
tubular heart 管狀心臟
心包囊 pericardial cavity 心包囊
dorsal diaphragm 背腦
ostia 心門
gut 胃腸道
perivisceral cavity 圍臟腔
haemocoel 血腔
ventral diaphragm 腹腦

heart 心臟
e.g. human 圖例：人類的心臟

aorta 主動脈
inferior 下腔
vena cava 靜脈
superior 上腔
vena cava 靜脈
right atrium 右心房
tricuspid valve 三尖瓣
right ventricle 右心室
pulmonary artery 肺動脈
pulmonary veins 肺靜脈
left atrium 左心房
atrioventricular valve (bicuspid valve) (二尖瓣)
septum 中隔
left ventricle 左心室

atrium (n) the region or chamber of the heart (↑) which receives the blood (p. 90). The heart of a mammal (p. 80) has a left and a right atrium which are the receivers of oxygenated (p. 126) blood from the lungs (p. 115) and deoxygenated (p. 126) blood from the body respectively. Also known as **auricle. atria** (pl.).

ventricle (n) a muscular (p. 143) region or chamber of the heart (↑) which by regular contractions pumps the blood (p. 90). The heart of a mammal (p. 80) has a left and a right ventricle which pump oxygenated (p. 126) blood to the body and deoxygenated (p. 126) blood to the lungs respectively.

cardiac cycle the cycle in which, by rhythmical muscular (p. 143) contractions, blood (p. 90) flows into the atria (↑) of the heart (↑) and is pumped out of the ventricles (↑).

systole (n) contraction phase of cardiac cycle (↑).

diastole (n) relaxation phase of cardiac cycle (↑).

心房(名) 心臟(↑)接受血液(第90頁)的區域或空腔。哺乳動物(第80頁)的心臟分左、右兩個心房，分別接受來自肺部(第115頁)的充氧(第126頁)血液和來自全身的脫氧(第126頁)血液。心房也稱為心耳。(複數形式為 atria)

心室(名) 心臟(↑)的一個由肌肉(第143頁)構成的區域或空腔。它藉有規律的收縮來汲送血液(第90頁)。哺乳動物(第80頁)的心臟分後左、右兩個心室，分別將充氧的(第126頁)血液泵送至全身和將脫氧的(第126頁)血液泵送至肺部。

心搏周期 血液(第90頁)藉心肌(第143頁)的節律性收縮，流入心(↑)房(↑)和流出心室(↑)所構成的一個周期。

心縮期(名) 心搏周期(↑)的心肌收縮階段。

心舒期(名) 心搏周期(↑)的心肌舒張階段。

position of heart in various invertebrates
不同無脊椎動物的心臟之位置

earthworm 蚯蚓

crustacean 甲殼動物

spider 蜘蛛

valve (*n*) a flap or pocket which only allows a liquid, e.g. blood (p. 90), to flow in one direction.

atrioventricular valve the valve which separates the left ventricle (↑) and atrium (↑) preventing blood (p. 90) from flowing back into the atrium by the closure of two membranous (p. 14) flaps. Also known as **mitral valve**.

bicuspid valve = atrioventricular valve (↑).

tricuspid valve the valve which separates the right ventricle (↑) and atrium (↑).

tendinous cords the tough connective tissue (p. 88) in the heart (↑) which prevents the atrioventricular (↑) and tricuspid valves(↑) from turning inside out during contraction.

pocket valves valves between the ventricles (↑) and the pulmonary artery (p. 128) and aorta (↓) which, when closed, prevent the return of blood to the ventricles.

semi-lunar valves = pocket valves.

aorta (*n*) the major artery (p. 127) carrying oxygenated (p. 126) blood (p. 90) from the heart (↑).

myogenic muscle (*n*) muscle (p. 143), e.g. cardiac muscle (p. 143), which may contract without nervous (p. 149) stimulation although its rate of contraction is controlled by such stimulation.

heartbeat (*n*) the rhythmic contraction of the myogenic muscle (↑) of the heart (↑). Also known as **cardiac rhythm**.

sino-atrial node a group of cells in the right atrium (↑) which is responsible for maintaining the heartbeat (↑) by nervous (p. 149) stimulation relayed by it.

pacemaker (*n*) = sino-atrial node (↑).

atrio-ventricular node a second group of cells in the right atrium (↑) which receives the nervous (p. 149) stimulation from the sino-atrial node (↑).

Purkinje tissue nervous (p. 149) tissue (p. 83) which conducts the nervous stimulation from the sino-atrical node (↑) to the tip of the ventricle (↑) ensuring that the ventricle contracts from its tip upwards to force blood (p. 90) out through the arteries (p. 127).

sympathetic nerve a motor nerve (p. 149) which arises from the spinal nerve and releases adrenaline (p. 152) into the heart muscle (p. 143) to increase the heartbeat (↑).

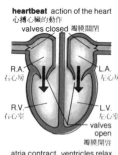

heartbeat action of the heart
心搏心臟的動作
valves closed 瓣膜關閉

R.A. 右心房　L.A. 左心房
R.V. 右心室　L.V. 左心室
valves open 瓣膜開啟

atria contract ventricles relax
心房收縮　心室鬆弛

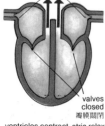

valves open 瓣膜開啓

valves closed 瓣膜關閉

ventricles contract atria relax
心室收縮　心房鬆弛

瓣膜、瓣（名） 只容許一種液體如血液（第90頁）朝一個方向流動的一種片狀物或囊形物。

房室瓣 將左心室（↑）和左心房（↑）隔開的瓣膜。它可藉兩個膜（第14頁）片的關閉作用，阻止血液（第90頁）由左心室返回左心房。又稱爲僧帽瓣。

二尖瓣 同房室瓣（↑）。

三尖瓣 將右心室（↑）和右心房（↑）隔開的瓣膜。

腱索 心臟（↑）中堅韌的結締組織（第88頁）。在心臟收縮的過程中，它可防止房室瓣（↑）和三尖瓣（↑）從裏向外翻出。

囊形瓣 位於心室（↑）與肺動脈（第128頁）之間，以及心室與主動脈（↓）之間的瓣膜。當其關閉時，可防止血液返回心室。

半月瓣 同囊形瓣。

主動脈（名） 運送來自心臟（↑）的充氧（第126頁）血液（第90頁）的主要動脈（第127頁）。

肌原性肌（名） 未經神經（第149頁）刺激即可收縮的肌肉（第143頁）如心肌（第143頁），但其收縮率受這種刺激控制。

心搏（名） 心臟（↑）的肌原性肌（↑）的節律性收縮。心搏也稱爲心節律。

竇房結 右心房（↑）內的一羣細胞。這羣細胞負責維持由其所傳導的神經（第149頁）興奮引起的心搏（↑）。

起搏點（名） 同竇房結（↑）。

房室結 右心房（↑）內的另一羣細胞。這羣細胞接受來自竇房結（↑）的神經（第149頁）興奮。

栢金氏組織 將來自竇房結（↑）的神經興奮傳導到心室（↑）頂端的神經（第149頁）組織（第83頁）。這種組織確保心室從其頂端處向上收縮，壓出血液（第90頁），使血液在動脈（第127頁）中保持流通。

交感神經 由脊神經發出的一種運動神經（第149頁）。這種運動神經可將腎上腺素（第152頁）釋入心肌（第143頁）中，使心搏（↑）增加。

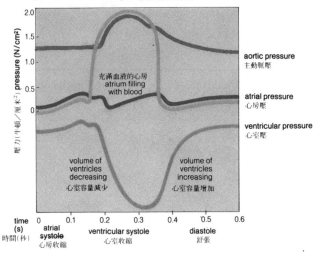

changes in volume and pressure 哺乳動物心搏周期的
during a mammalian cardiac cycle 容量和壓力之變化

壓力（牛頓／厘米²）pressure (N/cm²)

充滿血液的心房
atrium filling
with blood

aortic pressure
主動脈壓

atrial pressure
心房壓

ventricular pressure
心室壓

volume of
ventricles
decreasing
心室容量減少

volume of
ventricles
increasing
心室容量增加

time (s) 時間(秒)   0   0.1   0.2   0.3   0.4   0.5   0.6

atrial systole 心房收縮   ventricular systole 心室收縮   diastole 舒張

vagus nerve a motor nerve (p. 149) which arises from the medulla oblongata (p. 156) and releases acetylcholine (p. 152) into the heart muscle (p.143) to decrease the heartbeat (p.125).

pulse (n) a wave of increased blood (p. 90) pressure which passes through the arteries (↓) as the left ventricle (p. 124) pumps its contents into the aorta (p. 125).

pigment (n) a coloured substance. For example, *myoglobin* is a variety of haemoglobin (↓) found in muscle (p. 143) cells, and *chlorocruorin* is a respiratory pigment containing iron which is found in the blood (p. 90) of some polychaetes (p. 65). *See also* chlorophyll (p. 12).

haemoglobin (n) a red pigment (↑) and protein (p. 21) containing iron, which is found in the cytoplasm (p. 10) of the red blood cells (p. 91) of vertebrates (p. 74). It combines readily with oxygen to form oxyhaemoglobin and in this form oxygen is transported to the tissues (p. 83) from the lungs (p. 115).

oxyhaemoglobin (n) see haemoglobin (↑).

oxygenated (adj) containing or carrying oxygen.

deoxygenated (adj) not oxygenated (↑).

迷走神經　由延髓(第156頁)發出的一種運動神經(第149頁)。這種運動神經可將乙醯膽鹼(第152頁)釋入心肌(第143頁)中，使心搏(第125頁)減慢。

脈搏(名)　左心室(第124頁)將其中的血液壓入主動脈(第125頁)時，由血(第90頁)壓昇高形成的一種通過動脈(↓)的壓力波。

色素(名)　一種有色物質。例如，肌紅蛋白是一種肌肉(第143頁)細胞中含有的血色蛋白(↓)；血綠蛋白是一種在某些多毛綱的環節動物(第65頁)血液(第90頁)中所含的含鐵呼吸色素。參見「葉綠素」(第12頁)。

血紅蛋白、血紅素(名)　一種含鐵的紅色色素(↑)和蛋白質(第21頁)。它存在於脊椎動物(第74頁)的紅血細胞(第91頁)的細胞質(第10頁)內。血紅蛋白易於與氧結合生成氧合血紅蛋白；氧便以這種形式從肺(第115頁)部運送到各組織(第83頁)中去。

氧合血紅蛋白、氧化血紅素(名)見血紅蛋白(↑)。

充氧的(形)　含氧或攜氧的。

脫氧的(形)　非充氧的(↑)。

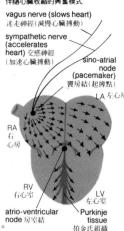

the pattern of excitation that accompanies contraction of the heart
伴隨心臟收縮的興奮模式

vagus nerve (slows heart)
迷走神經 (減慢心臟搏動)

sympathetic nerve (accelerates heart) 交感神經 (加速心臟搏動)

Sino-atrial node (pacemaker)
竇房結 (起搏點)

LA 左心房

RA 右心房

RV 右心室

LV 左心室

atrio-ventricular node 房室結

Purkinje tissue 柏金氏組織

a network of capillaries
微血管網絡

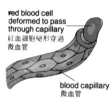

red blood cell
deformed to pass
through capillary
紅血細胞變形穿過
微血管

blood capillary
微血管

artery 動脈

artery
動脈

thick
muscular
wall 厚的肌壁

vein 靜脈

vein
靜脈

thin muscular
wall 薄的肌壁

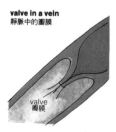

valve in a vein
靜脈中的瓣膜

valve
瓣膜

haemocyanin (*n*) a blue pigment (↑) and protein (p. 21) containing copper, which is found in the plasma (p. 90) of certain invertebrates (p. 75). It also combines with oxygen to transport it to the tissues (p. 83).

Bohr effect the effect of increasing the likelihood of the dissociation of oxygen from oxyhaemoglobin (↑) as the level of carbon dioxide is increased so that with increased activity more oxygen is passed to the body tissues (p. 83).

vascular system the system of vessels (↓) which transport fluid throughout the body of an organism.

capillary (*n*) any of the very large numbers of tiny blood vessels (↓) which form a network throughout the body. They present a large surface area and are thin-walled to aid gas exchange (p. 112).

sphincter muscle any of the muscles (p. 143) which, by contraction, close any of the hollow tubes, organs, or vessels (↓) in an organism.

vessel[a] (*n*) a channel or duct with walls e.g. blood (p. 90) flows through a blood vessel.

vein[a] (*n*) any one of the tubular vessels (↑) that conveys blood (p. 90) back to the heart (p. 124). Veins are quite large in diameter but thinner walled than arteries (↓) and the blood is carried under relatively low pressure. Veins have pocket valves (p. 125) which ensure that the blood is carried towards the heart only.

venule (*n*) a small blood vessel (↑) that receives blood from the capillaries (↑) and then, with other venules forms the veins (↑).

artery (*n*) any one of the tubular vessels (↑) that conveys blood (p. 90) from the heart (p. 124). They are smaller in diameter than veins (↑) but the walls are thicker and more elastic and the blood is carried at relatively high pressure. With the exception of the aorta (p. 125) and the pulmonary artery (p. 128), arteries have no pocket valves (p. 125).

arteriole (*n*) a small artery (↑).

sinus (*n*) any space or chamber, such as the sinus venosus which is a chamber found within the heart (p. 124) of some vertebrates (p. 74) especially amphibians (p. 77), and lies between the veins (↑) and the atrium (p. 124).

血藍蛋白（名）　一種含銅的藍色色素（↑）和蛋白質（第21頁）。某些無脊椎動物（第75頁）的血漿（第90頁）中含有血藍蛋白，它可與氧結合，將氧運送到各組織（第83頁）中去。

玻爾氏效應　二氧化碳含量增加時，使氧從氧合血紅蛋白（↑）中解離的可能性得以增加的效應，因此，隨着活動的增加，更多的氧運送到身體的各組織（第83頁）中。

血管系統、維管系統　整個生物體內運輸液體的導管（↓）系統。

微血管、毛細血管（名）　在全身形成網絡的數量極多的任何一根微細血管（↓）。它們的表面積很大，管壁較薄，有助於進行氣體交換（第112頁）。

括約肌　可藉收縮作用使生物體內的空心管、器官或脈管（↓）關閉的肌肉（第143頁）。

管、脈管[a]（名）　有壁的通管或導管。例如，血液（第90頁）經由血管流動。

靜脈[a]（名）　任何一根將血液（第90頁）送回心臟（第124頁）的管狀脈管（↑）。靜脈的直徑很大，但其管壁較動脈（↓）爲薄，在較低的壓力下輸送血液。靜脈有囊形瓣（第125頁），可確保血液只朝心臟方向輸送。

小靜脈（名）　接受來自微血管（↑）血液的小血管（↑）。爲數衆多的小靜脈即構成靜脈（↑）脈絡。

動脈（名）　任何一根輸送來自心臟（第124頁）血液（第90頁）的管狀脈管（↑）。動脈的直徑較靜脈（↑）小，但其管壁較厚，更富彈性，在較高的壓力下輸送血液。除主動脈（第125頁）和肺動脈（第128頁）之外，其他動脈都沒有囊形瓣（第125頁）。

小動脈（名）　小的動脈（↑）。

寶（名）　如靜脈寶之類的任何腔或室。某些脊椎動物（第74頁）、尤其是兩棲動物（第77頁）心臟（第124頁）內的一個室包括靜脈寶；它位於靜脈（↑）和心房（第124頁）之間。

pulmonary circulation the part of the double circulation (p. 123) in which deoxygenated (p. 126) blood (p. 90) is pumped from the heart (p. 124) to the lungs (p. 115).

pulmonary artery the artery (p. 127) which carries the deoxygenated (p. 126) blood (p. 90) pumped from the heart (p. 124) to the lungs (p.115).

pulmonary vein the vein (p. 127) which carries the oxygenated (p. 126) blood (p. 90) from the lungs (p. 115) back to the heart (p. 124).

systemic circulation the part of the double circulation (p. 123) in which blood (p. 90) is pumped from the heart (p. 124) throughout the body of the animal.

arterio-venous shunt vessel a small blood vessel (p. 127) which bypasses the capillaries (p. 127) and carries blood (p. 90) from the arteries (p.127) to the veins (p. 127) and therefore regulates the amount of blood which enters the capillaries.

lymph (n) a milky or colourless fluid which drains from the tissues (p. 83) into the lymphatic vessels (↓) and is not absorbed (p. 81) back into the capillaries (p. 127). It is similar to tissue fluid and contains bacteria (p. 42) but does not contain large protein (p. 21) molecules.

lymphatic vessel any one of the vein-like (p. 127) vessels (p. 127) that carry lymph (↑) from the tissues (p. 83) into the large veins that enter the heart (p. 124).

lymph node a swelling in the lymphatic vessel (↑), especially in areas such as the groin or armpits, which contain special white blood cells (p. 91) known as macrophages (p. 88).

platelet (n) any of the fragments of cells present in the blood (p. 90) plasma (p. 90) which are formed in the red bone marrow (p. 90) and which prevent bleeding by combining at the point of an injury and releasing a hormone (p. 130) which stimulates blood clotting (↓). They also release other substances which cause blood vessels (p. 127) to constrict so that they also prevent capillary (p. 127) bleeding.

coagulate (v) = clot (↓).

anticoagulant (n) a substance that stops blood (p. 90) clotting (↓).

肺循環　雙循環(第123頁)的一部份。這部份循環指的是將脫氧的(第126頁)血液(第90頁)從心臟(第124頁)壓出，送至肺(第115頁)部所進行的循環。

肺動脈　將心臟(第124頁)壓出的脫氧(第126頁)血(第90頁)運至肺(第115頁)的動脈(第127頁)。

肺靜脈　將來自肺(第115頁)部的充氧(第126頁)血液送回心臟(第124頁)的靜脈(第127頁)。

體循環　雙循環(第123頁)的一部份。這部份循環指的是血液(第90頁)從心臟(第124頁)壓出，送至動物體全身所進行的循環。

動靜脈分路管　一種小血管(第127頁)。它是為微血管(第127頁)所加的分路，可將來自動脈(第127頁)的血液(第90頁)輸送至靜脈(第127頁)，從而調節進入微血管的血量。

淋巴(名)　從組織(第83頁)流入淋巴管(↓)內的一種乳白色或無色液體。這種液體不再被吸收(第81頁)到微血管(第127頁)內。它類似於組織液，含有細菌(第42頁)，但不含大的蛋白質(第21頁)分子。

淋巴管　可將來自組織(第83頁)的淋巴(↑)運送到進入心臟(第124頁)的大靜脈內的任何一根靜脈狀(第127頁)管(第127頁)。

淋巴結　淋巴管(↑)內的隆突。它尤其存在於如腹股溝或腋窩這些區域，其內含有被稱為巨噬細胞(第88頁)的特殊白血細胞(第91頁)。

血小板(名)　血(第90頁)漿(第90頁)中含的任何細胞碎片。這些細胞碎片在紅骨髓(第90頁)中形成，它們能在損傷處相互結合，並釋放出一種刺激血液凝固(↓)的激素(第130頁)，以阻止出血。血小板還可釋放其他一些使血管(第127頁)收縮的物質，從而也可防止微血管(第127頁)出血。

凝結(動)　同凝固(↓)。

抗凝血劑(名)　可阻止血液(第90頁)凝固(↓)的一種物質。

lymphatic system 淋巴系統

| lymphatic vessel 淋巴管 | blood flow 血流 | lymph capillary 淋巴毛細管 |

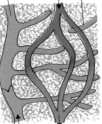

lymph flow 淋巴流　blood flow 血流

cells bathed in tissue fluid
浸在組織液中的細胞

lymph vessels
淋巴管

blood capillaries oxygenated blood
內含充氧血液的微血管

blood capillaries deoxygenated blood
內含脫氧血液的微血管

platelet 血小板

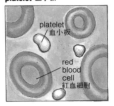

platelet 血小板

red blood cell 紅血細胞

clot (v) of liquids to become solid, for example, blood (p. 90) clots in air. **clot** (n).

blood groups in humans, there is a system of multiple alleles (p. 205) which gives rise to four main different blood groups with different antigens (p. 234) or proteins (p. 21) on the surface of the red blood cells (p. 91). The A allele, B allele and O allele (producing no antigens) may combine to give any of the following blood group combinations: AA, AO, BB, BO, AB or OO. A and B alleles are both dominant (p. 197) to O so that there are four groups, A, B, O and AB.

rhesus factor an antigen (p. 234) which is present in the blood (p. 90) of rhesus monkeys and most, but not all, humans. During pregnancy (p. 195) or following transfusion of blood containing the rhesus factor (Rh+) into blood lacking it (Rh–), breakdown of the red blood cells (p. 91) can occur with dangerous results.

凝固（動） 指液體變爲固體而言，例如血液（第90頁）在空氣中凝固。（名詞形式爲 clot）。

血型 人類中含有的一種複等位基因（第205頁）體系，該體系可衍生出在紅血細胞（第91頁）表面上具有不同抗原（第234頁）或蛋白質（第21頁）的四種不同的主要血型。等位基因 A、B 和 O（不產生抗原）可相互結合，產生下列任何一種血型組合：AA、AO、BB、BO、AB 或 OO。等位基因 A 和 B 兩者對等位基因 O 來說，均爲顯性（第197頁），所以有四種血型，即 A、B、O 和 AB。

Rh 因子、獼猴因子 一種抗原（第234頁）。獼猴和大多數（但不是全部）人的血液（第90頁）中含有抗原。在懷孕（第195頁）期間或在將含有 Rh 因子（Rh＋）的血液輸入到無 Rh 因子（Rh－）的血液之後，均有可能發生紅血細胞（第91頁）分解，帶來危險後果。

the four main blood groups
四種主要血型

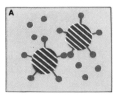
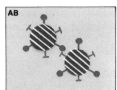

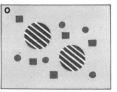

●━**A** antigen ━◼**B** antigen ◼**A** antibody ●**B** antibody
A 抗原 B 抗原 A 抗體 B 抗體

	blood group	血型	antigens on red cells	紅血細胞上的抗原	antibodies in serum	血清中的抗體	can receive blood type	可接受的血型	can donate blood to	可輸往的血型
	A	A	A	A	B	B	groups **A and O**	A 和 O 型	groups **A and AB**	A 和 **AB** 型
	B	B	B	B	A	A	groups **B and O**	B 和 O 型	groups **B and AB**	B 和 **AB** 型
universal recipients 全適受血者	**AB**	AB	A and B	A 和 B	none	A 和 B 均無	groups **A, B, AB and O**	A、B、AB 和 O 型	group **AB**	**AB** 型
universal donors 全適供血者	**O**	O	none	A 和 B 均無	A and B	A 和 B	groups **O only**	僅 O 型	groups **A, B, AB and O**	A、B、AB 和 O 型

homeostasis (*n*) the maintenance of constant internal conditions within an organism, thus allowing the cells to function more efficiently, despite any changes that might occur in the organism's external environment (p. 218).

endocrine system a system of glands (p. 87) in animals which produce hormones (↓). This system and the nervous system (p. 149) combine to control the functions of the body.

endocrine gland a gland (p. 87) which produces hormones (↓).

hormone (*n*) a substance made in very small amounts in one part of an organism and transported to another part where it produces an effect. (1) In plants, the hormones can be referred to as growth substances (p. 138). (2) In animals, hormones are secreted (p. 106) by the endocrine glands (↑) into the blood stream (p. 90), where they circulate to their destination.

adrenal glands in mammals (p. 80), a pair of endocrine (↑) glands (p. 87) near the kidneys (p. 136). They are divided into two parts; the *medulla*, the inner part which secretes (p. 106) adrenaline (p. 152) and noradrenaline (p. 152), and the *cortex*, the outer part which secretes various steroid (p. 21) hormones (↑).

homoiothermic (*adj*) of an organism which maintains its body temperature at a constant level in changing external circumstances. These organisms, including mammals (p. 80), for example, are usually regarded as 'warm blooded' because their body temperature is usually above that of the surroundings.

endothermic (*adj*) = homoiothermic (↑).

poikilothermic (*adj*) of an organism whose body temperature varies with and is roughly the same as that of the environment (p. 218). These organisms, not including birds and mammals (p. 80), are usually regarded as 'cold blooded' although their body temperature may be higher or lower than that of the environment depending on such factors as wind speed or the sun's radiation (↓). They have a lower metabolic rate (p. 32) as their body temperature falls.

exothermic (*adj*) = poikilothermic (↑).

體內平衡、內環境穩定（名） 生物體保持體內穩定的現象。這種現象使細胞不受生物體外部環境（第218頁）中可能發生的任何變化的影響，從而更有效地行使功能。

內分泌系統 動物體內產生激素（↓）的腺體（第87頁）系統。該系統和神經系統（第149頁）聯合控制身體的許多功能。

內分泌腺 產生激素（↓）的腺體（第87頁）。

激素、荷爾蒙（名） 在生物體內的某一部位產生的一種極少量的物質，當其被輸送到另一個部位時，才發揮作用。(1)在植物中，激素可稱為生長物質（第138頁）。(2)在動物中，激素由內分泌腺（↑）分泌（第106頁），進入血（第90頁）流，經血液循環到達其目的地。

腎上腺 哺乳動物（第80頁）體內靠近腎臟（第136頁）的一對內分泌（↑）腺（第87頁）。它們由兩個部份組成：裏面的部份為髓質，它分泌（第106頁）腎上腺素（第152頁）和去甲腎上腺素（第152頁），外面的部份為皮質，它分泌各種類固醇（第21頁）激素（↑）。

恒溫的（形） 指生物而言，在外部環境發生變化的情況下，其體溫可保持不變的。例如，這些生物包括哺乳動物（第80頁）在內，通常被認為是"溫血的"動物，因其體溫通常高於環境溫度。

吸熱的（形） 同恒溫的（↑）

變溫的的（形） 指生物而言，其體溫隨環境（第218頁）溫度而變化、並與環境溫度大致相同的。這些生物不包括鳥類和哺乳動物（第80頁），通常被認為是"冷血的"動物，其體溫可受風速或太陽輻射（↓）這類因素影響而高於或低於環境溫度。當它們的體溫下降時，其代謝率（第32頁）也隨之降低。

放熱的（形） 同變溫的（↑）。

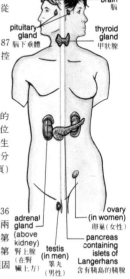

the main human endocrine glands 人體的主要內分泌腺

brain 腦
pituitary gland 腦下垂體
thyroid gland 甲狀腺
adrenal gland (above kidney) 腎上腺（在肝臟上方）
testis (in men) 睪丸（男性）
ovary (in women) 卵巢（女性）
pancreas containing islets of Langerhans 含有胰島的胰腺

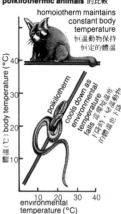

comparison of homoiothermic and poikilothermic animals 恒溫動物和變溫動物的比較

homoiotherm maintains constant body temperature 恒溫動物保持恒定的體溫

poikilotherm cools down as environmental temperature falls 變溫動物的體溫隨環境溫度下降而下降

body temperature (°C) 體溫（℃）

environmental temperature (°C) 環境溫度（℃）

heat gains and losses in a reptile (poikilothermic) animal
爬蟲動物（變溫動物）的得熱和失熱

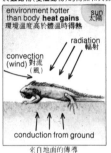

environment hotter than body **heat gains**
環境溫度高於體溫時得熱

radiation 輻射

sun 太陽

convection (wind) 對流（風）

conduction from ground
來自地面的傳導

environment cooler than body **heat losses**
環境溫度低於體溫時失熱

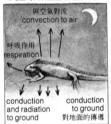

convection to air 與空氣對流

respiration 呼吸作用

conduction and radiation to ground
對地面的傳導與輻射

conduction to ground
對地面的傳導

radiation (*n*) the transfer of heat from a hot object, such as the sun, to a cooler object, such as the earth or the body of an organism, through space without increasing the temperature of the space.

evaporation (*n*) the change from a liquid to a vapour or gas that takes place when the liquid is warmed to a temperature at or below its boiling point.

conduction (*n*) the transfer of heat through a solid.

convection (*n*) the transfer of heat in a fluid as the warmed portion of the fluid rises and the cool portion falls.

hair (*n*) a single-celled or many celled outgrowth from the dermis (↓) of a mammal (p. 80) made up of dead material and including the substance keratin. Among other functions, a coat of hair insulates the mammal's body from excessive warming or cooling especially if the hairs are raised to trap a layer of insulating air around the body. Hair is a characteristic of mammals.

skin (*n*) the outer covering of an organism which insulates it from excessive warming or cooling, prevents damage to the internal organs, prevents the entry of infection, reduces the loss of water, protects it from the sun's radiation (↑) and it also contains sense organs which make the organism aware of its surroundings.

dermis (*n*) = skin (↑).

epidermis (*n*) the outer layer of the skin (↑). The epidermis is made up of three main layers of cells: the continuous *Malpighian layer* is able to produce new cells by division and so replace epidermal layers as they are worn away; the *granular layer* grades into the harder, outer *cornified layer* which is composed of dead cells only and forms the main protective part of the skin. This cornified layer may become very hard and thick in areas that are constantly subject to wear, such as the soles of the feet.

sebaceous gland one of the many glands (p. 87) contained within the skin (↑) that open into the follicles (p. 132) and which secrete (p. 106) an oily antiseptic substance which repels water and keeps the skin flexible.

輻射（名）　熱量從一個熱的物體（例如太陽），通過空間，在不升高該空間溫度的情況下，向一個較冷的物體（例如地球或一個生物體）進行的傳遞。

蒸發（名）　當一種液體加熱到其沸點或低於其沸點的溫度時，該液體變成爲蒸氣體的變化。

傳導（名）　熱量通過固體所進行的傳遞。

對流（名）　當一種流體中熱的部份上升，冷的部份下沉時，在該流體中所進行的熱量傳遞。

毛髮（名）　哺乳動物（第80頁）眞皮（↓）內長出的單細胞或多細胞贅疣，由無生命的物質包括角蛋白所組成。除了其他一些功能之外，毛髮的覆蓋層還能使哺乳動物的身體避免過熱或過冷，尤其是當毛髮長到在身體四周形成一個空氣隔絕層時，情況更是如此。具有毛髮是哺乳動物的一個特徵。

皮膚（名）　生物體外表的被覆物。它可使體避免過熱或過冷，防止內部器官受損，防止傳染病菌侵入，減少水份散失，保護生物體免受太陽的輻射（↑）；它也具有一些使生物能察覺其周圍環境的感覺器官。

眞皮（名）　同皮（膚）（↑）。

表皮（名）　皮膚（↑）的外層。表皮由三個主要的細胞層構成：連續的馬氏層，可藉分裂產生新的細胞，因此，當表皮層磨損時得以替換；顆粒層漸次變化爲較堅硬的、外面的角質層；角質層僅由死的細胞構成，並形成皮膚的主要保護部份。這種角質層在經常受磨損的區域，例如足底，可變硬而厚。

皮脂腺　位於皮膚（↑）內的諸多腺體（第87頁）。這些腺體通入毛囊（第132頁），可分泌（第106頁）一種能排斥水份、保持皮膚柔潤的油質抗菌物質。

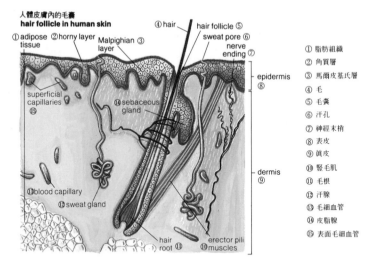

人體皮膚內的毛囊
hair follicle in human skin

④ hair　　hair follicle ⑤
　　　　　sweat pore ⑥
①adipose ②horny layer Malpighian ③　　　nerve
tissue　　　　　layer　　　　　ending ⑦

superficial
capillaries
⑮

⑭sebaceous
gland

epidermis
⑧

⑬blood capillary

⑫sweat gland

dermis
⑨

hair
root ⑪

erector pili
⑩muscles

① 脂肪組織
② 角質層
③ 馬爾皮基氏層
④ 毛
⑤ 毛囊
⑥ 汗孔
⑦ 神經末梢
⑧ 表皮
⑨ 眞皮
⑩ 竪毛肌
⑪ 毛根
⑫ 汗腺
⑬ 毛細血管
⑭ 皮脂腺
⑮ 表面毛細血管

hair follicle (*n*) a pit within the Malphigian and granular layers of the epidermis (p. 131) which contains the root of a hair and from which the hair grows by cell division. Sebaceous glands (p. 131) open into the hair follicles and also contain muscles (p. 143) to erect the hairs for additional insulation as well as nerve (p. 149) endings for sensitivity.

sweat gland a coiled tubular gland (p. 87) within the epidermis (p. 131) which absorbs moisture containing some salts and minerals from surrounding cells and releases it to the surface through a tube causing the skin (p. 131) to be cooled as the moisture evaporates into the atmosphere.

hibernation (*n*) a process by which certain organisms respond to very low external temperatures in a controlled fashion. The core temperature of the animal falls to near that of the environment (p. 218) and there is a resulting drop in the metabolic rate (p. 32) although the nervous system (p. 149) continues to operate so that should the temperature fall near to the lethal temperature, the animal increases its metabolic rate to cope with it.

毛囊（名）　表皮（第131頁）的馬爾皮基氏層和顆粒層內的小凹。其內包含毛根；毛髮經由細胞分裂從該處長出。皮膚腺（第131頁）通入毛囊，還具有肌肉（第143頁），可使毛髮竪起，以輔助保溫，也可使神經（第149頁）末梢勃起，以增強感受性。

汗腺　表皮（第131頁）內捲曲的管狀腺體（第87頁）。該腺體可從周圍的細胞吸收含有某些鹽份和礦物質的水份，並通過一導管將其排放到皮膚表面，這樣，由於水份蒸發到大氣中，從而使皮膚（第131頁）變涼。

冬眠（名）　某些生物以可控制的方式對非常低的外部氣溫作出反應的一個過程。動物的核心溫度降至接近於環境（第218頁）溫度，代謝率（第32頁）也隨之下降，但神經系統（第149頁）繼續起作用。因此，如果動物的核心溫度降至接近於致死溫度時，動物便會提高其代謝率，以對付這種情況。

hyperthermia (*n*) overheating. The condition in which, as a result of vigorous activity, disease, or heating by radiation (p. 131), the body temperature of an organism rises above its normal level. As a result of nerve impulses (p. 150) sent to the hypothalmus (p. 156) counter measures are taken including dilation of the blood vessels (p. 127) to allow greater heat loss by radiation, convection (p.131) and conduction (p. 131), and sweating to cool the body surface as moisture evaporates.

hypothermia (*n*) overcooling. The condition in which the temperature of an organism falls below its normal level. As a result of nerve impulses (p. 150) sent to the hypothalamus (p. 156) counter measures are taken including a reduction in sweating, constriction of the blood vessels (p. 127) to reduce the amount of heat being lost by radiation (p. 131), conduction (p. 131) and convection (p. 131), rapid spasmodic contraction of the muscles (p. 143) to cause shivering, and an increase in metabolic rate (p.32).

aestivation (*n*) (1) the condition of inactivity or torpor into which some animals enter during periods of drought or high temperatures. Lung fish, for example, bury themselves in mud at the start of the dry season and re-emerge when the rain begins to fall again. (2) the arrangement of parts in a flower bud.

leaf fall the condition into which some plants enter during periods of extreme water shortage by losing some of their leaves to reduce water losses by transpiration (p. 120).

osmoregulation (*n*) the process by which an organism maintains the osmotic potential (p.118) in its body fluids at a constant level e.g. freshwater fish take in large volumes of water through the gills (p.113) by osmosis (p.118) which are then excreted (p.134) as urine (p.135) from the kidneys (p.135). Marine fish, either drink sea water (bony fish (p.76)) so that salts are absorbed (p. 81) by the gut (p.98) and water then follows by osmosis, with the salts eliminated by the gills, or retain urea (p.134) so that their fluids are hypertonic (p. 118) to sea water and then, like freshwater fish, they take in more water through their gills.

osmoregulation in bony fish 硬骨魚的滲透調節
drinks large amount of water 吸入大量的水

salt water fish 海水魚
salt 鹽　water 水　small volume of urine (moderate salt) 少量尿 (鹽份適中)

salt in foods drinks small amount of water 食物中的鹽份 吸入少量的水
salt 鹽
water 水

freshwater fish 淡水魚
large volume of urine (little salt) 大量尿 (鹽份很少)

體溫過高 (名)　即過熱。生物體由於劇烈運動、生病或輻射 (第131頁) 受熱而引起體溫高於其正常水平時所出現的狀況。由於神經衝動 (第150頁) 傳遞給下丘腦 (第156頁)，結果身體即採取對抗措施：包括擴張血管 (第127頁)，使大量的熱藉輻射、對流 (第131頁) 和傳導 (第131頁) 而散失；以及出汗，藉水份的蒸發使身體表面變涼。

體溫過低 (名)　即過冷。生物體的體溫降至其正常水平以下時所出現的狀況。由於神經衝動 (第150頁) 傳遞給下丘腦 (第156頁) 的結果，身體即採取對抗措施：包括減少出汗；收縮血管 (第127頁)，以減少由輻射 (第131頁)、傳導 (第131頁) 和對流 (第131頁) 而引起的熱量散失；肌肉 (第143頁) 快速痙攣性收縮，以產生顫栗；增加代謝率 (第32頁)。

夏眠；花被捲疊式 (名)　(1)某些生物在乾旱或高溫期間所進入的一種不活動或假死狀態。例如，肺魚在旱季開始時即蟄匿於泥中，等到再度開始下雨時方重新出現；(2)花芽內各組成部份的一種排列方式。

落葉　某些植物在環境極度缺水期間所進入的一種狀態。它們藉失去一些葉，以降低由於蒸騰作用 (第120頁) 而引起的水份損失。

滲透調節 (名)　生物將其體液內的滲透勢 (第118頁) 保持在恒定水平所經歷的過程。例如，淡水魚可藉滲透作用 (第118頁) 經由鰓 (第113頁) 吸入大量水，然後將它以尿 (第135頁) 的形式由腎臟 (第135頁) 排泄 (第134頁) 出去。海水魚 (硬骨魚 (第76頁)) 或者吸入海水，藉消化道 (第98頁) 吸收 (第81頁) 其中的鹽份，隨後再藉滲透作用讓水滲入，而將鹽份由鰓排除；或者將尿素 (第134頁) 保留下來，使其體液對海水是高滲的 (第118頁)，隨後像淡水魚那樣，由鰓吸入更多的水。

carotid body a small oval structure in the carotid artery (p. 127) containing nerves (p. 149) which respond to the oxygen and carbon dioxide content of the blood (p. 90) and so control the level of respiration (p. 112).

carotid sinus a small swelling in the carotid artery (p. 127) containing nerves (p. 149) which respond to blood (p. 90) pressure and so control circulation (p. 123).

excretion (n) the process by which the waste and harmful products of metabolism (p. 26) such as water, carbon dioxide, salts and nitrogenous compounds, are eliminated from the organism.

含氮廢物的化學式 **chemical formulae of nitrogenous wastes**

| ammonia 氨 | NH_3 |

$$H_2N-\underset{\underset{O}{\|}}{C}-NH_2$$
urea 尿素

urea (n) a nitrogenous organic compound (p. 15) which is soluble in water and which is the main product of excretion (↑) from the breakdown of amino acids (p. 21) in certain animals.

uric acid a nitrogenous organic compound (p. 15), insoluble in water, which is the main product of excretion (↑) from the breakdown of amino acids (p. 21) in certain animals. Because it is insoluble, uric acid is not toxic and can be excreted without losing large amounts of water.

ammonia (n) a nitrogenous inorganic compound (p. 15) which is very toxic and may only be found as a product of excretion (↑) in organisms where large amounts of water are available for its removal, such as in aquatic animals.

contractile vacuole a vacuole (p. 11) present in the endoplasm (p. 44) of the Protozoa (p. 44) which is important in the osmoregulation (p. 133) of these organisms. In hypotonic (p.118) solutions (p. 118) the vacuole swells and then seems to contract, releasing to the exterior, water which has entered the cell with food or from the surroundings. The water passes into the solution. In hypertonic (p. 118) and isotonic (p. 118) solutions the vacuole disappears.

頸動脈體 頸動脈(第127頁)中含有很多神經(第149頁)的一種細小的卵形結構。頸動脈體中的神經可感受血液(第90頁)的氧和二氧化碳變化,從而控制呼吸(第112頁)的程度。

頸動脈竇 頸動脈(第127頁)中含有很多神經(第149頁)的一種小的隆起物。頸動脈竇中的神經可感受血(第90頁)壓的變化,從而控制血液循環(第123頁)。

排泄(作用)(名) 新陳代謝(第26頁)的廢物和有害物質,例如水、二氧化碳、鹽和含氮化合物從生物體內被排除出去的過程。

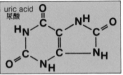

含氮廢物的化學式

尿素(名) 一種含氮的有機化合物(第15頁)。它溶於水;在某些動物中,是氨基酸(第21頁)分解的主要排泄(↑)物。

尿酸 一種含氮的有機化合物(第15頁)。它不溶於水中;在某些動物中,是氨基酸(第21頁)分解的主要排泄(↑)物。正因爲尿酸不溶於水,所以它是無毒的,並可在不失去大量水份的情況下被排泄出去。

氨(名) 一種含氮的無機化合物(第15頁)。它的毒性很大,也許僅在像水生動物之類生物中,才作爲一種排泄(↑)物而存在。在這些生物中,有大量水可將之除去。

伸縮泡 原生動物(第44頁)內質(第44頁)中具有的一種液泡(第11頁)。它在這些生物體的滲透調節(第133頁)中起重要作用。在低滲(第118頁)溶液(第118頁)中,伸縮泡先膨脹,然後似乎收縮,將已經和食物一起進入細胞的水份或來自周圍環境的水份排放到外部。這樣,水便進入該溶液中。在高滲(第118頁)和等滲(第118頁)溶液中,伸縮泡則消失。

無脊椎動物的排泄器官
excretory organs in invertebrates
Malpighian tubules of insect

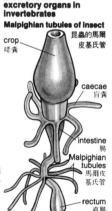

crop 昆蟲的馬爾皮基氏管
crop 嗉囊
caecae 盲囊
intestine 腸
Malpighian tubules 馬爾皮基氏管
rectum 直腸

contractile vacuole of Amoeba 變形蟲的伸縮泡

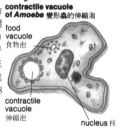

food vacuole 食物泡
contractile vacuole 伸縮泡
nucleus 核

Malpighian tubule one of a number of narrow, blind tubes which are the main organs of osmoregulation (p. 133) and excretion (↑) in insects (p. 69) and other members of the Arthropoda (p. 67). They arise from the gut (p. 98) and in them, uric acid (↑) crystals are produced which can be eliminated with little water loss. Although insects do possess a thin, waterproof cuticle (p. 145) water is still lost through the joints (p. 146) and by respiration (p. 112).

馬爾皮基氏管 許多狹窄的盲管，它們是節肢動物門（第67頁）中的昆蟲（第69頁）和其他成員進行滲透調節（第133頁）與排泄（↑）主要器官。它們由腸（第98頁）部長出；產生於其內的尿酸（↑）結晶，以極少的水份損失即可予以排除。雖然昆蟲確實具有一層薄而防水的表皮（第145頁），但是水份仍然經由關節（第146頁）和呼吸作用（第112頁）散失。

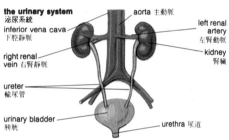

the urinary system 泌尿系統
inferior vena cava 下腔靜脈
right renal vein 右腎靜脈
aorta 主動脈
left renal artery 左腎動脈
kidney 腎臟
ureter 輸尿管
urinary bladder 膀胱
urethra 尿道

ureter (*n*) the tube or duct which carries urine (↓) from the kidney (p. 136) to the bladder (↓).

bladder[a] (*n*) an extensible sac into which the ureter (↑) passes which fills with urine (↓) secreted (p. 106) continuously by the kidneys (p. 136). When full, it is opened by a sphincter muscle (p. 127), contracts, and releases the urine.

urine (*n*) the fluid which is finally expelled from the kidneys (p. 136). It contains urea (↑), or uric acid (↑) together with other materials and water.

anti-diuretic hormone a hormone (p. 130) synthesized by the hypothalamus (p. 156) and secreted (p. 106) by the pituitary gland (p.157). It increases the reabsorption of water in the kidney (p. 136) tubules thus increasing the concentration of the urine (↑).

aldosterone (*n*) a hormone (p. 130) secreted (p.106) by the adrenal glands (p. 130) which stimulates the reabsorption of sodium from the kidneys (p. 136) and increases the excretion (↑) of potassium thus tending to increase the concentration of sodium in the blood (p. 90) while decreasing the concentration of potassium.

輸尿管（名） 將尿（↓）從腎臟（第136頁）運送至膀胱（↓）的管子或導管。

膀胱（名） 一種有輸尿管（↑）通入其內的可伸展的囊。該囊內充有由腎臟（第136頁）不斷分泌（第106頁）的尿液（↓）。當它充滿尿液時，即藉括約肌（第127頁）的作用開口，同時收縮，將尿液排除。

尿（名） 由腎臟（第136頁）最後排出的液體。它含有尿素（↑）、尿酸（↑），以及其他物質和水。

抗利尿激素 由下丘腦（第156頁）合成並由腦下垂體（第157頁）分泌（第106頁）的一種激素（第130頁）。這種激素可增加腎（第136頁）管中水份的再吸收，從而提高尿（↑）的濃度。

醛固酮、醛甾酮（名） 由腎上腺（第130頁）分泌（第106頁）的一種激素（第130頁）。這種激素可促進腎臟（第136頁）中鈉的再吸收，並增加鉀的排泄（↑），從而增高血液（第90頁）中鈉濃度，降低鉀濃度。

kidney (*n*) in mammals (p. 80), one of a pair of organs which form the main site of excretion and may also be involved in osmoregulation (p. 113). Blood (p. 90) is pumped by the heart (p. 124) under pressure through the kidneys and, by reabsorptions and secretions (p. 106), useful substances are returned to the blood while wastes are eliminated in the urine. (↑).

nephron (*n*) one of the main structures of excretion (p. 134) in the kidneys (↑). It is a microscopic (p. 9) tubule which is made up of a Malpighian corpuscle and a drainage duct. It is in the nephrons that the processes of filtration and reabsorption take place.

腎(臟)(名) 哺乳動物(第80頁)體內形成主要排泄部位的一對器官。腎臟也可參與滲透調節(第113頁)。由心臟(第124頁)泵送的血液(第90頁)，在壓力下流經腎臟，藉再吸收和分泌作用(第106頁)，可使有用的物質回到血液內，而將廢物排入尿(↑)中。

腎元(名) 腎臟(↑)內的主要排泄(第134頁)結構之一。它是由馬爾皮基氏體和排泄組成的一根微型(第9頁)小管。過濾和再吸收的過程就是在這些腎元內進行。

kidney 腎臟

cortex 皮質
pyramid 腎錐體
medulla 髓質
renal artery 腎動脈
renal pelvis 腎盂
renal vein 腎靜脈
ureter 輸尿管

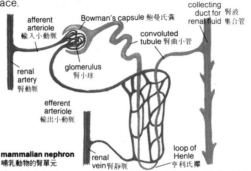

collecting duct for renal fluid 腎液集合管
Bowman's capsule 鮑曼氏囊
afferent arteriole 輸入小動脈
convoluted tubule 腎曲小管
renal artery 腎動脈
glomerulus 腎小球
efferent arteriole 輸出小動脈
mammalian nephron 哺乳動物的腎單元
renal vein 腎靜脈
loop of Henle 亨利氏襻

Bowman's capsule part of the Malpighian corpuscle in the nephron (↑). It is a cup-shaped swelling surrounding the glomerulus (↓) forming part of the structure through which blood (p. 90) is forced under pressure and by which it is 'purified' by a process known as ultra-filtration (↓).

loop of Henle a U-shaped tubule in the nephron (↑) into which isotonic (p. 118) renal fluid (↓) is pumped. As the fluid passes in the counter direction along the other arm of the tubule, sodium ions are actively transported (p. 122) across into the first arm so that there is a high concentration of sodium ions in the bend of the tube. From here there are collecting ducts which open into the renal pelvis from where water is drawn out by osmosis (p. 118). Thus, in the collecting ducts, the renal fluid becomes hypertonic (p. 118).

鮑曼氏囊、腎小球囊 腎元(↑)中的馬爾皮基氏體的一部份。鮑曼氏囊是包圍着腎小球(↓)的一種杯狀隆起，並形成這種結構的一部份，血液(第90頁)在壓力下通過這種結構，並藉這種結構經由超濾作用(↓)過程而被"純化"。

亨利氏襻、髓襻 腎元(↑)中的一根 U 型小管。亨利氏襻內注入有等滲的(第118頁)腎液(↓)。當腎液沿着 U 型小管的另一側朝相反方向流動時，鈉離子即隨之運輸(第122頁)到達管子的前半部，因此管子彎曲部份的鈉離子濃度較高。從此處發出許多通向腎盂的集合管；水藉滲透作用(第118頁)從腎盂中排出。因此，在集合管內，腎液便變成高滲的(第118頁)。

mesophyte e.g. beech
中生植物圖例：山毛櫸

sheds leaves in autumn
在秋天落葉

xerophyte 旱生植物
e.g. cactus 圖例：仙人掌

thick succulent　　hot
stems with　　　　dry
thick cuticle　　　air
具有厚　　　　　　空氣乾熱
角質層
的肉　　　　　　spines
質莖　　　　　　　刺

dry desert sand
乾燥的荒沙

glomerulus (*n*) a knot of capillaries (p. 127) which form part of the Malpighian corpuscle and into which blood (p. 90) is pumped via the renal artery (p. 127) and arterioles (p. 127). Blood is forced through the walls of the capillaries into the Bowman's capsule (↑).

renal fluid a fluid consisting mostly of blood plasma (p. 90) and the soluble materials contained within it, which, the process of ultrafiltration (↓) passes through the kidneys (↑).

ultrafiltration (*n*) the process by which a large proportion of the blood plasma (p. 90) and the soluble materials contained within it are forced under pressure through the walls of the glomerulus (↑), through the walls of the Bowman's capsule (↑) and into the lumen (↓) of the nephron (↑).

lumen (*n*) a space within a tube or sac.

腎小球（名）　構成馬爾皮基氏體一部份的毛細血管（第127頁）結。血液（第90頁）經由腎動脈（第127頁）和小動脈（第127頁）送入其內，並穿過這些微血管壁再進入鮑曼氏囊（↑）。

腎液　主要由血漿（第90頁）及其中含有可溶性物質組成的一種液體。腎液經由超濾作用（↓）過程而通過腎臟（↑）。

超濾作用（名）　大部份血漿（第90頁）及其中含有的可溶性物質，在壓力下被迫穿過腎小球（↑）的壁，穿過鮑曼氏囊（↑）的壁，而進入腎元（↑）的腔（↓）內所經歷的過程。

腔（名）　管或囊內的間隙。

hydrophyte e.g. *Nuphar* 水生植物圖例：萍蓬草

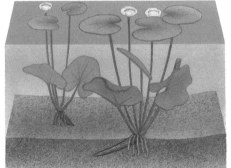

hydrophyte (*n*) a plant which is adapted to grow in water or in very wet conditions. Leaves and stems often contain air spaces to aid the flotation of the whole plant or part of it.

mesophyte (*n*) a plant which is adapted to grow in habitats (p. 217) with normal supplies of water. Typically, they have large, flattened leaves that are lost during leaf fall (p. 133).

xerophyte (*n*) a plant which is adapted to grow in very dry environments (p. 218).

水生植物（名）　適於在水中或在十分潮濕的環境生長的植物。其葉和莖內常含有氣隙，可使整個植物或其一部份漂浮起來。

中生植物（名）　適於在供水正常的生境（第217頁）生長的植物。其特徵是有大而扁平的葉；在落葉（第133頁）期間，這些葉會脫落。

旱生植物（名）　適於在十分乾燥的環境（第218頁）生長的植物。

growth substance (*n*) a hormone (p. 130) which, in very small quantities, can increase, decrease, or otherwise change the growth of a plant or part of a plant.

indole-acetic acid IAA. A growth substance (↑) which causes plant cells to grow longer and causes cells to divide. IAA is the most common of the group of growth substances called auxins.

auxin (*n*) *see* indole-acetic acid (↑).

生長物質（名） 一種激素（第130頁）。這種激素以很少的量就能促進、減緩或者用其他方法改變一株植物或其中一部份的生長。

吲哚乙酸 即 IAA。一種可使植物細胞伸長並引起細胞分裂的生長物質（↑）。IAA 是植物生長素之類生長物質中最普通的一種。

生長素（名） 見吲哚乙酸↑）

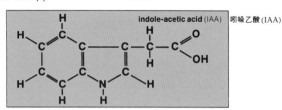

indole-acetic acid (IAA) 吲哚乙酸 (IAA)

gibberellin (*n*) a growth substance (↑) which may effect the elongation of cells in stems and which may cause the area of leaves to increase. They also promote a variety of effects in plant growth such as seed germination (p. 168), flowering and the setting of fruit.

赤霉素（名） 一種生長物質（↑）。這種生長物質可促使莖細胞伸長並增加葉面面積，也促進植物的生長例如種子發芽（第168頁）、開花和座果。

gibberellin 赤霉素
e.g. gibberellic acid 1 圖例：赤霉酸1

cytokinin (*n*) a growth substance (↑) which, in association with IAA, affects the rate of cell division, promotes the formation of buds, and is essential for the growth of healthy leaves. Also known as **kinin**.

細胞分裂素（名） 一種與 IAA 有關的生長物質（↑）。這種生長物質可影響細胞分裂的速度，促進芽的形成，也是葉子健康長所必不可少的。它也稱爲激動素。

cytokinins 細胞分裂素
e.g. kinetin 圖例：激動素

abscisin (*n*) a growth substance (↑) which inhibits plant growth, prevents germination (p. 168), and tends to promote buds to become dormant. Abscisin seems to work against normal growth substances by preventing the manufacture of proteins (p. 21) etc.

脫落素（名） 一種生長物質（↑）。這種生長物質可抑制植物生長，阻止發芽（第168頁），並促使芽休眠。脫落素似乎藉阻止蛋白質（第21頁）等的製造與植物正常的生長物質的作用相對抗。

abscisin
脫落素

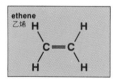

ethene (*n*) a growth substance (↑) which is produced as the result of normal metabolism (p. 26) in plants and which may cause leaves to fall and fruit to ripen. Also known as **ethylene**.

florigen (*n*) a growth substance (↑) which, although it has never been isolated, is believed to promote the production of flowers.

tropism (*n*) the way in which the direction of growth of a plant responds to external stimuli. **tropic (adj).**

geotropism (*n*) a tropism (↑) in which the various parts of a plant grow in response to the pull of the Earth's gravity. For example, primary roots grow downwards and are referred to as being positively geotropic while the main stems grow upwards and are referred to as being negatively geotropic.

乙烯（名） 一種生長物質（↑）。這種生長物質是植物正常新陳代謝（第26頁）所產生的，可導致脫葉，並催熟果實。亦可拼寫為 ethylene。

成花激素（名） 一種生長物質（↑）。雖然還未分離出這種物質，但可認為它能促進花的形成。

向性（名） 植物的生長方向相應於外部刺激而變化的方式。（形容詞形式為 tropic）

向地性（名） 一種向性（↑）。這種向地性使植物各個不同部份的生長，均隨重力的作用而定。例如，初生根向下生長，具有正向地性；而主莖向上生長，具有負向地性。

geotropism 向地性
gravity 重力
stem grows upwards 莖向上生長
root grows downwards 根向下生長

statolith (*n*) a large grain of starch (p. 18) which is found in plant cells and which is thought to respond to the effects of gravity causing the effects of geotropism (p. 139).

phototropism (*n*) a tropism (p. 139) in which various parts of the plant grow in response to the direction from which light is falling on the plant. Stems tend to grow towards the light and are referred to as being positively **phototropic** . The roots of some plants e.g. those of climbers grow away from the source of light and are referred to as being negatively **phototropic** .

平衡石（名）　植物細胞中含有的一種大顆粒澱粉（第18頁）。它能對引起向地性（第139頁）效應的重力作用起反應。

向光性（名）　一種向性（第139頁）。這種向性使植物各個不同部份生長，隨照射在植物上的光線方向而定。莖趨光生長，具有正向光性。某些植物例如攀緣植物的根，離光生長，具有負向光性。

phototropism 向光性

light
光

auxin in phototropism 生長素的向光性

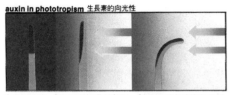

1 shoot tip in dark, auxin evenly concentrated　**2** exposed to light from one side, auxin concentration increases on dark side and decreases on light side　**3** increased relative auxin concentration on dark side causes cells on dark side to elongate, and the shoot bends towards the light

1. 無光照時，苗尖部的生長素均勻分佈
2. 從一側照射光時，無光側苗尖部的生長素濃度增加，而有光側苗尖部的生長素濃度降低
3. 無光側苗尖部的生長素濃度的相對增加，使無光側的細胞伸長，導致苗朝光照方向彎曲

hydrotropism (*n*) a tropism (p. 139) in which the roots of plants grow towards a source of water. Hydrotropism will usually override the effects of geotropism (p. 139). **hydrotropic (adj).**

chemotropism (*n*) a tropism (p. 139) in which the roots of a plant or the hyphae (p. 46) of a fungus (p. 46) may grow towards a source of food materials. **chemotropic** (*adj*).

thigmotropism (*n*) a tropism (p. 139) in which, by the stimulus of touch, certain parts of particular plants, such as the stems of climbing plants, may coil around a support. **thigmotropic (adj).**

nastic movements growth movements, such as the opening and closing of flowers, which although they occur as a result of external stimuli, such as the presence or absence of light, do not take place in a particular direction.

photonasty (*n*) a nastic movement (↑) which is a response to the presence or absence of light or even to light levels e.g. the flowers of daisies close at night and only open during the daylight.

向水性（名）　一種向性（第139頁）。這種向性使植物的根朝向水源生長。向水性的效應通常都超過向地性（第139頁）的效應。（形容詞形式為 hydrotropic）

向化性（名）　一種向性（第139頁）。這種向性使植物的根或真菌（第46頁）的菌絲（第46頁）朝食物源生長。（形容詞形式為 chemotropic）

向觸性（名）　一種向性（第139頁）。這種向性使一些特殊植物的某些部份，例如攀緣植物的莖，由於受到接觸刺激，可以纏繞一支持物生長。（形容詞形式為 thigmotropic）

感性運動　花朵的開放或閉合之類的生長運動。雖然這類生長運動都是由外部的刺激（例如光線的存在與否）所致，但其並不朝一特定方向進行。

感光性（名）　一種感性運動（↑）。這種感性運動是對有無光線，甚至是對光線的強弱所作出的一種反應。例如，雛菊的花朵在晚上閉合，而僅在白天開放。

thigmonastic movements of the leaves of a 'sensitive plant' after it has been touched
"含羞草"葉子受觸摸後所產生的感觸運動

apical dominance 頂端優勢

apex 頂端

axillary buds 腋芽

intact plant: auxin translocated from apex inhibits growth of axillary buds into lateral shoots
完整的植物：由頂端輸送的植物生長素抑制着腋芽長成側枝

apex removed: lateral shoots grow
除去頂端後，側枝生長

thermonasty (*n*) a nastic movement (↑) which is a response to the surrounding temperature. For example, the flowers of some plants will open when the weather is warm.

thigmonasty (*n*) a nastic movement (↑) in which the response is to touch. For example, the leaves of the South American plant, commonly known as the 'sensitive plant', fold back when touched.

taxic movements the movement of an organism in which the response takes place in relation to the direction of the stimulus.

phototaxis (*n*) a taxic movement (↑) in which the movement may be away from or towards the direction from which the light is coming. For example, certain insects (p. 69) may hide from the light and are referred to as negatively phototaxic while many algae (p. 44) will move towards the light and are described as positively phototaxic.

thermotaxis (*n*) a taxic movement (↑) in which the movement may be away from or towards regions of higher or lower temperature. For example, a mammal (p. 80) may seek the shade of a tree during the heat of the day to prevent overheating.

chemotaxis (*n*) a taxic movement (↑) in which an organism may move towards a chemical stimulus. For example, a spermatozoid (p. 178) may swim towards a female organ which secretes (p. 106) a substance, such as sucrose (p. 18).

hygroscopic movements movements which take place as the parts of organisms dry out and thicker parts move differently from thinner parts.

autonomic movements movements which take place in an organism without any external stimulus. The stimulus comes from within the organism itself and may include movements such as the coiling of the tendrils of climbing plants such as peas.

apical dominance the state which may occur in plants in which the bud at the tip of a plant stem grows but the lateral ones do not. If the apical bud is removed, lateral branches may grow.

vernalization (*n*) the process whereby certain plants, such as cereal crops, need to be subjected to low temperatures, such as that which occurs through overwintering, during an early part of their growth before they will be induced to flower.

感熱性（名） 一種感性運動（↑）。這種感性運動是植物對環境溫度作出的一種反應。例如，某些植物的花在天氣暖和時才開放。

感觸性（名） 植物對接觸作出反應的一種感性運動（↑）。例如，通常稱爲"含羞草"的一種南美植物葉子，被觸及時即折叠起來。

趨性運動 生物體受到刺激而作出運動反應的一種方式，這種反應方式與刺激方向有關。

趨光性（名） 一種趨性運動（↑）。生物體作這種趨性運動時，可以離開光照方向進行，也可以朝着光照方向進行。例如，某些昆蟲（第69頁）可以迴避光線而運動，具有負趨光性；而許多藻類（第44頁）可以朝光照方向運動，具有正趨光性。

趨溫性（名） 一種趨性運動（↑）。生物體作這種趨性運動時，可以離開高溫或低溫區域進行，也可以朝向高溫或低溫區域進行。例如，哺乳動物（第80頁）在天熱時，能尋找樹蔭，以防止過熱。

趨化性（名） 一種趨性運動（↑）。生物體作這種趨性運動時，朝着化學刺激的方向進行。例如，游動精子（第178頁）會朝向分泌（第106頁）蔗糖（第18頁）之類物質的雌性器官游動。

吸濕運動 生物體的各組成部份在乾燥時所發生的運動。較厚組成部份的運動不同於較薄組成部份的運動。

自發運動 生物體在無任何外部刺激情況下所發生的運動。刺激來自於生物體本身。自發運動可包括像攀緣植物例如豌豆捲鬚捲曲之類的運動。

頂端優勢 植物中存在的一種狀態。處於這種狀態時，植物莖頂端的芽生長，而側面的芽則不生長。如果將頂芽除去，則側枝就能生長。

春化作用（名） 指穀類作物的某些植物的生長過程，它們在早期生長階段，需要經受低溫，以促使其開花，例如在越冬期間所經受的那種低溫過程。

phytochrome (*n*) a light-sensitive pigment (p. 126) present in plant leaves which exists in two forms that can be converted from one to the other. One absorbs red light, the other far red light. In the absence of light, the latter slowly changes back to the former. Phytochromes initiate the formation of hormones (p. 130).

etiolation (*n*) plant growth which takes place in the absence of light. The plants may lack chlorophyll (p. 12) so that they will be yellow or even white in colour. The leaves will be reduced in size and the stems tend to grow much longer.

photoperiodism (*n*) the process in which certain activities, such as flowering or leaf fall (p. 133), respond to seasonal changes in day length.

long-day plants plants, such as cucumber, which only usually flower during the summer months in temperate climates when the hours of daylight exceed about fouteen in twenty-four.

short-day plants plants, such as chrysanthemums, which only usually flower during the spring or autumn months in temperate climates when the hours of daylight are less than about fourteen in twenty-four.

day-neutral plants plants, such as the pea, in which the hours of daylight have no influence on the flowering period.

光敏色素（名） 植物葉子中的一種感光色素（第126頁）。有兩種可以相互轉化的形式。一種形式的光敏色素能吸收紅光；另一種形式的光敏色素能吸收遠紅光。在缺乏光照情況下，後者可緩慢地逆變成前者。光敏色素可引發形成多種激素（第130頁）。

黃化（現象）（名） 在缺乏光照情況下所發生的生長現象。這些植物可能缺乏葉綠素（第12頁），致使它們的顏色變黃甚至變白。其葉子會縮小，而其莖則往往長得更長。

光周期現象（名） 植物的某些活動，例如開花和落葉（第133頁），隨晝長的季節性變化而作出反應的過程。

長日照植物 如黃瓜之類植物，這些植物通常僅在溫帶氣候條件下的春季月份，當一天24小時中的日照時間超過約14小時的情況下才開花。

短日照植物 如菊類植物，這些植物通常僅在溫帶氣候條件下的春季和秋季月份，當一天24小時的日照時間少於約14小時的情況下才開花。

中性日照植物、光期鈍感植物 日照時間對開花期沒有影響的一些植物，例如豌豆。

兩種形式的光敏色素的相互轉化
interconversion of two forms of phytochrome

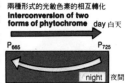

etiolation 黃化現象

etiolated young plant grown in dark ④
hooked apex ④
no chlorophyll ②
little leaf development ③
long shoot ⑤

normal ⑥
young plant grown in light
leaves well developed with chlorophyll ⑦
upright ⑧ apex
shorter shoot ⑨

photoperiodism 光周期現象

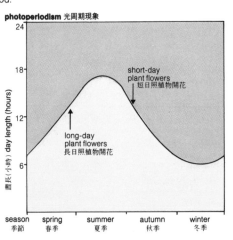

day length (hours) 晝長（小時）
24
18
short-day plant flowers 短日照植物開花
12
long-day plant flowers 長日照植物開花
6
season 季節 spring 春季 summer 夏季 autumn 秋季 winter 冬季

① 在黑暗中長大的黃化植物幼株
② 無葉綠素
③ 幾乎無莖生成
④ 具鉤的頂端
⑤ 長枝
⑥ 在光照條件下長大的正常植物幼株
⑦ 葉子生長正常，並具葉綠素
⑧ 直立的頂端
⑨ 較短的枝

long-day plant 長日照植物

kept under short days 保持短日照
kept under long days 保持長日照

short-day plant 短日照植物

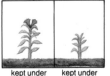

kept under short days 保持短日照
kept under long days 保持長日照

**fibre of voluntary muscle
隨意肌的肌纖維**

striped
band
條紋帶

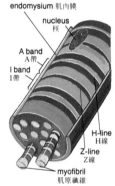

**structure of striated muscle
橫紋肌的結構**

endomysium 肌肉膜

nucleus
核

A band
A帶

I band
I帶

H-line
H線

Z-line
Z線

myofibril
肌原纖維

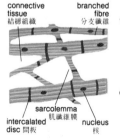

**structure of cardiac muscle
心肌的結構**

connective
tissue
結締組織

branched
fibre
分支纖維

sarcolemma
肌纖維膜

intercalated
disc 間板

nucleus
核

locomotion (*n*) the ability of an organism to move all or part of its body independent of any outside force. An animal can usually move its whole body whereas a plant may only be able to move certain parts, such as petals (p. 179) or leaves, in response to changes in the environment (p. 218).

muscle (*n*) tissue (p. 83) which is made up of cells or fibres that are readily contracted.

fibre (*n*) a thread-like structure.

skeletal muscle muscle (↑) tissue (p. 83) consisting of elongated cells with many nuclei (p. 13) and cross striations in the cytoplasm (p. 10). It usually occurs in bundles and is under voluntary control of the central nervous system (p. 149) so that it contracts when stimulated to do so. These muscles are attached to parts of the skeleton (p. 145) and their contractions cause these parts to move. Skeletal muscle which has a striped look is known as striated muscle. This consists of long, narrow muscle fibres bounded by a membrane (p. 14) and containing many nuclei. The muscle fibres are bound together into bundles. They contract when stimulated.

voluntary muscle = skeletal muscle (↑).

striated muscle = skeletal muscle (↑). Also known as **striped muscle**.

unstriated muscle = involuntary muscle (↓).

involuntary muscle the muscle (↑) which is found in the internal organs and blood vessels (p. 127) and consists of simple tubes or sheets. It is under the involuntary control of the autonomic nervous system (p. 155). Also known as **smooth muscle**.

visceral muscle a smooth or unstriated (↑) muscle (↑) tissue (p. 83) made up of elongated cells held together by connective tissue (p. 88) and activated involuntarily. It is found in all internal organs as well as blood vessels (p. 127) with the exception of the heart (p. 124).

cardiac muscle muscle (↑) tissue (p. 83) found only in the heart (p. 124) walls. It consists of fibres containing cross striated (↑) myofibrils (p. 144). It contracts rhythmically and automatically (i.e. without nervous (p. 149) stimulation).

運動、行動(名) 生物在不依賴任何外力情況下移動其全部或部份身體的能力。動物通常能夠根據環境(第218頁)的變化移動其整個身體,而植物卻只能移動其某些部份,例如移動其花瓣(第179頁)或葉。

肌、肌肉(名) 由易收縮的一些細胞或纖維構成的組織(第83頁)。

纖維(名) 一種線狀結構。

骨骼肌 由一些長形細胞構成的肌肉(↑)組織(第83頁)。這些長形細胞有許多細胞核(第13頁),其細胞質(第10頁)內還有諸多交叉條紋。骨骼肌通常以束狀形式存在,並受中樞神經系統(第149頁)隨意控制,因此,當其受到刺激時,便會收縮。這些肌肉附著在骨骼(第145頁)的各組成部份上;骨骼肌收縮使這些骨骼的組成部份移動。具條紋的骨骼肌稱為橫紋肌。橫紋肌由許多細長的肌纖維構成,這些肌纖維外面包覆著一層膜(第14頁),並含有許多細胞核。這些肌纖維聚合在一起,形成許多肌束;當受到刺激時,便會收縮。

隨意肌 同骨骼肌(↑)。

橫紋肌 同骨骼肌(↑)。其另一英文名稱為 striped muscle。

無橫紋肌 同不隨意肌(↓)。

不隨意肌 內臟器官和血管(第127頁)中具有的肌肉(↑)。這種肌肉由單管或單片構成。它處於自主神經系統(第155頁)的不隨意控制之下。亦稱平滑肌。

內臟肌 由長形細胞構成的一種平滑肌(↑)組織(第83頁)。平滑肌組織由結締組織(第88頁)結合在一起,並可不隨意收縮。除心臟(第124頁)之外的所有內臟器官和血管(第127頁)中都有內臟肌。

心肌 僅存在於心(第124頁)壁中的肌肉(↑)組織(第83頁)。心肌由肌纖維構成,這些肌纖維含有許多交叉橫紋的(↑)肌原纖維(第144頁)。它能有節律地和自動地(即未經神經(第149頁)刺激的情況下)收縮。

muscle fibre the elongated cells which make up striated muscles (p. 143) and which consist of a number of myofibrils (↓).

myofibril (*n*) the very fine threads which make up the muscle fibres (↑) and are found in smooth, striated (p. 143), and cardiac muscles (p. 143). They contain the contractile proteins (p. 21) myosin (↓), actin (↓) and tropomyosin (↓).

sarcomere (*n*) the part of the myofibril (↑) which is responsible for the contraction. It is made up of a dark central A band composed of myosin (↓) on either side of which are I bands composed of actin (↓). Each sarcomere is joined to the next by the Z membrane (p. 14). During contraction the I band shortens while the A bands stay more or less the same length so that the muscle filaments slide between one another.

thick filaments the filaments of a myofibril (↑) which are composed of myosin (↓).

thin filaments the filaments of a myofibril (↑) which are composed of actin (↓).

actin (*n*) the contractile protein (p. 21) which comprises one of the main elements in muscle (p. 143) myofibrils (↑). When stimulated actin and myosin (↓) join together to form actomyosin (↓).

myosin (*n*) the contractile protein (p. 21) which comprises the most abundant element in muscle (p. 143) myofibrils (↑). When stimulated actin (↑) and myosin join together to form actomyosin (↓).

actomyosin (*n*) a complex of the two proteins (p. 21) actin (↑) and myosin (↑) which, when they interact to form the complex, result in the contraction of the muscle (p. 143).

tropomyosin (*n*) the third protein found in myofibrils (↑) which may be responsible for controlling the contractions of muscle (p.143).

sliding filament hypothesis the theory (p. 235) which suggests that when a muscle (p. 143) contracts, the individual filaments do not shorten but that they slide between one another because it can be seen under the electron microscope (p. 9) that the pattern of striations (p. 143) changes during the contraction.

肌纖維　指一些長形細胞。橫紋肌(第143頁)即由其構成。這些長形細胞含有許多肌原纖維(↓)。

肌原纖維(名)　在平滑肌、橫紋肌(第143頁)和心肌(第143頁)中組成肌纖維(↑)的一些細絲。這些細絲含有收縮蛋白(第21頁)，肌球蛋白(↓)、肌動蛋白(↓)和原肌球蛋白(↓)。

肌原纖維節(名)　肌原纖維(↑)中擔負收縮功能的那一部份。由位於中央的暗色的 A 帶和位於 A 帶兩側的 I 帶組成。A 帶由肌球蛋白(↓)組成；I 帶由肌動蛋白(↓)組成。每一肌原纖維節均由Z膜(第14頁)與相鄰的肌原纖維節連結。在收縮過程中，I 帶縮短，而 A 帶近乎保持同樣的長度，致使肌絲相互產生滑動。

粗絲　由肌球蛋白(↓)組成的肌原纖維(↑)絲。

細絲　由肌動蛋白(↓)組成的肌原組織(↑)絲。

肌動蛋白(名)　一種收縮蛋白(第21頁)。它係肌肉(第143頁)肌原纖維(↑)的主要成份之一。受到刺激時，肌動蛋白和肌球蛋白(↓)結合形成肌動球蛋白(↓)。

肌球蛋白(名)　一種收縮蛋白(第21頁)。它係肌肉(第143頁)的肌原纖維(↑)中含量最大的成份。受到刺激時，肌動蛋白(↑)和肌球蛋白結合形成肌動球蛋白(↓)。

肌動球蛋白(名)　肌動蛋白(↑)和肌球蛋白(↑)這兩種蛋白(第21頁)的複合物。當這兩種蛋白相互作用，形成複合物時，便會導致肌肉(第143頁)收縮。

原肌球蛋白(名)　肌原纖維(↑)中含有的第三種蛋白。它對調節肌肉(第143頁)收縮起主要作用。

滑絲假說　這種學說(第235頁)認為，肌肉(第143頁)收縮時，各單條肌絲並不縮短而是彼此之間產生滑動，因為在電子顯微鏡(第9頁)下可以看到，在收縮過程中，條紋(第143頁)圖案發生了變化。

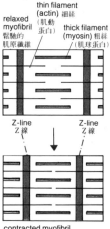

滑絲假說
sliding filament hypothesis

thin filament
(actin) 細絲
(肌動蛋白)

thick filament
(myosin) 粗絲
(肌球蛋白)

relaxed
myofibril
鬆弛的
肌原纖維

Z-line
Z 線

Z-line
Z 線

contracted myofibril
收縮的肌原纖維

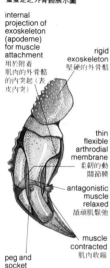

claw of crab showing exoskeleton
蟹螯足之外骨骼展示圖

internal projection of exoskeleton (apodeme) for muscle attachment
用於附着肌肉的外骨骼的內突起（表皮內突）

rigid exoskeleton
堅硬的外骨骼

thin flexible arthrodial membrane
柔韌的動關節膜

antagonistic muscle relaxed
頡頏肌鬆弛

muscle contracted
肌肉收縮

peg and socket joint 釘狀關節

sarcoplasmic reticulum a smooth endoplasmic reticulum (p. 11) which is responsible for absorbing (p. 81) the calcium that is necessary for muscle (p. 143) contraction.

muscle spindle a modified muscle fibre (↑) which is receptive to stimulation and controls the way in which a muscle (p. 143) contracts.

skeleton (n) a supporting structure. It can be jointed (p. 146) and the joints (p. 146) are connected by muscles (p. 143) that, when they contract against the limbs as levers, enable the animal to operate the limbs, e.g. the legs, allowing movement on land.

exoskeleton (n) the external skeleton (↑) of organisms, such as insects (p. 69), which provides protection for the internal organs and is the structure to which the muscles (p. 143) are attached. For example, the shell of a cockle would be referred to as an exoskeleton.

apodeme (n) one of a number of projections on the inside of the exoskeleton (↑) where there are joints (p. 146) and to which the muscles (p. 143) for the movement of those joints are attached.

cuticle[a] (n) the outer layer of the endoskeleton (↓) which in animals, such as insects (p. 69), acts as the skeleton (↑) itself. It may be composed of chitin (p. 49) but in shellfish may be hardened with lime-rich salts. It is secreted (p. 106) by the epidermis (p. 131) and is non-cellular. The outer epicuticle is waxy and waterproof in insects and other arthropods (p. 67).

hydrostatic skeleton a form of skeleton (↑) found in soft-bodied animals such as earthworms (p. 66) in which the body fluids themselves provide the structure against which the muscles (p. 143) act.

endoskeleton (n) a bony or cartilaginous (p. 90) structure contained within the body of vertebrates (p: 74) which is usually jointed (p. 146) to allow movement and to which the muscles (p. 143) are attached to provide the mechanisms for movement

musculo-skeletal system the system which enables the animal to move by providing a jointed (p. 146) skeleton (↑) against which the muscles (p. 143) can act to cause operation of the joints using the limbs as levers.

肌質網、肌漿網　一種擔負鈣質吸收（第81頁）功能的光滑型內質網（第11頁）。鈣質對肌肉（第143頁）收縮是必不可少的。

肌梭　一種變形的肌纖維（↑）。這種肌纖維能感受刺激並控制肌肉（第143頁）收縮的方式。

骨骼（名）　一種支持結構。骨骼能形成關節（第146頁），而各關節（第146頁）則由肌肉（第143頁）連結。當骨骼肌收縮而牽動作爲槓桿的肢體時，即能使動物的肢體如腿動作，以便在陸地上運動。

外骨骼（名）　各種生物如昆蟲（第69頁）的外部骨骼（↑）。外骨骼對內部器官起保護作用；也是附着肌肉（第143頁）的結構。例如，蛤的外殼可稱爲一種外骨骼。

表皮內突（名）　外骨骼（↑）裏面的許多突出物。在這些突出物附近有關節（第146頁）存在，而在這些突出物上，則附有肌肉（第143頁），以使關節得以活動。

角質層[a]（名）　內骨骼（↓）的外層，在如昆蟲（第69頁）之類動物中，相當於骨骼（↑）本身。角質層可由幾丁質（第49頁）組成，但在有殼水生動物中，角質層可由富含石灰質的鹽類硬化而成。角質層是由表皮（第131頁）分泌（第106頁）的，而不是由細胞構成的。昆蟲其他節肢動物（第67頁）的外上表皮是蠟質的，可以防水。

水壓骨骼　蚯蚓（第66頁）之類軟體動物中具有的一種骨骼（↑）形式。在這些軟體動物中，其體液本身形成肌肉（第143頁）藉以動作的結構。

內骨骼（名）　脊椎動物（第74頁）體內具有的一種硬骨或軟骨（第90頁）結構。這種結構通常連接（第146頁）在一起，以進行活動；其上附着肌肉（第143頁），以形成運動機構。

肌肉骨骼系統　藉形成一副有關節（第146頁）的骨骼（↑），使動物能夠進行運動的系統。骨骼肌（第143頁）能倚着該骨骼動作，以肢體爲槓桿，使關節得以活動。

swimming in fish waves of lateral undulations pass from the head back along the body
魚的游泳魚左右擺動所形成的波浪從其頭部沿着身體向後部傳遞

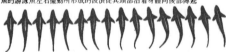

swimming (*n*) the process by which an organism such as a fish propels itself through or on the surface of the water by the action of fins (p. 75) or flexing of the whole body. In fish, when the muscles (p. 143) contract, the body cannot shorten so that it moves from side to side to provide propulsion.

caudal fin the tail fin. The main organ by which a fish propels itself through the water. It is a membrane (p. 14) supported by fin rays attached to the vertebral column (p. 74) of the fish.

myotome muscle one of a number of blocks of striated muscle (p. 143) that completely enclose the vertebral column (p. 74) of fish.

joint (*n*) the region at which any two or more bones of a skeleton (p. 145) come into contact. For example, the elbow joint in a human.

ball and socket joint a movable joint (↑) between limbs in which one bone terminates in a knob-shaped structure which fits into a cup-shape in the meeting bone allowing some movement in all directions, e.g. the joint between the femur and socket of the pelvic girdle (↓).

hinge joint a movable joint (↑) between bones in which the movement can take place in one plane or direction only e.g. the knee joint.

pivot hinge a movable joint (↑) between bones in which movement can take place in all directions by a twisting or rotating movement. For example, the joints in the neck.

ligament (*n*) the strong, elastic connective tissue (p. 88) which, for example, holds together the limb bones at a joint (↑) and which helps to control the movement of the joint.

tendon (*n*) the thick cord of connective tissue (p. 88) which connects a muscle (p. 143) to a bone. It is non-elastic so that when the muscle contracts, it pulls against the bone forcing it to move at the joint (↑).

游泳（名） 某種生物如魚，藉鰭（第75頁）動作或整個身體屈曲，在水中或水面上前進的過程。就魚而言，當其肌肉（第143頁）收縮時，身體不會縮短，因此，它可左右搖擺着前進。

尾鰭 尾部的鰭。魚在水中前進所依靠的主要器官。尾鰭是一層膜（第14頁），由附着魚的脊柱（第74頁）上的一些鰭條支持。

肌節肌、生肌節 將魚的脊柱（第74頁）完全包圍住的諸多橫紋肌（第143頁）塊。

關節（名） 骨骼（第145頁）中任何兩塊或兩塊以上的骨相接觸的部位，例如人的肘關節。

球窩關節 肢體之間的可動關節（↑）。在這種關節中，一塊骨的球形端嵌入另一塊相接骨的杯形窩內，使關節朝各個方向都能作某種轉動，例如骨盆帶（↓）的股骨與骨臼之間的關節。

鉸鏈關節、屈成關節 骨與骨之間的一種動關節（↑）。在這種關節中，其轉動只能在一個平面上或朝一個方向進行，例如膝關節。

樞軸關節 骨與骨之間的一種動關節（↑）。這種關節藉扭轉或旋轉能朝所有向轉動，例如頸部的關節。

韌帶（名） 一種強有力的、富有彈性的結締組織（第88頁），例如它可在關節（↑）處將肢骨結合在一起，並不幫助控制關節運動。

腱（名） 可將肌肉（第143頁）連接到骨上的粗索狀結締組織（第88頁）。腱沒有彈性，所以當肌肉收縮時，即牽引骨頭，使其在關節（↑）處移動。

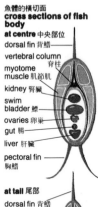

魚體的橫切面
cross sections of fish body
at centre 中央部位
dorsal fin 背鰭
vertebral column 脊柱
myotome muscle 肌節肌
kidney 腎臟
swim bladder 鰾
ovaries 卵巢
gut 腸
liver 肝臟
pectoral fin 胸鰭

at tail 尾部
dorsal fin 背鰭
vertebral column 脊柱
myotome muscle 肌節肌
caudal fin 尾鰭

structure of a joint 關節的結構
synovial capsule 關節囊
bone 骨
ligament 韌帶
cartilage 軟骨

joints 關節
ball and socket joint 球窩關節
socket 窩
ball 球
hinge joint 屈成關節

哺乳動物後腿的肌肉
muscles of hind leg of a mammal

pelvic girdle 腰帶
ball and 球窩關
socket joint 節
head of femur 股骨頭
femur 股骨
hinge joint 屈戍關節
tibia-fibula 脛腓骨
tarsus 跗骨
foot 足

muscles 肌肉
abductor 外展肌
rotator 旋肌
retractor 牽縮肌
protractor 牽引肌
extensor 伸肌
flexor 屈肌
extensor 伸肌
flexor 屈肌

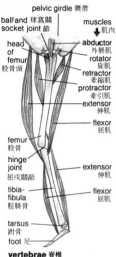

vertebrae 脊椎
vertebral column (backbone) 脊柱（背脊骨）
skull 顱骨

椎骨之間的關節
joint between vertebrae
centrum 椎體
vertebrae 椎骨
cartilage disc 軟骨片

a vertebra 椎骨
neural spine 神經棘
neural canal 椎管
centrum 椎體
transverse processes 橫突

protractor muscle a muscle (p. 143) which on contraction draws the limb bone forwards.

retractor muscle a muscle (p. 143) which on contraction draws the limb bone backwards.

adductor muscle a muscle (p. 143) which on contraction draws the limb bone inwards.

abductor muscle a muscle (p. 143) which on contraction draws the limb bone outwards.

rotator muscle a muscle (p. 143) which on contraction rotates a limb bone outwards or inwards.

flexor muscle a muscle (p. 143) which on contraction draws two limb bones together.

extensor muscle a muscle (p. 143) which on contraction draws two limb bones apart.

vertebra (n) one of a number of bones, or, in some cases, segments of cartilage (p. 90), which make up the vertebral column (p. 74). Each vertebra is usually hollow and has muscles (p. 143) attached to it.

pelvic girdle the part of the skeleton (p. 145) of a vertebrate (p. 74) to which the hind limbs are attached. It is rigid and provides the main support for the weight of the body.

limb (n) any part of the body of an animal, apart from the head and the trunk, including, for example, the arms, the legs or the wings.

flight (n) that form of locomotion (p. 143) such as is found in most birds and many insects, whereby the animal is borne through the air either by gliding on the wind using outstretched membranes (p. 14) or by using the lift generated by the special shape and the power provided by wings.

feather (n) one of a very large number of structures which provide the body covering of birds. They distinguish birds from all other animals. Feathers insulate the bird's body, repel water, streamline it, and aid in the power of flight.

down (n) the soft, fluffy feathers (↑) which form the initial covering of very young birds and which are also found on the undersides of adult birds. The individual barbs (p. 148) are not joined together so that down provides better insulation than flight feathers (p. 148) by trapping a layer of air close to the bird's body.

牽引肌　一種在收縮時牽引肢骨向前的肌肉（第143頁）。

牽縮肌　一種在收縮時牽引肢骨向後的肌肉（第143頁）。

內縮肌　一種在收縮時牽引肢骨向內的肌肉（第143頁）。

外展肌　一種在收縮時牽引肢骨向外的肌肉（第143頁）。

旋肌　一種在收縮時使肢骨向外或向內旋轉的肌肉（第143頁）。

屈肌　一種在收縮時將兩塊肢骨牽引在一起的肌肉（第143頁）。

伸肌　一種在收縮時將兩塊肢骨拉開的肌肉（第143頁）。

椎骨、脊椎（名）　構成脊柱（第74頁）的許多骨塊，在某些情況下爲一些軟骨（第90頁）節段。每一椎骨通常是中空的，其上附有肌肉（第143頁）。

骨盆帶、腰帶　脊椎動物（第74頁）骨骼（第145頁）的一部份。後（下）肢附著在這一部份上。骨盆帶是堅硬的，是身體重量的主要支持物。

肢（名）　動物身體中除了頭和軀幹之外的任何部份，例如臂、腿或翅等。

飛行（名）　大多數鳥和許多昆蟲所具有的那種行動（第143頁）形式。依靠這種行動形式，動物或者藉助於伸展開的膜（第14頁），在風中滑翔，或者藉助於特殊體型所產生的昇力和由雙翅提供的動力，在空氣中前進。

羽、羽毛（名）　覆蓋鳥類身體的大量結構物。這些結構物使鳥類區別於所有其他動物。衆多的羽毛可使鳥體保溫，防水，形成流線型結構和提高飛行能力。

絨羽　構成雛鳥身體初始覆蓋物的柔軟、呈絨毛狀的羽（↑）。這種羽也見於成鳥身體下面。絨羽的各個羽支（第148頁）不連接在一起，因此，在靠近鳥體處藉助所截留的一層空氣，使其保溫性優於飛羽（第148頁）。

flight feather one of a number of feathers (p. 147) giving birds their streamlined effect and which are elongated to provide the flight surface.

shaft (*n*) the central rod-like but flexible stem of a feather (p. 147) which is the continuation of the quill (↓) that is attached to a feather (like hair) follicle (p. 132) in the skin of a bird. The skin of a bird has no sweat glands (p. 132).

quill (*n*) the hard tube-like part of the feather (p. 147) that is attached to the feather follicle (p. 132) and which is also connected by muscles (p. 143) that are able to alter the angle at which the feather lies in relation to the body of the bird. For example, in cold weather, the bird would raise its feathers to trap extra layers of air for more efficient insulation.

vane (*n*) the flat, blade-like part of the feather (p. 147) which is composed of the shaft (↑) and its attached web of barbs (↓) and barbules (↓).

barb (*n*) a hook-like process which projects from the shaft (↑) of a feather (p. 147). Barbs are interlocked by the barbules (↓).

barbule (*n*) one of the tiny barbs (↑) attached to the barb of a feather (p. 147) and which link the barbs by a system of hooks and troughs to make up the web or vane (↑) of the feather.

pectoralis muscle one of the large, powerful muscles (p. 143) which pull on the wings of a bird to force it upwards and downwards providing the power for flight. Pectoralis muscles are attached to the sternum (↓) of the bird. The pectoralis major is the muscle which depresses the wing while the pectoralis minor is responsible for raising it.

飛羽 使鳥具有流線型體形的許多羽(第147頁)。這些羽毛長得較長,以形成飛行的翼面。

羽幹(名) 羽毛(第147頁)中具有韌性的中央桿狀體。羽幹是羽根(↓)的延長部份。羽根附着在鳥皮膚中的羽(像毛)囊(第132頁)上。鳥的皮膚沒有汗腺(第132頁)。

羽根(名) 羽毛(第147頁)堅硬的管狀部份。它附着在羽囊(第132頁)之上,也由肌肉(第143頁)相連接,而這些肌肉能改變羽毛與鳥體之間的角度。例如,在寒冷的天氣,鳥會將其羽毛堅起,截留額外的空氣層,以便更有效地保溫。

羽片(名) 羽毛(第147頁)的扁平片狀部份。它由羽幹(↑)和附着羽幹的羽支(↓)與羽小支(↓)形成的網所組成。

羽支(名) 羽毛(第147頁)之羽幹(↑)上的一種鈎狀突出物。羽支由為數眾多的羽小支(↓)縱橫交叉地聯系在一起。

羽小支(名) 附着於羽毛(第147頁)羽支上的小羽支(↑)。小羽支以一個具有鈎和槽的組織體系將羽支連接在一起,形成羽支連接在一起,形成羽毛的網或羽片(↑)。

胸肌 能牽拉鳥翅、迫使其上、下運動、提供飛行動力的強有力的大塊肌肉(第143頁)。胸肌附着在鳥的胸骨(↓)上。胸大肌是能將鳥翅下拉的肌肉,而胸小肌的功能則是提昇鳥翅。

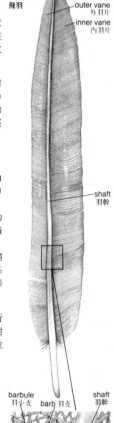

flight feather 飛羽

outer vane 外羽片
inner vane 內羽片
shaft 羽幹
barbule 羽小支
barb 羽支
shaft 羽幹

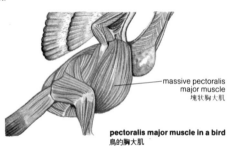

massive pectoralis major muscle
塊狀胸大肌

pectoralis major muscle in a bird
鳥的胸大肌

sternum (*n*) the breast bone of tetrapods.

gliding (*n*) the process of flight (p. 147) whereby the animal holds the wings outstretched so that they function as aerofoils and the animal soars on a cushion of supporting air.

irritability (*n*) the ability of an organism to respond to changes in its environment (p. 218) e.g. the movement of animals in response to noise or being touched.

nervous system the system within the body of an organism which permits the transmission of information through the body so that its various parts are able rapidly to respond to any stimuli.

central nervous system CNS. That part of the nervous system (↑), which in vertebrates (p. 74) includes the brain (p. 155) and the spinal cord (p. 154) which receives nerve impulses (p. 150) from all parts of the body, internal and external, and responds by delivering the appropriate commands to the various organs and muscles (p. 143) to react accordingly.

peripheral nervous system that part of the nervous system (↑) excluding the CNS (↑). It consists of a network of nerves (↓) running through the body of the organism and connected with the CNS.

neurone (*n*) one of the many specially modified cells which make up the nervous system (↑). Each neurone is connected via synapses (p. 151) to others by a single thread-like axon (↓) or nerve fibre and numerous dendrons (↓) which transmit nerve impulses (p. 150) from neurone to neurone.

nerve cell = neurone (↑).

cell body the part of the neurone (↑) with the nucleus (p. 13). Also known as **centron**.

dendron (*n*) a branching process of cytoplasm (p. 10) from the body of a neurone (↑) which ends in a synapse (p.151). They may branch into dendrites.

Nissl's granules granules found in the cytoplasm (p. 10) of a neurone (↑). They are rich in RNA (p. 24)

axon (*n*) the long process of a neurone (↑) filled with axoplasm which normally conducts nerve impulses (p. 150) away from the cell body (↑). The axon is enclosed in a myelin (p. 150) sheath which is bounded by the thin membrane (p. 14), the *neurilemma*, of the *Schwann cell*.

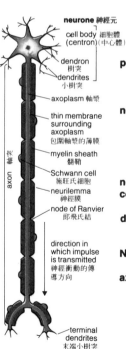

neurone 神經元

cell body 細胞體
(centron)(中心體)

dendron
樹突

dendrites
小樹突

axoplasm 軸漿

thin membrane
surrounding
axoplasm
包圍軸漿的薄膜

myelin sheath
髓鞘

Schwann cell
施旺氏細胞

neurilemma
神經膜

node of Ranvier
郎飛氏結

axon 軸突

direction in
which impulse
is transmitted
神經衝動的傳
導方向

terminal
dendrites
末端小樹突

胸骨（名）　指四足動物的胸骨。

滑翔（名）　動物使翅保持展狀態所利用的飛行（第147頁）過程。因此，翅可起機翼作用，使動物能在支承氣墊上翔翔。

感應性、應激性（名）　生物隨環境（第218頁）變化作出反應的能力，例如一些動物隨噪聲或受觸而產生的運動。

神經系統　生物體內使信息在身體中傳遞的系統。這樣，身體的各個部份便能夠迅速地對任何刺激作出反應。

中樞神經系統　即CNS。包括脊椎動物（第74頁）的腦（第155頁）和脊髓（第154頁）在內的那部份神經系統（↑）。這部份神經系統接受來自身體所有各部份、起源於身體內、外的神經衝動（第150頁），而對這些神經衝動作出響應的方式為向各種不同的器官和肌肉（第143頁）發出適當的指令，使之作出相應的反應。

周圍神經系統　除CNS（↑）之外的那部份神經系統（↑）。周圍神經系統係一種遍佈生物體、並與CNS相連的神經（↓）網。

神經元（名）　組成神經系統（↑）的許多特殊的變形細胞。每一個神經元藉單個線狀軸突（↓）或神經纖維和為數眾多的樹突（↓），經由突觸（第151頁），與其他一些神經元相連。這些樹突將神經衝動（第150頁）從一個神經元傳導到另一個神經元。

神經細胞　同神經元（↑）。

細胞體　神經元（↑）中具有細胞核（第13頁）的那一部份。它也稱為中心體。

樹突（名）　神經元（↑）細胞體上的細胞質（第10頁）的分枝突起，其端部是突觸（第151頁）。樹突又可分枝成為小樹突。

尼氏粒　神經元（↑）細胞質（第10頁）中含有的一些顆粒。這些顆粒富含RNA（第24頁）。

軸突　充滿軸漿的神經元（↑）的長形突起。這種突起通常傳導離開細胞體的神經衝動（第150頁）。軸突被髓（磷脂）（第150頁）鞘所包繞，髓鞘外是一層施旺氏細胞的薄膜（第14頁）——神經膜。

myelin (*n*) a fatty substance which insulates the axon (p. 149) and speeds up the transmission of nerve impulses (↓). In vertebrates (p. 74) not all axons are myelinated. The myelin sheath is broken at intervals by constrictions called *nodes of Ranvier*.

neuroglia (*n*) specialized cells which protect and support the central nervous system (p. 149).

nerve impulse one of an interspaced succession of impulses or signals that are carried between the neurones (p. 149) via the exchange of sodium ions and changes in the electrical state of the neurone. The impulses travel at a constant speed throughout the nervous system (p. 149) and the energy for the impulse is not provided by the stimulus itself.

髓磷脂、髓鞘質（名） 一種可將軸突（第149頁）隔開、並加速神經衝動（↓）傳導的脂質。在脊椎動物（第74頁）中，並非所有軸突都有髓鞘。髓鞘每相隔一段距離，均有稱爲郎飛氏結的縮窄部份所形成的節束。

神經膠質 保護和支持中樞神經系統（第149頁）的特化細胞。

神經衝動 一系列有間隔的衝動或信號。這些衝動或信號借助於鈉離子的交換和神經元帶狀態的變化，在神經元（第149頁）之間傳遞。神經衝動在整個神經系統（第149頁）中以恒定速度傳導；而用於傳導神經衝動的能量並不是由刺激本身所提供的。

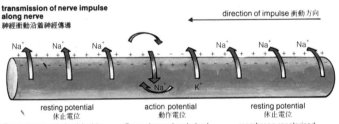

transmission of nerve impulse along nerve
神經衝動沿着神經傳導

direction of impulse 衝動 方向

resting potential
休止電位
① membrane polarized: inside negative, outside positive. Sodium ions expelled by sodium pump mechanism

action potential
動作電位
② membrane depolarized: sodium ions enter axon; inside positive, outside negative

resting potential
休止電位
membrane repolarized
膜再極化

① 膜極化：
細胞內帶負電，
細胞外帶正電。
鈉離子由鈉泵機制排出

② 膜去極化：
鈉離子進入軸突；
細胞內帶正電，
細胞外帶負電

resting potential the state which occurs when a neurone (p. 149) is inactive so that the neurone carries a greater negative charge within the cell and a greater positive charge outside.

action potential the state in which an electrical charge moves along the membrane (p. 14) of the axon (p. 149).

sodium pump mechanism the mechanism by which sodium ions are pumped out of a neurone (p. 149) as soon as the nerve impulse (↑) has passed.

polarization (*n*) the process in which sodium ions are pumped out of the neurone (p. 149) by the sodium pump mechanism (↑) so that the inside of the cell is restored to its resting potential (↑).

休止電位 神經元（第149頁）不活動時出現的狀態，結果神經元在細胞內帶較多負電荷，在細胞外帶較多正電荷。

動作電位 電荷沿着軸突（第149頁）的膜（第14頁）移動時出現的狀態。

鈉泵機制 神經衝動（↑）一通過，即將鈉離子泵送出神經元（第149頁）的機制。

極化 藉鈉泵機制（↑）將鈉離子泵送出神經元（第149頁）的過程，結果細胞內部恢復到其休止電位（↑）狀態。

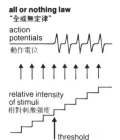

all or nothing law
"全或無定律"

action
potentials
動作電位

relative intensity
of stimuli
相對刺激強度

threshold
intensity of
stimulation
刺激的閾強度

depolarization (*n*) the process in which the membrane (p. 14) of the neurone (p. 149) becomes permeable to the passage of sodium ions which then enter the cell so that the cell becomes positively charged.

stimulus (*n*) any change in the external environment (p. 218) or the internal state of an organism which, (via the nervous system (p. 149) in animals), provokes a response to that change without supplying the energy for it.

threshold intensity the level of stimulus below which there is no nervous (p. 149) response of the stimulated organism.

'all or nothing law' the law which states that an organism will respond to a stimulus in only two ways: that is, either no nervous (p. 149) response at all or a response which is of a degree of intensity which does not vary with the intensity of the stimulus.

refractory period the length of time which passes between the passage of a nervous impulse (↑) through a neurone (p. 149) and its return to the resting potential (↑). During this period the neurone cannot further be stimulated.

absolute refractory period a refractory period (↑) in which a further stimulus of any intensity will result in the passage of no further nerve impulse (↑).

relative refractory period a refractory period (↑) in which another, unusually intense stimulus will result in the passage of a further nerve impulse (↑).

transmission speed the speed at which a nervous impulse (↑) travels and which is dependent upon the diameter of the neurone (p. 149).

synapse (*n*) the gap which exists between neurones (p. 149) and which is bridged during the action potential (↑) by a substance secreted (p. 106) by the neurone.

synaptic transmission the process by which nervous impulses (↑) are transmitted between neurones (p. 149) via the synaptic knob (p. 152). The action potential (↑) stops at the synapse (↑) but it causes a substance to be released which travels across the synapse and generates a new action potential in the neighbouring neurone.

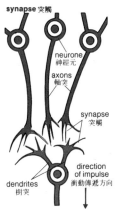

synapse 突觸

neurone
神經元

axons
軸突

synapse
突觸

direction
of impulse
衝動傳遞方向

dendrites
樹突

去極化（名）　神經元（第149頁）的膜（第14頁）變成可透過鈉離子的過程，其後鈉離子進入細胞，使細胞帶正電。

刺激（名）　有機體的外部環境（第218頁）或內部狀態的任何變化。這種變化（在動物中，經由神經系統（第149頁））激發出對該變化的反應，但並不為之提供能量。

閾強度、臨界強度　一種刺激量值。低於該量值時，受到刺激的生物體不會產生神經（第149頁）反應。

"全或無定律"　這個定律認為，一個生物體對某種刺激作出反應時僅有兩種方式：即完全沒有神經（第149頁）反應，或者產生了具有一定強度的反應，而這種反應強度並不隨刺激強度而變化。

不應期、休復期　在神經衝動（↑）通過神經元（第149頁）與該神經元回復到休止電位（↑）之間所經歷的一段時間。在這段時間內，神經元不能再引起興奮。

絕對不應期　一種不應期（↑）。在絕對不應期中，任何強度的第二次刺激都不能導致第二次神經衝動（↑）的通過。

相對不應期　一種不應期（↑）。在相對不應期中，另一次不尋常的較強刺激可導致第二次神經衝動（↑）的通過。

傳導速度　指神經衝動（↑）傳導的速度。傳導速度取決於神經元（第149頁）的直徑大小。

突觸（名）　神經元（第149頁）與神經元之間的間隙。在出現動作電位（↑）時，該間隙由神經元分泌（第106頁）的一種物質所填補。

突觸傳遞　神經衝動（↑）經由突觸小結（第152頁）在神經元（第149頁）之間傳遞所經歷的過程。動作電位（↑）中止於突觸（↑）處；但它可使一種物質釋放出來，該物質能穿越突觸，並在相鄰的神經元中產生新的動作電位。

synaptic knob (*n*) the knob-like ending of the axon (p. 149) which projects into the synapse (p. 151).

acetylcholine (*n*) one of the substances that are released as the action potential (p. 150) in a neurone (p. 149) arrives at the synapse (p. 151). It is specifically produced between a neurone and a muscle (p. 143) cell. There is a special enzyme (p. 28) called acetylcholine esterase which breaks it down so its effect does not continue.

atropine (*n*) a substance which is found in the plant, deadly nightshade, and which acts as a poison by preventing nerve impulses (p. 150) being transmitted from the neurone (p. 149) to the body tissues (p. 83).

strychnine (*n*) a substance which is obtained from the seed of an east Indian tree and which has a powerful stimulating effect on the central nervous system (p. 149), so much so, that in greater than minute quantities it acts as a poison.

adrenaline (*n*) a substance, similar to noradrenaline (↓), released by the adrenal glands (p. 130), which increases the metabolic rate (p. 32) and other functions when it is released into the bloodstream (p. 90) during stress or in preparation for action.

noradrenaline (*n*) one of the substances which is released as the action potential (p. 150) in a neurone (p. 149) arrives at the synapse (p. 151). It is produced in the autonomic nervous system (p. 155). It is also secreted (p. 106) by the adrenal glands (p. 130) and affects cardiac muscle (p. 143) and involuntary muscle (p. 143) etc.

summation (*n*) the process in which the additive effect of nerve impulses (p. 150) arriving at different neurones (p. 149) stimulates the impulse in another neurone while the arrival of just one of the impulses produces no effect.

facilitation (*n*) the process in which the stimulation of a neurone (p. 149) is increased by summation (↑).

reflex action the fundamental and innate (p. 164) response of an animal to a stimulus. For example, the automatic escape reaction away from a source of threat or pain, such as withdrawing the hand from a hot object.

突觸小結(名)　軸突(第149頁)伸入突觸(第151頁)內的球形末端。

乙醯膽鹼(名)　神經元(第149頁)中的動作電位(第150頁)到達突觸(第151頁)時所釋放的物質之一。乙醯膽鹼的特定釋放地點是在神經元與肌肉(第143頁)細胞之間。有一種稱為乙醯膽鹼酯酶的特殊酶(第28頁)可將乙醯膽鹼分解，所以乙醯膽鹼不會連續起作用。

阿托品(名)　毒性茄屬植物中含的一種物質。這種物質作為一種毒物，可以阻止神經衝動(第150頁)從神經元(第149頁)傳遞至身體組織(第83頁)。

馬錢子鹼(名)　一種從東印度喬木的種子中獲得的物質。這種物質對中樞神經系統(第149頁)有強烈的激作用，以致極少量即起毒物的作用。

腎上腺素(名)　由腎上腺(第130頁)釋放的一種類似於去甲腎上腺素(↓)的物質。在緊張狀態或為動作作準備期間，腎上腺素釋入血流(第90頁)中，即可增強代謝率(第32頁)和其他功能。

去甲腎上腺素(名)　神經元(第149頁)中的動作電位(第150頁)到達突觸(第151頁)時所釋放的物質之一。去甲腎上腺素產生於自主神經系統(第155頁)之中。它也是腎上腺(第130頁)分泌(第106頁)的物質，能影響心肌(第143頁)和不隨意肌(第143頁)等的作用。

總和、累積作用(名)　所謂累積作用，是指到達不同神經元(第149頁)的一些神經衝動(第150頁)所產生的加性效應，可激發另一個神經元的衝動，而僅僅一個神經衝動到達時就沒有這種效果。

促進作用(名)　對某一神經元(第149頁)的刺激，由累積作用(↑)得到增強的過程。

反射動作　動物對刺激所產生的固有和先天(第164頁)反應，例如離border威嚇或疼痛源的自動逃避反應，例如手從熱物體處縮回。

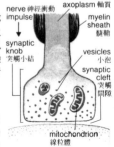

synaptic knob 突觸小結

nerve 神經衝動 impulse / synaptic knob 突觸小結 / axoplasm 軸漿 / myelin sheath 髓鞘 / vesicles 小泡 / synaptic cleft 突觸間隙 / mitochondrion 線粒體

membrane of post-synaptic dendrite
突觸後樹突膜

simple reflex arc 簡單反射弧

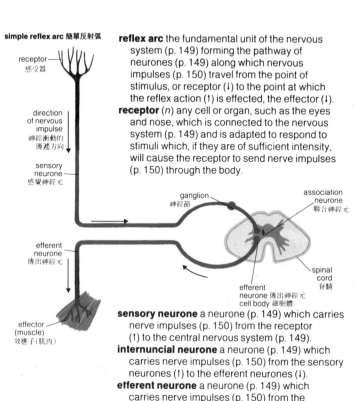

receptor
感受器

direction
of nervous
impulse
神經衝動的
傳遞方向

sensory
neurone
感覺神經元

ganglion
神經節

association
neurone
聯合神經元

efferent
neurone
傳出神經元

spinal
cord
脊髓

efferent
neurone 傳出神經元
cell body 細胞體

effector
(muscle)
效應子(肌肉)

reflex arc the fundamental unit of the nervous system (p. 149) forming the pathway of neurones (p. 149) along which nervous impulses (p. 150) travel from the point of stimulus, or receptor (↓) to the point at which the reflex action (↑) is effected, the effector (↓).

receptor (n) any cell or organ, such as the eyes and nose, which is connected to the nervous system (p. 149) and is adapted to respond to stimuli which, if they are of sufficient intensity, will cause the receptor to send nerve impulses (p. 150) through the body.

sensory neurone a neurone (p. 149) which carries nerve impulses (p. 150) from the receptor (↑) to the central nervous system (p. 149).

internuncial neurone a neurone (p. 149) which carries nerve impulses (p. 150) from the sensory neurones (↑) to the efferent neurones (↓).

efferent neurone a neurone (p. 149) which carries nerve impulses (p. 150) from the internuncial neurones (↑) to the effectors (↓). Also known as **motor neurone**.

effector (n) any cell or organ, such as a muscle (p. 143), which responds in some way to stimulus from a nervous impulse (p. 150).

conditioned reflex a reflex action (↑) which has been modified by learning or experience and which needs to be reinforced periodically to be retained.

taxis (n) a reflex action (↑) in which the movements in response to the stimulus are determined by the direction of the stimulus. For example, the movement towards or away from bright light or the force of gravity. See also taxic movement (p. 141). **taxes** (pl.).

反射弧　構成神經元(第149頁)通路的神經系統(第149頁)的基本單元。神經衝動(第150頁)沿着該通路從刺激點或感受器(↓)傳遞到產生反射動作(↑)的處所——效應子(↓)。

感受器(名)　與神經系統(第149頁)相連、適合於對各種刺激產生反應的任何細胞或器官,例如眼和鼻。如果這些刺激具有足夠的強度,它們就會使感受器將神經衝動(第150頁)傳遞到整個身體。

感覺神經元　一種將神經衝動(第150頁)從感受器(↑)傳遞到中樞神經系統(第149頁)的神經元(第149頁)。

聯絡神經元　一種將神經衝動(第150頁)從感覺神經元(↑)傳遞到傳出神經元(↓)的神經元(第149頁)。

傳出神經元　一種將神經衝動(第150頁)從聯絡神經元(↑)傳遞到效應子(↓)的神經元(第149頁)。它也稱爲**運動神經元**。

效應子(名)　以某種方式對來自神經衝動(第150頁)刺激產生反應的任何細胞或器官,例如肌肉(第143頁)。

條件反射　一種經學習或經驗調整了反射動作(↑)。條件反射的維持需要定期强化。

趨性(名)　一種反射動作(↑)。產生該反射動作時,各種隨刺激產生的運動均由刺激方向決定。例如朝光或離光的運動;朝向重力或離開重力的運動。參見"趨性運動"(第141頁)。(複數形式爲 taxes)

kinesis (*n*) a reflex action (p. 152) in which the rate of movement is affected by the intensity of the stimulus and which is unaffected by its direction. For example, woodlice move faster in drier surroundings than in damp ones.

spinal cord the part of the central nervous system (p. 149) in vertebrates (p. 74) which is contained within a hollow tube running the length of the vertebral column (p. 74) and runs from the medulla oblongata (p. 156). It consists of neurones (p. 149) and nerve fibres with a central canal containing fluid. Pairs of spinal nerves (↓) leave the spinal cord to pass into the body. The spinal cord carries nerve impulses (p. 150) to and from the brain (↓) and the body.

動態、運動、動作（名） 一種反射動作（第152頁）。動作運動速度受刺激強度影響，而不受刺激方向影響。例如，地鼈在乾燥的環境下比在潮濕環境下運動得更快。

脊髓 脊椎動物（第74頁）中樞神經系統（第149頁）的一部份，脊柱（第74頁）內一根與脊柱等長的空管中含有脊髓，其前端連接延髓（第156頁）。脊髓由神經元（第149頁）、神經纖維以及一根含有液體的中央管構成。脊髓發出成對的脊神經（↓），分佈到身體各部，脊髓傳遞來往於腦（↓）和身體的神經衝動（第150頁）。

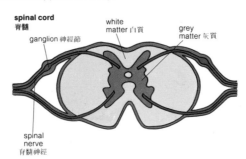

spinal cord 脊髓

ganglion 神經節 · white matter 白質 · grey matter 灰質

spinal nerve 脊髓神經

meninges (*n.pl.*) the three membranes (p. 14) which protect the central nervous system (p. 149) of vertebrate (p. 74) animals.

pia mater one of the meninges (↑). The soft, delicate inner membrane (p. 14) next to the central nervous system (p. 149) which is densely packed with blood vessels (p. 127).

arachnoid mater the central of the three meninges (↑) and separated from the pia mater (↑) by fluid-filled spaces.

dura mater one of the meninges (↑). The stiff, tough outer membrane which is in direct contact with the arachnoid mater (↑) and which contains blood vessels (p. 127).

spinal nerve one of the pairs of nerves (p. 149) which arise from the spinal cord (↑) in segments.

腦〔脊〕膜（名、複） 脊椎動物（第74頁）中樞神經系統（第149頁）的三層保護膜（第14頁）。

軟〔脊〕膜 腦〔脊〕膜（↑）中的一層膜。軟〔脊〕膜系一層柔軟而精細的內膜（第14頁），緊貼於中樞神經系統（第149頁），其上佈滿血管（第127頁）。

蛛網膜 腦脊膜（↑）三層膜中的中層膜。充有液體的間隙將蛛網膜與軟腦脊膜（↑）隔開。

硬〔脊〕膜 腦脊膜（↑）中的一層膜。硬腦脊膜系一層厚而韌的外膜，直接與蛛網膜（↑）接觸，上面佈有血管（第127頁）。

脊髓神經 由椎骨內的脊髓（↑）兩側發出的成對神經（第149頁）。

meninges 腦脊膜

① skull (bone) — dura mater ④
② arachnoid mater — cerebro- ⑤ spinal fluid
③ pia mater — brain ⑥

① 顱（骨）
② 蛛網膜
③ 軟腦脊膜
④ 硬腦脊膜
⑤ 腦脊髓液
⑥ 腦

grey matter nervous tissue (p. 91) that is grey in colour and is found in the centre of the spinal cord (↑) as well as in parts of the brain (↓). It contains large numbers of synapses (p. 151) and consists mainly of nerve (p. 149) cell bodies (p. 149).

white matter nervous tissue (p. 91) that is whitish in colour and is found on the outer region of the spinal cord (↑) as well as in parts of the brain (↓). It connects different parts of the central nervous system (p.149) and consists mainly of axons (p.149).

autonomic nervous system the part of the central nervous system (p. 149) in vertebrates (p. 74) which carries nerve impulses (p. 150) from receptors (p. 153) to the smooth muscle fibres (p. 144) of the heart (p. 124), gut (p. 98) and other internal organs.

sympathetic nervous system the part of the autonomic nervous system (↑) which increases the heart (p. 124) rate and breathing (p. 112) rate, the secretion (p. 106) of adrenaline (p. 152), the blood (p. 90) pressure, and slows the digestion (p. 98) so that the vertebrate's (p. 74) body is prepared for emergency action in response to stimuli.

parasympathetic nervous system the part of the autonomic nervous system (↑) which effectively works in opposition to the sympathetic nervous system (↑) slowing down the heart (p. 124) beat etc. Both systems act in co-ordination to control the rates of action.

ganglion (*n*) a bundle of nerve (p. 149) cell bodies (p. 149) contained within a sheath which, in invertebrates (p. 75) may form part of the central nervous system (p. 149), and in vertebrates (p. 74) are generally found outside the central nervous system. Also, in the brain (↓) some of the masses of grey matter (↑) are referred to as ganglia (*pl.*).

nerve net an interconnecting network of nerve (p.149) cells found in the bodies of some invertebrates (p. 75) to form a simple nervous system (p. 149).

brain (*n*) the part of the central nervous system (p. 149) which effectively co-ordinates the reactions of the whole body of the organism. It forms as an enlargement of the spinal cord (↑) and is situated at the anterior end of the body.

灰質　脊髓（↑）中央和腦（↓）的各部份含有的灰色神經組織（第91頁）。灰質含有大量突觸（第151頁），主要由神經（第149頁）細胞體（第149頁）構成。

白質　脊髓（↑）外部和腦（↓）的各部份含有的呈白色神經組織（第91頁）。白質與中樞神經系統（第149頁）的不同部份相連接，主要由軸突（第149頁）構成。

自主神經系統　脊椎動物（第74頁）中樞神經系統（第149頁）的一部份。它將神經衝動（第150頁）從感受器（第153頁）傳遞至心臟（第124頁）、腸（第98頁）和其他內部器官的平滑肌纖維（第144頁）。

交感神經系統　自主神經系統（↑）的一部份。它可增加心（第124頁）搏率、呼吸（第112頁）速率、腎上腺素（第152頁）分泌（第106頁）和血（第90頁）壓，減慢消化（第98頁），從而使脊椎動物（第74頁）的身體對刺激做好應急動作準備。

副交感神經系統　自主神經系統（↑）的一部份。其作用正好與交感神經系統（↑）相反，它可減慢心（第124頁）搏率等。交感神經系統和副交感神經系統協同作用，控制動作的速度。

神經節（名）　鞘內的一叢神經（第149頁）。含的細胞體（第149頁）。在無脊椎動物（第75頁）中，神經節構成中樞神經系統（第149頁）的一部份；在脊椎動物（第74頁）中，神經節通常在中樞神經系統之外。此外，在腦（↓）中，某些灰質（↑）團塊也可稱爲神經節。

神經網　某些無脊椎動物（第75頁）體內構成簡單神經系統（第149頁）的一種神經（第149頁）細胞互連網絡。

腦（名）　中樞神經系統（第149頁）的一部份。它可有效地調整個生物體的各種反應。腦構成脊髓（↑）的一個膨大部份，位於身體的前端。

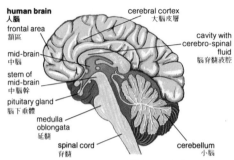

**human brain
人腦**
frontal area
額區
mid-brain
中腦
stem of
mid-brain
中腦幹
pituitary gland
腦下垂體
medulla
oblongata
延髓
spinal cord
脊髓

cerebral cortex
大腦皮層
cavity with
cerebro-spinal
fluid
腦脊髓液腔
cerebellum
小腦

cerebral hemispheres the paired masses of grey matter (p. 155), beneath which is white matter (p. 155), that occur at the front end of the forepart of the brain (p. 155) and by which many of the animal's activities are controlled. Each hemisphere controls actions on the opposite side of the body from which it is situated.

cerebral cortex the highly convoluted grey matter (p. 155) that forms part of the cerebral hemispheres (↑).

corpus callosum the band of nerve (p. 149) fibres which connects the cerebral hemispheres (↑) allowing their action to be co-ordinated.

medulla oblongata the continuation of the spinal cord (p. 154) with the hind region of the brain (p. 155). It contains centres of grey matter (p. 155) which are responsible for controlling many of the major functions and reflexes (p. 153) of the body, for example, the medulla oblongata contains the respiratory centre (p.117).

cerebellum (n) the part of the brain (p. 155) lying between the medulla oblongata (↑) and the cerebral hemispheres (↑) which is deeply convoluted and is responsible for controlling voluntary muscle (p. 143) action which is stimulated by the cerebral hemispheres.

hypothalamus (n) that region of the forepart of the brain (p. 155) which is responsible for monitoring and regulating metabolic (p. 26) functions such as body temperature, eating, drinking and excretion (p. 134). It controls the activity of the pituitary gland (↓).

大腦半球 腦(第155頁)前部的前端包含的成對灰質(第155頁)團塊,其下是白質(第155頁)。動物的許多活動都由灰質控制。每一個大腦半球各控制身體中與其相反一側的各種動作。

大腦皮層 構成大腦半球(↑)一部份的、呈深度皺摺狀的灰質(第155頁)。

胼胝體 連接兩個大腦半球(↑)、使它們協同作用的神經(第149頁)纖維束。

延髓 包括腦(第155頁)後段在內的脊髓(第154頁)延續部份。延體包含灰質(第155頁)中負責控制身體的許多主要功能和反射(第153頁)的一些中樞,例如呼吸中樞(第117頁)。

小腦(名) 位於延髓(↑)和大腦半球之間的那一部份腦(第155頁)。小腦呈深度皺摺狀;負責控制由大腦半球所刺激的隨意肌(第143頁)運動。

下丘腦、丘腦下部(名) 指腦(第155頁)前部的那個區域。丘腦下部擔負監控和調節諸如體溫、吃喝及排泄(第134頁)等的代謝(第26頁)功能。它控制腦下垂體(↓)的活動。

thalamus (*n*) the part of the brain (p. 155) which carries and co-ordinates nerve impulses (p. 150) from the cerebral hemispheres (↑).

pituitary gland a gland (p. 87) in the brain (p. 155) which secretes (p. 106) a number of hormones (p. 130) that in turn stimulate the secretion of hormones from other glands to affect such metabolic (p. 26) processes as growth, secretion of adrenaline (p. 152), milk production, and so on.

pineal body a gland (p. 87) found as an outgrowth on the top of the brain (p. 155) and which may be responsible for secreting (p. 106) a hormone (p. 130) associated with colour change.

丘腦（名） 負責傳遞和協調來自大腦半球（↑）的神經衝動（第150頁）的那一部份腦（第155頁）。

腦下垂體 腦（第155頁）中的一種腺體（第87頁）。這種腺體能分泌（第106頁）多種激素（第130頁），而這些激素又可刺激其他一些腺體分泌激素，以影響諸如生長、腎上腺素（第152頁）分泌及產乳等代謝（第26頁）過程。

松果體 以贅疣形式存在於腦（第155頁）頂的一種腺體（第87頁）。其功能可能是分泌（第106頁）一種與體色改變有關的激素（第130頁）。

human ear
人耳

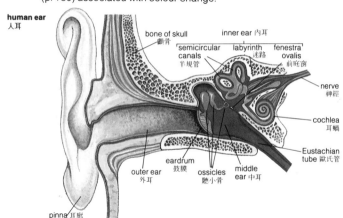

bone of skull 顱骨 ／ inner ear 內耳
semicircular canals 半規管 ／ labyrinth 迷路 ／ fenestra ovalis 前庭窗
nerve 神經
cochlea 耳蝸
Eustachian tube 歐氏管
eardrum 鼓膜 ／ ossicles 聽小骨 ／ middle ear 中耳
outer ear 外耳
pinna 耳廓

ear (*n*) one of the pair of sense organs, situated on either side of the head in vertebrates (p. 74), which are used for hearing (p. 159) and balance (p. 159).

outer ear the tube which leads from the outside of the head to the eardrum (p. 158). In amphibians (p. 77) and some reptiles (p. 78), it is not present because the eardrum is situated at the skin surface.

pinna (*n*) the part of the outer ear (↑), present in mammals (p. 80), situated on the outside of the head and consisting of a flap of skin and cartilage (p. 90), which helps to direct sound into the ear(↑).

耳（名） 位於脊椎動物（第74頁）頭部兩側的一對感覺器官，用於聽覺（第159頁）和平衡（第159頁）感覺。

外耳 從頭的外側通至鼓膜（第158頁）的管道。兩棲動物（第77頁）和某些爬蟲動物（第78頁），由於鼓膜位於皮膚表面，因而沒有外耳。

耳殼、耳廓（名） 哺乳動物（第80頁）中位於頭之外側的那一部份外耳（↑），由皮瓣和軟骨（第90頁）構成。耳廓有助於將聲音引入耳（↑）內。

inner ear the innermost part of the ear (p. 157) which is situated within the skull and which detects sound as well as the position of the animal in relation to gravity and acceleration, so enabling the animal to balance. It is fluid filled and is connected to the brain (p. 155) by an auditory nerve (p. 149) so that it is able to convert sound waves into nervous impulses (p. 150). It is made up of a labyrinth of membraneous (p. 14) tubes contained within bony cavities.

middle ear an air-filled cavity situated between the outer ear (p. 157) and the inner ear (↑). It is separated from the outer ear by the eardrum (↓) and, in mammals (p. 80), contains three small bones or ossicles (↓).

eardrum (n) a thin, double membrane (p. 14) of epidermis (p. 131) separating the outer ear (p. 157) from the middle ear (↑) and which is caused to vibrate by sound waves. These vibrations are then transmitted through the middle ear, where their force is amplified, into the inner ear (↑). Also known as **tympanic membrane**.

fenestra ovalis a small, oval, membraneous (p. 14) window which connects the middle ear (↑) with the inner ear (↑) allowing vibrations from the eardrum (↑) to be transmitted to the inner ear. It is twenty times smaller than the eardrum so that the force of the vibrations is increased.

fenestra rotunda a small, round, membraneous (p. 14) window which connects the inner ear (↑) with the middle ear (↑) and which bulges into the middle ear as vibrations of the fenestra ovalis (↑) cause pressure increases in the inner ear.

ear ossicle one of usually three small bones that occur in the middle ear (↑) of mammals (p. 80) and which, by acting as levers, transmit and increase the force of vibrations produced by the eardrum (↑) and carry them to the inner ear (↑).

malleus (n) a hammer-shaped ear ossicle (↑) which is connected with the eardrum (↑).

incus (n) an anvil-shaped ear ossicle (↑) situated between the malleus (↑) and the stapes (↓).

stapes (n) a stirrup-shaped ear ossicle (↑) which is connected with the fenestra ovalis (↑).

內耳 耳(第157頁)的最裏面部份。內耳位於顱骨內。它可感受聲音以及與重力和加速度有關的動物體位置，從而使動物能保持身體平衡。內耳充有液體，經由一根聽神經(第149頁)與腦(第155頁)相通，故它能將聲波轉變成神經衝動(第150頁)。內耳亦稱爲迷路，而迷路是由骨腔內的一些膜質(第14頁)管構成的。

中耳 外耳(第157頁)和內耳(↑)之間的一個充氣腔。耳鼓(↓)將中耳與外耳隔開。哺乳物物(第80頁)的中耳含有三塊小骨或聽小骨(↓)。

耳鼓(名) 將外耳(第157頁)與中耳(↑)隔開的一張薄的表皮(第131頁)雙層膜(第14頁)。聲波使耳鼓振動，隨後，這些振動便傳到中耳，其強度經放大後，即傳入內耳(↑)。也稱爲鼓膜。

前庭窗 連接中耳(↑)與內耳、使聲波振動從鼓膜(↑)傳至內耳的一種橢圓形膜質(第14頁)小窗。耳鼓爲前庭窗的20倍，所以聲波振動的強度得以增大。

正圓窗、蝸窗 連接內耳(↑)與中耳(↑)的一種圓型膜質(第14頁)小窗。當前庭窗(↑)的振動導致內耳中的壓力增加時，蝸窗便鼓脹起來，凸入中耳之內。

聽小骨 哺乳動物(第80頁)的中耳內通常有的三塊小骨。聽小骨藉其槓桿作用，傳播鼓膜(↑)所產生的振動，增大振動強度，並將之傳至內耳(↑)

錘骨(名) 與鼓膜(↑)相連的一塊錘狀聽小骨(↑)。

砧骨(名) 位於錘骨(↑)和鐙骨(↓)之間的一塊砧狀聽小骨(↑)。

鐙骨(名) 與前庭窗(↑)相連的一塊鐙狀聽小骨(↑)。

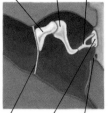

ear ossicles 聽小骨
malleus 錘骨 incus 砧骨 stapes 鐙骨

eardrum　　fenestra　　fenestra
耳鼓　　　　ovalis　　　rotunda
　　　　　　前庭窗　　　蝸窩

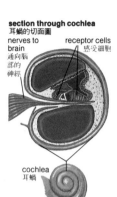

section through cochlea
耳蝸的切面圖

nerves to brain 通向腦部的神經

receptor cells 感受細胞

cochlea 耳蝸

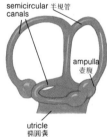

semicircular canals 半規管

semicircular canals 半規管

ampulla 壺腹

utricle 橢圓囊

Eustachian tube a tube which connects the middle ear (↑) with the back of the throat. It is normally closed but opens during yawning and swallowing to balance the pressure on either side of the eardrum (↑) thus preventing the eardrum from bursting.

vestibular apparatus the apparatus contained in a cavity in the inner ear (↑) immediately above and behind the fenestra ovalis (↑) and which contains the organs concerned with the sense of balance and posture.

cochlea (*n*) a spirally coiled tube, which is a projection of the saccule (p. 160), and found within the inner ear (↑). It is concerned with sensing the pitch (↓) of the sound waves entering the ear (p. 157).

hearing (*n*) the sense whereby sound waves or vibrations enter the outer ear (p. 157) and cause the eardrum (↑) to vibrate. In turn, these vibrations are transmitted through the middle ear (↑) and into the inner ear (↑) where they are converted into nervous impulses (p. 150) and transmitted to the brain (p. 155).

intensity (*n*) the degree of loudness or softness of sound. If a sound entering the ear (p. 157) is very loud, muscles (p. 143) attached to the ear ossicles (↑) prevent them from vibrating too much.

pitch (*n*) the degree of height or depth of a sound which depends on the frequency of the sound waves - those of a high frequency are referred to as high and vice versa. Different parts of the cochlea (↑) respond to sounds of different pitch.

balance (*n*) the ability of the animal to orient itself properly in relation to the force of gravity. Animals rely on information received by apparatus within the inner ear (↑) and by the eyes and other senses to achieve balance and posture.

semicircular canal one of three looped tubes positioned at right angles to one another within the inner ear (↑). They contain fluid which flows in response to movement of the head. The movement of the fluid is detected by the sensory hairs in ampullae (p. 160) at the ends of the canals.

耳咽管、歐氏管、咽鼓管 溝通中耳(↑)與咽後部的一根管道。耳咽管通常呈關閉狀態，但在張開口和吞嚥時開放，以使鼓膜(↑)兩側的壓力保持平衡，從而防止鼓膜破裂。

前庭器官 緊靠前庭窗(↑)後上方的內耳(↑)腔中的器官。前庭器官內含有與平衡覺和姿勢覺有關的器官。

耳蝸(名) 內耳(↑)中的一個螺旋形管道。耳蝸是球囊(第160頁)的一個突出物。它感覺進入耳(第157頁)中的聲波音調(↓)。

聽覺(名) 聲波或振動進入外耳(第157頁)、使鼓膜(↑)振動引起的感覺。這些振動又通過中耳(↑)傳至內耳(↑)，並在內耳轉變成神經衝動(第150頁)，再傳給腦(第155頁)。

聲強(名) 聲音的響亮或柔和程度。如果進入耳(第157頁)內的聲音非常響亮，附著在聽小骨(↑)上的肌肉(第143頁)會阻止其振動得太厲害。

音調(名) 隨聲波頻率而定的聲音高低程度──高頻率聲波稱為高音，反之亦然。耳蝸(↑)的不同部份分別對不同音調的聲音產生感應。

平衡(名) 動物使自身適應於重力的能力。動物依靠內耳(↑)中的器官、依靠眼睛和其他感官所接收的信息，以實現身體的平衡和作出姿勢。

半規管 內耳(↑)中互相垂直的三個環形管。半規管內含有隨頭部運動而流動的液體。該液體的流動由在半規管端部壺腹(第160頁)內的感覺毛感知。

ampulla (*n*) a swelling at the end of each semicircular canal (p. 159). It bears a gelatinous cupula (↓), sensory hairs, and receptor (p. 153) cells which are responsible for transmitting information to the brain (p. 155) via the auditory nerve (p. 149). **ampullae** (*pl.*).

utricle (*n*) a fluid-filled sac within the inner ear (p. 158) from which arise the semicircular canals (p. 159). Within the fluid of the utricle are otoliths (↓) of calcium carbonate. If the head is tilted, the otoliths are pulled downwards by gravity and in turn pull on sensory fibres attached to the wall of the utricle.

saccule (*n*) a lower, fluid-filled cavity in the inner ear (p. 158) from which arises the cochlea (p. 159). Like the utricle (↑), it also contains otoliths (↓) which respond to the orientation of the head with respect to the force of gravity.

cupula (*n*) the gelatinous body which forms part of the ampulla (↑) and which is displaced by fluid that moves in response to the position of the head.

otolith (*n*) one of the granules of calcium carbonate present in the fluid of the utricle (↑) and saccule (↑) which respond to tilting movements of the head by force of gravity.

eye (*n*) the sense organ of sight which is sensitive to the direction and intensity of light and which, in vertebrates (p. 74), is also able to form complex images of the outside world which are transmitted to the brain (p. 155) via the optic nerve (p. 149). The eyes of most animals are roughly spherical in shape, and in vertebrates are contained in depressions within the skull, to which they are attached by muscles (p. 143).

retina (*n*) the light-sensitive inner layer of the eye (↑) which contains rod-shaped receptor (p. 153) cells and cone-shaped receptor cells. Nerve (p. 149) fibres leave the retina and join together to form the optic nerve.

choroid layer a layer of tissue surrounding the eye (↑) between the retina (↑) and the sclerotic layer (↓). It contains pigments (p. 126) to reduce reflections within the eye and blood vessels (p. 127) which supply oxygen to the eye.

壺腹(名) 每個半規管(第159頁)端部的隆起部份。壺腹有一個膠質蝸頂(↓)，以及許多感覺毛和感受(第153頁)細胞。這些感覺細胞負責將信息通過聽神經(第149頁)傳遞給腦(第155頁)。(複數形式 ampullae)

橢圓囊(名) 內耳(第158頁)中的一個充液囊，由此處生出半規管(第159頁)。橢圓囊的液體內含有碳酸鈣耳石(↓)。如果頭部傾斜，則耳石由於重力的作用被向下拉，進而又牽拉附着在橢圓囊壁上的感覺纖維。

球囊(名) 內耳(第158頁)的一個下部充液腔，由此處生出耳蝸(第159頁)。與橢圓囊(↑)一樣，球囊也含有耳石(↓)。這些耳石能根據重力對頭部的定向作出反應。

蝸頂(名) 構成壺腹(↑)一部份的膠質體。隨頭部位置的不同而流動的液體，可使蝸頂發生位移。

耳石、耳沙(名) 橢圓囊(↑)和球囊(↑)液體內含有的碳酸鈣顆粒。耳石由於重力作用，可對頭部傾斜運動作出反應。

眼、眼睛(名) 能感受光方向和強度的視覺器官。脊椎動物(第74頁)的眼能形成外部世界的複雜圖像，並藉視神經(第149頁)將之傳遞給腦(第155頁)。大多數動物的眼基本上爲球形。脊椎動物的眼在顱骨的凹窩內，由肌肉(第143頁)將它們附着於其上。

視網膜(名) 眼(↑)的感光內層。它含有桿狀感受(第153頁)細胞和錐狀感受細胞。神經(第149頁)纖維離開們視網膜後，結合在一起構成視神經。

脈絡膜 在視網膜(↑)和鞏膜(↓)之間包圍眼睛(↑)的一層組織。脈絡膜含有多種色素(第126頁)，可減少眼睛內的反射作用，並富含向眼睛供氧的血管(第127頁)。

ampulla 壺腹

gelatinous cupula 膠質蝸頂

sensory hairs 感覺毛

hair cells 毛細胞

sensory neurones 感覺神經元

human eye 人的眼睛

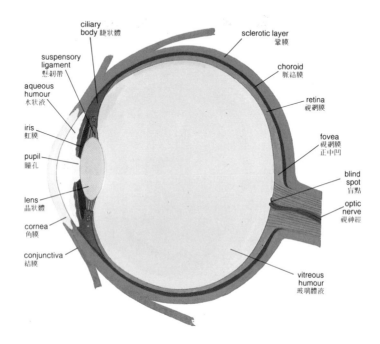

ciliary body 睫狀體

suspensory ligament 懸韌帶

aqueous humour 水狀液

iris 虹膜

pupil 瞳孔

lens 晶狀體

cornea 角膜

conjunctiva 結膜

sclerotic layer 鞏膜

choroid 脈絡膜

retina 視網膜

fovea 視網膜正中凹

blind spot 盲點

optic nerve 視神經

vitreous humour 玻璃體液

sclerotic layer the tough, fibrous (p. 143) non-elastic layer which surrounds and protects the eye (↑) and is continuous with the cornea (↓).

cornea (n) the disc-shaped area at the front of the eye (↑) which is continuous with the sclerotic layer (↑) and which is transparent to light. It is curved so that light passing through it is refracted and begins to converge before reaching the lens (p. 162). Indeed, in land-living mammals (p. 80) this is the main element of the eye's focusing power.

鞏膜　包圍和保護眼睛 (↑) 一層強韌、非彈性的纖維 (第 143 頁) 膜。它與角膜 (↓) 相連接。

角膜 (名)　眼睛 (↑) 前方的圓盤狀區域。該區域與鞏膜 (↑) 相連接，光線可以透過。角膜成曲面狀，所以，透過它的光線在到達晶狀體 (第162頁) 之前，便產生折射，並開始聚焦。陸生哺乳物 (第80頁) 的角膜是實際上使眼睛具有聚焦能力的主要元件。

lens accommodation 晶狀體調節

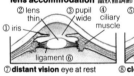

② lens thin ③ pupil wide ④ ciliary muscle ⑤ lens thick
① iris
ligament ⑥
⑦**distant vision** eye at rest　⑧**close vision**

① 虹膜　　　　⑤ 晶狀體增厚
② 晶狀體變薄　⑥ 懸韌帶
③ 瞳孔放大　　⑦ 遠距視覺處於休息狀態時的眼睛
④ 睫狀肌　　　⑧ 近距視覺

lens (n) a transparent disc which is convex on both faces and which is attached to the ciliary body (↓) by suspensory ligaments (↓). It consists of an elastic, jelly-like material held in a skin. When the ciliary muscles (↓) contract, the convexity of the lens is increased so that light rays entering the eye can be focused on the retina (p. 160). This allows for the fine focusing power of the eye (p. 160).

refraction (n) the change in direction of light which takes place as it crosses a boundary between two substances.

convex (adj) of a lens which causes light passing through it to move closer together.

concave (adj) of a lens which causes light passing through it to move further apart rather than closer together. See also convex (↑).

suspensory ligament the fibrous (p. 143) ligament (p. 146) which holds the lens (↑) in position.

ciliary body the circular, thickened outer edge of the choroid (p. 160) at the front of the eye which contains the ciliary muscles and to which the lens (↑) is attached. It also contains glands (p. 87) which secrete (p. 106) the aqueous humour (↓).

ciliary muscles see ciliary body (↑).

iris (n) a ring of opaque tissue (p. 83) that is continuous with the choroid (p. 160) and which has a hole or pupil (↓) in the centre through which light can pass. There are circular muscles (p.143) and radial muscles surrounding the pupil which increase or decrease the size of the pupil in accordance with the intensity of light.

pupil (n) the hole in the centre of the iris (↑) through which light enters the eye (p. 160). It is usually circular but may be other shapes in some animals.

晶狀體（名）　一種兩面均為凸面的透明圓片。晶狀體由懸韌帶（↓）繫於睫狀體（↓）上。它由固定在皮膚內一種富有彈性的膠狀物質組成。當晶狀體（↓）收縮時，晶狀體的凸度增加，使進入眼睛的光線得以聚焦在視網膜（第160頁）上。這使眼睛（第160頁）具有精密的聚焦能力。

折射（名）　光線穿過兩種物質之間的邊界時所發生的方向變化。

凸的、凸面的（形）　指透鏡而言，該透鏡可使光線通過之後相互靠近。

凹的、凹面的（形）　指透鏡而言，該透鏡可使光線通過之後彼此遠離，而不是相互靠近。參見"凸的"（↑）。

懸韌帶　指將晶狀體（↑）固定在適當位置的纖維（第143頁）韌帶（第146頁）。

睫狀體　眼睛前方的脈絡膜（第160頁）的環形增厚外緣。睫狀體含有睫狀肌；其上繫有晶狀體（↑）。它還含有可分泌（第106頁）水狀液（↓）的多個腺體（第87頁）。

睫狀肌　見睫狀體（↑）。

虹膜（名）　與脈絡膜（第160頁）相連的一圈不透明組織（第83頁）。其中央有一個光線可通過的圓孔，稱為瞳孔（↓）。瞳孔周圍有環肌（第143頁）和輻射狀肌。這些肌肉可根據光線的強度使瞳孔放大或縮小。

瞳孔（名）　虹膜（↑）中央的圓孔。光線通過該孔進入眼睛（第160頁）。瞳孔通常為圓形，但有些動物的瞳孔可能為其他形狀。

refraction 折射

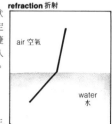

air 空氣

water 水

凸透鏡 **convex lens**

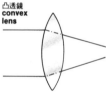

凹透鏡 **concave lens**

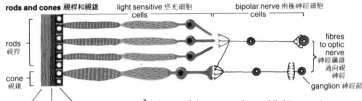

rods and cones 視桿和視錐　　light sensitive 感光細胞 cells　　bipolar nerve 兩極神經細胞 cells

rods 視桿

cone 視錐

fibres to optic nerve 神經纖維通向視神經

ganglion 神經節

sclerotic layer 鞏膜　choroid 脈絡膜　epithelium 上皮

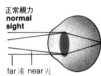

正常視力
normal sight

far 遠　near 近

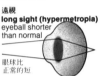

遠視
long sight (hypermetropia)
eyeball shorter than normal
眼球比正常的短

convex 凸透鏡 lens

近視
short sight (myopia)
eyeball longer than normal
眼球比正常的長

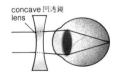

concave 凹透鏡 lens

cone[a] (*n*) one of the cone-shaped light receptors (p. 153) present in the retina (p. 160). Cones contain three different pigments (p. 126) which are in turn sensitive to red, green, and blue light so that cones are primarily responsible for colour vision. Cones are concentrated mainly in and around the fovea (↓) and are not present at the edge of the retina. They are also receptors of light of high intensity.

fovea (*n*) a small, central depression in the retina (p. 160) in which most of the receptor (p. 153) cells, especially the cones (↑) are concentrated. It is directly opposite the lens and provides the main area for acute and accurate daylight vision (↓).

daylight vision vision of great sharpness which takes place in bright light since most of the light entering the eye (p. 160) falls on the fovea (↑).

rod (*n*) one of the rod-shaped light receptors (p. 153) present in the retina (p. 160) which are much more sensitive to light of low intensity but are not sensitive to colour. They are not present in the fovea (↑) and increase in numbers towards the edges of the retina. They are also sensitive to movements.

night vision vision which takes place in light of low intensity using the rods (↑).

aqueous humour a watery fluid that fills the space between the cornea (p. 161) and the vitreous humour (↓) and in which lie the lens (↑) and the iris (↑). It is secreted (p. 106) by glands (p. 87) in the ciliary body (↑).

vitreous humour the jelly-like fluid that fills the space behind the lens (↑).

blind-spot the area of the retina (p. 160) from which the optic nerve (p. 149) leaves the eye. It contains no rods (↑) or cones (↑) so that no image is recorded on this part of the retina.

視錐[a]（名）　視網膜（第160頁）中包含的錐狀光感受器（第153頁）。視錐含有三種不同的色素（第126頁），分別感受紅光、綠光和藍光，所以，視錐主要擔負色覺的功能。視錐主要集中在視網膜正中凹（↓）內及其周圍，而不存在於視網膜邊緣處。視錐還是強光的感受器。

視網膜正中凹（名）　位於視網膜（第160頁）中央的一個小凹陷。大多數感受器（第153頁）細胞，尤其是視錐（↑）皆集中於該處。視網膜正中凹正好對準晶狀體，是提供敏銳而準確的白晝視覺（↓）的主要區域。

白晝視覺、明視覺　在明亮光線照射下，由於進入眼睛（第160頁）的大部份光線落在視網膜正中凹（↑）處而產生極清晰的視覺。

視桿（名）　視網膜（第160頁）中包含的桿狀光感受器。它們特別能感弱光，但不感顏色。視桿在視網膜正中凹（↑）內並不存在，而朝着視網膜邊緣方向，視桿的數量則增多。視桿也易感受運動。

夜間視覺、暗視覺　在弱光下，由視桿（↑）所產生的視覺。

水狀液　充填在角膜（第161頁）和玻璃體液（↓）之間空隙處的一種水樣液體。晶狀體（↑）和虹膜（↑）都位於這種液體之中。水狀液是由睫狀體（↑）內的腺體（第87頁）分泌（第106頁）的。

玻璃體液　充填在晶狀體（↑）後部的空隙內的膠狀液體。

盲點　視網膜（第160頁）上的一個區域，視神經（第149頁）從該區域離開眼睛。盲點處無視桿（↑）和視錐（↑），所以，視網膜的這一部份不顯示圖像。

behaviour (n) all observable activities carried out by an animal in response to its external and internal environment (p. 218).

ethology (n) the study or science of the behaviour (↑) of animals in their natural environment (p. 218).

instinctive behaviour behaviour (↑) which is believed to be controlled by the genes (p. 196) and which is unaffected by experience, e.g. the courtship behaviour of many animals, such as fish and birds, is stimulated by a particular signal provided the animal is sexually mature and has the appropriate level of sex hormones (p. 130) in its body.

innate behaviour behaviour (↑) that does not have to be learned. *See also* instinctive behaviour (↑).

learned behaviour behaviour (↑) in which the response to stimuli is affected by the experience of the individual animal to gain the best advantage from a situation.

habituation (n) a learned behaviour (↑) in which the response to a stimulus is reduced by the constant repetition of the stimulus.

associative learning a learned behaviour (↑) in which the animal learns to associate a stimulus with another that normally produces a reflex action (p. 152). For example, dogs respond to the sight of food by salivating. Pavlov's dogs learned to associate the sight of food with the ringing of a bell and would then salivate on hearing the bell without seeing the food.

imprinting (n) a learned behaviour (↑) which takes place during the very early stages of an animal's life so that e.g. the animal continues to follow the first object on which its attention is fixed by sight, sound, smell or touch. This is usually the animal's parent.

exploration (n) the process through which animals learn about their environment (p. 218) while they are young by play and contact with other animals.

orientation (n) the reflex action (p. 152) in which the animal changes the position of part or the whole of its body in response to a stimulus. For example, an animal might turn its head or prick up its ears in response to a sudden or unusual sound.

releaser (n) a stimulus which releases instinctive behaviour (↑) in an animal.

行為（名） 動物為響應其外部和內部環境（第218頁）而進行的一切可以觀察到的活動。

行為學（名） 研究動物在其自然環境（第218頁）中的行為（↑）的科學。

本能行為 由基因（第196頁）控制、不受經驗影響的行為（↑）。例如，許多動物諸如魚和鳥類的求偶行為就是受特殊信號刺激的，只要動物達到性成熟的階段，且其體內具有適量的性激素（第130頁）即可。

先天行為 並非必須通過學習獲得的行為（↑）。參見"本能行為"（↑）。

習得行為 一種行為（↑）。在這種行為中，動物個體對各種刺激所作出的反應是受其經驗影響的，以便從情境中獲得最大利益。

習慣化（名） 一種習慣行為（↑）。在這種行為中，對某種刺激出的反應，因刺激的不斷重複而減少。

聯合學習、聯想學習 一種習得行為（↑）。在這種行為中，動物學會將一種刺激與另一種通常產生反射動作（第152頁）的刺激聯系起來。例如，狗看到食物的反應是流涎。巴甫洛夫的狗學會將看到食物與鈴聲聯系起來，以致後來聽到鈴聲而未看到食物就會流涎。

印刻、銘記（名） 動物在其生命的最初的階段中所學到的一種習得行為（↑）。例如，動物會繼續追隨其注意力所關注的第一個目標，這種關注是通過視覺、聽覺、嗅覺和觸覺而固定下來的。這第一個目標通常為該動物的雙親。

探究、探索（名） 動物年幼時，藉玩耍和與其他動物接觸來瞭解其環境（第218頁）的過程。

定向（名） 一種反射動作（第152頁）。在這種反射動作中，動物隨刺激而改變其身體的部份位置或整體位置。例如，對突然發出的聲音或異乎尋常的聲音，動物可能會扭轉其頭部或豎起耳朵。

釋放體（名） 使動物的本能行為（↑）得以釋放的一種刺激。

incomplete metamorphosis
e.g. locust
不完全變態 圖例：蝗蟲

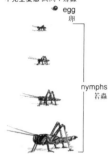

egg
卵

nymphs
若蟲

adult/imago 成蟲

complete metamorphosis
e.g. butterfly
完全變態 圖例：蝴蝶

egg
卵

larva
幼蟲

pupa
蛹

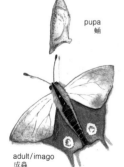

adult/imago
成蟲

growth (*n*) the permanent increase in size and dry mass of an organism which takes place as the cells absorb (p. 81) materials, expand, and then divide. The temporary take-up of water cannot be considered as growth.

growth rate the amount of growth which takes place in a given unit of time.

incomplete metamorphosis the change which takes place from young to adult form in which the young closely resembles the adult.

instar (*n*) a stage through which an insect (p. 69) passes during incomplete metamorphosis (↑).

ecdysis (*n*) the periodic moulting or shedding of the external cuticle (p. 145) allowing growth, which takes place between instars (↑) during incomplete metamorphosis (↑).

nymph (*n*) the early instar (↑) or young of an insect (p. 69) which is small, sexually immature, and unable to fly.

complete metamorphosis the change which takes place from young to adult form in which the young do not resemble the adult and which can occur through a pupal (↓) stage.

larva (*n*) the immature stage in the life cycle of an animal, for example, an insect (p. 69) which undergoes metamorphosis (p. 70). The larva is usually different in structure and appearance from the adult. It hatches from the egg and is able to fend for itself. **larvae** (*pl.*), **larval** (*adj*).

pupa (*n*) the stage between larva (↑) and adult in an insect (p. 69) in which movement and feeding stop and metamorphosis (p. 70) takes place. **pupae** (*pl.*), **pupate** (*v*).

imago (*n*) the sexually mature, adult stage of an insect's (p. 69) development.

corpora allata a pair of glands (p. 87) in the head of an insect (p. 69) which secrete (p. 106) a hormone (p. 130) that encourages the growth of larval (↑) structures and discourages that of adult structures.

ecdysial glands a pair of glands (p. 87) in the head of an insect (p. 69) which secrete (p. 106) a hormone (p. 130) that stimulates ecdysis (↑) and growth.

morphogenesis (*n*) the process in which the overall form of the organs of an organism is developed, leading to the development of the whole organism.

生長（名） 生物體的大小和乾物質的永久性增加。當生物體的細胞吸收（第81頁）各種物質，進行擴大，然後分裂時，即發生生長。暫時性的水份吸收不能視作生長。

生長率 在一定單位時間內所發生的生長量。

不完全變態 從幼蟲形態變成成蟲形態時所產生的一種變化。在這種變化過程中，幼蟲形態與成蟲形態非常相似。

齡（蟲）（名） 昆蟲（第69頁）在不完全變態（↑）中所經歷的一個時期。

蛻皮（名） 昆蟲爲生長而進行的外表皮（第145頁）的周期性脫落。它發生在不完全變態（↑）過程的兩個齡（↑）期之間。

若蟲（名） 較早齡（↑）期昆蟲（第69頁）或昆蟲的幼體。若蟲軀體小，性未成熟，且不會飛。

完全變態 從幼蟲形態變成成蟲形態時所產生的一種變化。在這種變化過程中，幼蟲形態與成蟲形態是不相似的，而且要經歷一個蛹（↓）期。

幼蟲、幼體（名） 指在動物如經歷變態（第70頁）的昆蟲（第69頁）生活史中的未成熟期。幼蟲的結構和外形通常不同於成蟲，它是由卵孵化而來的，且能夠自營生活。（複數形式爲 larvae，形容詞形式爲 larval。）

蛹（名） 昆蟲（第69頁）的幼蟲（↑）和成蟲之間的一個時期。在這個時期，昆蟲停止運動和攝食，發生變態（第70頁）。（複數形式爲 pupae，動詞形式爲pupate）

成蟲（名） 指昆蟲（第69頁）發育中處於性成熟的成體期。

咽側體（名） 昆蟲（第69頁）頭部的一對腺體（第87頁）。它們能分泌（第106頁）一種激素（第130頁），促進幼體（↑）結構生長，阻止形成成體結構。

蛻皮腺 昆蟲（第69頁）頭部的一對腺體（第87頁）。它們能分泌（第106頁）一種刺激蛻皮（↑）和生長的激素（第130頁）。

形態發生（名） 生物體各種器官全面形成的過程。它可導致整個生物體發育。

differentiation (*n*) the process in which cells (unspecialized) change in form and function, during the development of the organism, to give all of the different types of specialized cells that characterize that organism.

neotony (*n*) the retention in some animals of larval (p. 165) or embryonic (↓) features, either temporarily or permanently beyond the stage at which they would normally be lost. It is thought to be important in evolutionary (p. 208) development. For example, humans retain certain resemblances to young apes.

embryology (*n*) the study of embryos (↓).

embryo (*n*) the stage in the development of an organism between the zygote (↓) and hatching, birth, or germination (p. 168). **embryonic** (*adj*).

zygote (*n*) the diploid (p. 36) cell which results from the fusion of a haploid (p. 36) male gamete (p. 175) or spermatozoon (p. 188) and a haploid female gamete or ovum (p. 178 and p. 190).

cleavage (*n*) the process in which the nuclei (p. 13) and cytoplasm (p. 10) of the fertilized (p.175) zygote (↑) divide mitotically (p. 37) to form separate cells. Also known as **segmentation**.

blastula (*n*) the embryonic (↑) structure or mass of small cells which results from cleavage (↑).

blastocoel (*n*) the cavity which occurs in the centre of a blastula (↑) during the final stages of cleavage (↑).

gastrulation (*n*) the process which follows cleavage (↑) in which cell movements occur to form a gastrula (↓) that will eventually lead to the formation of the main organs of the animal. In simple instances, part of the blastula (↑) wall folds inwards to form a hollow gastrula.

gastrula (*n*) the stage in the development of an animal embryo (↑) which comprises a two-layered wall of cells surrounding a cavity known as the archenteron.

ectoderm (*n*) the external germ layer (↓) of an embryo (↑) which develops into hair, various glands (p. 87), CNS, the lining of the mouth etc.

endoderm (*n*) the internal germ layer (↓) of the embryo (↑) which develops into the lining of the gut (p. 98) and its associated organs.

分化（名）　生物個體發育過程中，細胞（非特化的）的形態和功能發生變化，以產生表徵該生物體所有不同類型的特化細胞的過程。

幼態持續、幼期性熟（名）　指某些動物的幼體（第165頁）特徵或胚胎（↓）特徵，在它們通常應該消失的階段之後產生的暫時性或永久性保留現象。幼態持續在進行（第208頁）進程中是重要的。例如，人保留了某些與幼猿相似之處。

胚胎學（名）　關於胚胎（↓）的學科。

胚、胚胎（名）　生物體處在合子（↓）與孵化、出生或發芽（第168頁）之間的發育階段。（形容詞形式為 embryonic）

合子、受精卵（名）　由一個單倍體（第36頁）雄配子（第175頁）或精子（第188頁）和一個單倍體雌配子或卵子（第178頁和第190頁）融合而產生的二倍體（第36頁）細胞。

卵裂（名）　受精（第175頁）合子（↑）的細胞核（第13頁）和細胞質（第10頁）進行有絲分裂（第37頁），形成多個獨立細胞的過程。它也稱為分裂。

囊胚（名）　由卵裂（↑）產生的胚胎（↑）結構或少數細胞構成的團塊。

囊胚腔（名）　在卵裂（↑）的最後階段，囊胚（↑）中央的空腔。

原腸胚形成（名）　卵裂（↑）後發生的一個過程。所謂原腸胚形成，指的是為形成原腸胚（↓）而發生的細胞運動。原腸胚將最終導致形成動物主要器官。在情況較為簡單的例子中，囊胚（↑）壁的一部份向內折疊，形成一個中空的原腸胚。

原腸胚（名）　動物胚胎（↑）發育的一個階段。在該階段，動物胚胎由兩層分別由細胞排列成的壁圍繞着一個稱為原腸的空腔所構成。

外胚層（名）　胚胎（↑）的外部胚層（↓）。它發育成毛髮、各種腺體（第87頁）、CNS 和口腔黏膜上皮等。

內胚層（名）　胚胎（↑）的內部胚層（↓）。它發育成腸（第98頁）及其關聯器官的黏膜上皮。

stages of cleavage in *Amphioxus*
文昌魚卵裂的各個時期

single cell zygote
單細胞
受精卵

2-cell stage
二細胞期

nucleus
細胞核

blastomere
分裂球

8-cell stage
八細胞期

blastula
囊胚

blastula in section
囊胚的切面

blastoderm (one cell thick)
囊胚層（一個細胞的厚度）

blastocoel
囊胚腔

gastrulation 原腸胚形成

1.

gastrula
原腸胚

infolding of blastoderm
囊胚層的折疊

ectoderm
外胚層

2.

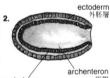

endoderm
內胚層

archenteron
原腸

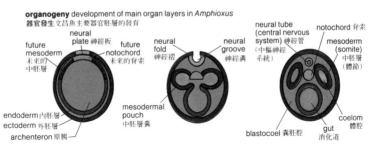

organogeny development of main organ layers in *Amphioxus*
器官發生 文昌魚主要器官胚層的發育

future mesoderm 未來的中胚層

neural plate 神經板

future notochord 未來的脊索

neural fold 神經褶

neural groove 神經溝

neural tube (central nervous system) (中樞神經系統) 神經管

notochord 脊索

mesoderm (somite) 中胚層 (體節)

endoderm 內胚層
ectoderm 外胚層
archenteron 原腸

mesodermal pouch 中胚層囊

blastocoel 囊胚腔

gut 消化道

coelom 體腔

organogeny (*n*) the formation of the organs during growth. **organogenesis** (*n*).

notochord (*n*) the flexible skeletal (p. 145) rod which is present at some stage in the development of all chordates (p. 74). It stretches from the central nervous system (p. 149) to the gut (p. 98) and, in vertebrates (p. 74), it persists as remnants in the backbone throughout the life of the animal although it is present primarily during the development of the embryo (↑).

neural tube the part of the brain (p. 155) and vertebral column (p. 74) which forms first during the growth of the embryo (↑).

mesoderm (*n*) the germ layer (↓) lying between the ectoderm (↑) and endoderm (↑) and which gives rise to connective tissue (p. 88), blood (p. 90), and muscles (p. 143), etc.

germ layer one of the two or three main layers of cells which can be seen in an embryo (↑) after gastrulation (↑). The endoderm (↑), ectoderm (↑) and mesoderm (↑) are all germ layers.

coelom (*n*) a fluid-filled cavity in the mesoderm (↑) of triploblastic (p. 62) animals which, in higher animals, forms the main body cavity in which the gut (p. 98) and other organs are suspended so that their muscular (p. 143) contractions may be independent of those of the body wall.

somite (*n*) any one of the blocks of mesoderm (↑) tissue (p. 83) that flank the notochord (↑) as parallel strips and which develop into blocks of muscle (p. 143), parts of the kidneys (p. 136) and parts of the axial skeleton (p. 145).

myotome (*n*) the part of a somite (↑) that develops into striped muscle (p. 143) tissue (p. 83).

器官發生（名） 指在生長過程中各種器官的形成。亦稱器官形成。

脊索（名） 所有脊索動物（第74頁）某一發育階段具有的堅韌柱狀骨骼（第145頁）。脊索從中樞神經系統（第149頁）延伸到消化道（第98頁）。在脊椎動物（第74頁）中，雖然它主要存在於胚胎（↑）的發育過程中，但是它作為脊柱的殘留物將存在於動物的一生中。

神經管 在胚胎（↑）發育過程中，首先形成的那一部份腦（第155頁）和脊柱（第74頁）。

中胚層（名） 位於外胚層（↑）和內胚層（↑）之間的胚層（↓）。它可形成結締組織（第88頁）、血液（第90頁）和肌肉（第143頁）等。

胚層 在原腸胚形成（↑）之後，可以在胚胎（↑）中見到的兩個或三個主要細胞層。內胚層（↑）、外胚層（↑）和中胚層（↑）都是胚層。

體腔（名） 三胚層（第62頁）動物的中胚層（↑）內的充液空腔。在高等動物中，體腔形成為主體腔，其內懸浮着消化道（第98頁）和其他一些器官，所以，它們的肌肉（第143頁）收縮與體壁的肌肉收縮無關。

體節（名） 構成中胚層（↑）組織（第83頁）的一系列區段中的任何一段。這些區段呈平行條帶狀，位於脊索（↑）兩側，可發育成各種肌肉（第143頁）、腎臟（第136頁）的各個部份和中軸骨骼（第145頁）的各個部份。

肌節（名） 體節（↑）中可發育成橫紋肌（第143頁）組織（第83頁）的部份。

germination (*n*) the first outward sign of the growth of the seeds or spores (p. 178) of a plant which takes place when the conditions of moisture, temperature, light, and oxygen are suitable. **germinate** (*v*).

hydration phase the stage of germination (↑) in which the seed absorbs (p. 81) water and the activity of the cytoplasm (p. 10) begins.

metabolic phase the stage of germination (↑) in which, under enzyme (p. 28) control, the water absorbed (p. 81) during the hydration phase (↑) hydrolyzes (p. 16) the stored food materials into the materials needed for growth.

plumule (*n*) the first apical (↓) leaves and stem which form part of the embryonic (p. 166) shoot of a spermatophyte (p. 57).

radicle (*n*) the first root in an embryonic (p. 166) spermatophyte (p. 57) which later develops into the plant's rooting system.

cotyledon (*n*) the first, simple, leaf-like structure that forms within a seed. Plants, e.g. grasses and cereals, have only one and are called monocotyledons (p. 58) while other flowering plants have two and are called dicotyledons (p.57). Cotyledons contain no chlorophyll (p.12) at first and may function as food reserves for the germinating (↑) plant but, in most dicotyledons, the cotyledons emerge above ground, turn green and photosynthesize (p. 93).

endosperm (*n*) the layer of tissue (p. 83) which surrounds the embryo (p. 166) in some spermatophytes (p. 57). It supplies nourishment to the developing embryo but, in some plants e.g. peas and beans, it has been absorbed (p.81) by the cotyledons (↑) by the time the seed is fully developed while, in others such as wheat, it is not absorbed until the seed germinates (↑).

testa (*n*) a hard, tough, protective coat which surrounds the seed and shields it from mechanical damage or the invasion of fungi (p. 46) and bacteria (p. 42). Also known as **seed coat**.

epicotyl (*n*) the part of the plumule (↑) which lies above the attachment point of the cotyledons (↑).

hypocotyl (*n*) the part of the plumule (↑) which lies below the attachment point of the cotyledons (↑).

萌發、發芽（名）　在濕度、溫度、光照和氧氣等條件皆適當時，植物的種子或孢子（第178頁）表現出向外生長的最初迹象。（動詞形式爲 germinate）

水合期　發芽（↑）的一個階段。此階段中，種子吸收（第81頁）水份，細胞質（第10頁）開始活動。

代謝期　發芽（↑）的一個階段。在這個階段於水合期（↑）吸收（第81頁）的水份，在酶（第28頁）的控制下，將貯存的養料水解（第16頁）成生長所需要的物質。

胚芽（名）　形成種子植物（第57頁）胚（第166頁）枝一部份的最初長出的幾片頂（↓）葉和莖。

胚根（名）　胚性（第166頁）種子植物（第57頁）中的最初長出的根，它後來發育成爲該植物的根系。

子葉（名）　種子內最初形成的簡單葉狀結構。諸如禾本科草類和禾谷類的一些植物，僅有一片子葉，稱爲單子葉植物第58頁）；而其他一些有花植物具有二片葉子，稱爲雙子葉植物（第57頁）。子葉最初不含葉綠素（第12頁），可起到爲發芽（↑）植物貯存養料的用；但是，在大多數雙子葉植物中，子葉長出地面之後，即變成綠色、並進行光合作用（第93頁）。

胚乳（名）　某些種子植物（第57頁）中包圍胚（第166頁）的那層組織（第83頁）。胚乳爲發育中的胚提供養料；但是，在豌豆和其他某些豆類植物中，到種子充分發育時，胚乳已爲子葉（↑）吸收（第81頁），而在如小麥的其他一些植物中，則待種子發芽（↑）時，胚乳才被子葉吸收。

外種皮（名）　包圍種子的一層硬而堅韌的保護性外皮。這層外皮可使種子免於機械損傷或眞菌（第46頁）和細菌（第42頁）的侵入。也稱爲種皮。

上胚軸（名）·　胚芽（↑）中位於子葉（↑）着生點上面的那一部份。

下胚軸（名）　胚芽（↑）中位於子葉（↑）着生點下面的那一部份。

types of seed
種子的類型

cotyledonous 具子葉的
food stored in cotyledons 養料貯存於子葉中
e.g. bean
e.g. 豆

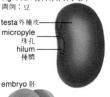

testa 外種皮
micropyle 珠孔
hilum 種臍
embryo 胚

plumule 胚芽
epicotyl 上胚軸
hypocotyl 下胚軸
radicle 胚根

cotyledons 子葉

endospermic 具胚乳的
most food stored in endosperm 大部份養料貯存於胚乳中
e.g. maize
e.g. 玉米

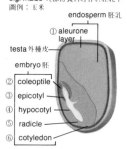

endosperm 胚乳
① aleurone layer
testa 外種皮
embryo 胚
② coleoptile
③ epicotyl
④ hypocotyl
⑤ radicle
⑥ cotyledon

① 糊粉層
② 胚芽鞘
③ 上胚軸
④ 下胚軸
⑤ 胚根
⑥ 子葉

hypogeal germination germination (↑) in which the cotyledons (↑) remain below the surface of the soil, e.g. in broad beans.

epigeal germination germination (↑) in which the cotyledons (↑) emerge above the soil surface and form the first photosynthetic (p. 93) seed leaves e.g. in lettuces.

留土式萌發　子葉（↑）保留在土壤表面以下的發芽（↑），例如蠶豆的發芽。

出土式萌發　子葉（↑）長出土地表面，生成最初幾片光合（第93頁）子葉的發芽（↑），例如萵苣的發芽

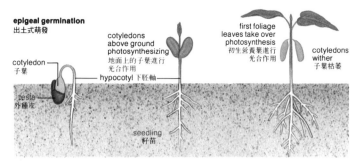

epigeal germination
出土式萌發

cotyledon
子葉

testa
外種皮

cotyledons above ground photosynthesizing
地面上的子葉進行光合作用

hypocotyl 下胚軸

seedling
籽苗

first foliage leaves take over photosynthesis
初生營養葉進行光合作用

cotyledons wither
子葉枯萎

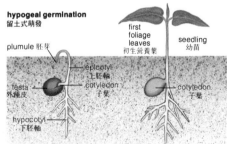

hypogeal germination
留土式萌發

plumule 胚芽

testa
外種皮

epicotyl
上胚軸

cotyledon
子葉

hypocotyl
下胚軸

first foliage leaves
初生營養葉

seedling
幼苗

cotyledon
子葉

meristem (n) the part of an actively growing plant where cells are dividing and new permanent plant tissue (p. 83) is formed.

apical meristem a meristem (↑) which occurs at shoot and root tips. The division of these cells which are small and contain granular cytoplasm (p. 10) with small vacuoles (p. 11), results in the growth of stems and roots.

apex (n) the top or pointed end of an object.
　apical (adj).

分生組織（名）　植物體中持續進行生長的部份，在該處，細胞不斷分裂，形成新的永久性植物組織（第83頁）。

頂端分生組織　苗端和根尖的一種分生組織（↑）。該組織中含有顆粒狀細胞質（第10頁）和小液泡（第11頁）的小細胞分裂，可導致莖和根的生長。

頂端、尖端（名）　指物體的頂端或尖端。（形容詞形式爲 apical）

lateral meristem a meristem (p. 169), including the vascular cambium (p. 86) and phellogen (p. 172), which occurs along the roots and stems of dicotyledons (p. 57) plants and which is made up of long, thin cells that give rise to xylem (p. 84) and phloem (p. 84).

lateral (adj) on, at or about the side of something.

ground meristem the part of the apical meristem (p. 169) from which pith (p. 86), cortex (p. 86), medullary rays (p. 86) and mesophyll (p. 86) are formed.

tunica (n) one of the two layers of tissue (p. 83) comprising the apical meristem (p. 169). It is the outer layer of tissue and may itself be made up of one or more layers of cells in which the division takes place at right angles to the surface of the plant (anticlinally).

corpus (n) one of the two layers of tissue (p. 83) comprising the apical meristem (p. 169). It is the inner layer of tissue and the division of cells occurs irregularly.

zone of cell division the part of the apex (p. 169) of a root or shoot which includes the apical meristem (p. 169) and the leaf primordium (↓) or the root cap (p. 81).

zone of expansion the part of the apex (p. 169)· of a root or shoot which lies behind the zone of cell division (↑) and in which the cells elongate and expand.

zone of differentiation the part of the apex (p. 169) of a root or shoot which lies behind the zone of expansion (↑) and in which the cells differentiate into the form and function of parts of the plant that characterize it.

primordium (n) the group of cells in the apex of a shoot or root which differentiates into a leaf etc. **primordia** (pl.).

primary growth the growth which takes place only in the meristems (p. 169) which were present in the embryo (p. 166). These include the apical meristems (p. 169) and primary growth results largely in the increase in length.

secondary growth the growth which takes place in the lateral meristems (↑) and which results in increase in girth rather than in length.

側生分生組織　包括維管形成層（第86頁）和木栓形成層（第172頁）在內的一種分生組織（第169頁）。這種分生組織縱貫於雙子葉植物（第57頁）的根和莖，由產生木質部（第84頁）和韌皮部（第84頁）的細長細胞組成。

側的、側面的（形）　在某物之側，側面之旁的或側面附近的。

基本分生組織　頂端分生組織（第169頁）中構成髓（第86頁）、皮膚（第86頁）、髓射線（第86頁）和葉肉（第86頁）的那一部份。

原套、外囊（名）　構成頂端分生組織（第169頁）的兩層組織（第83頁）之一。原套是組織的外層，本身可由一層或多層細胞構成，這些細胞的分裂，是在與植物表面垂直的方向進行的（即為垂周分裂）。

原體、內體（名）　構成頂端分生組織（第169頁）的兩層組織（第83頁）之一。原體是組織的內層，其細胞分裂不規則。

細胞分裂區　指根端（第169頁）或苗端中包括頂端分生組織（第169頁）和葉原基（↓）或根冠（第81頁）在內的那一部份。

伸長區　指根端（第169頁）或苗端中位於細胞分裂區（↑）後面的那一部份。細胞在這一部份內伸長、增大。

分化區　指根端（第169頁）或苗端中位於伸長區（↑）後面的那一部份。細胞在這一部份內分化成表徵植物一定形態和功能的各組成部份。

原基（名）　在苗端或根端，可分化成葉子等的那一羣細胞。（複數形式為 primordia）

初生生長　指僅發生胚（第166頁）的各分生組織（第169頁）的生長。這些分生組織包括頂端分生組織（第169頁）在內。初生生長主要是增加長度。

次生生長　指發生於各側生分生組織（↑）的生長。這種生長增加圍長而增加長度。

meristems of shoot 苗和根的分生組織 **and root**

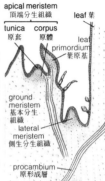

apical meristem
頂端分生組織

tunica　corpus
原套　原體

leaf 葉

leaf primordium
葉原基

ground meristem
基本分生組織

lateral meristem
側生分生組織

procambium
原形成層

shoot tip 苗尖

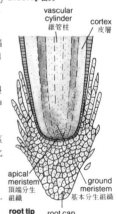

vascular cylinder
維管柱

cortex
皮層

apical meristem
頂端分生組織

ground meristem
基本分生組織

root tip
根尖

root cap
根冠

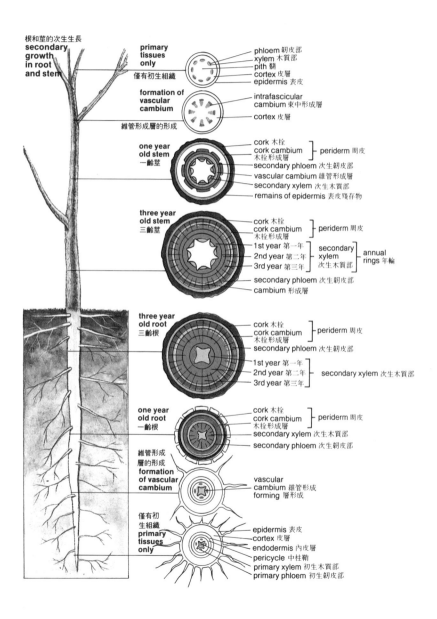

根和莖的次生生長
secondary growth in root and stem

primary tissues only
僅有初生組織
- phloem 韌皮部
- xylem 木質部
- pith 髓
- cortex 皮層
- epidermis 表皮

formation of vascular cambium
維管形成層的形成
- intrafascicular cambium 束中形成層
- cortex 皮層

one year old stem
一齡莖
- cork 木栓 ┐
- cork cambium 木栓形成層 ┘ periderm 周皮
- secondary phloem 次生韌皮部
- vascular cambium 維管形成層
- secondary xylem 次生木質部
- remains of epidermis 表皮殘存物

three year old stem
三齡莖
- cork 木栓 ┐
- cork cambium 木栓形成層 ┘ periderm 周皮
- 1st year 第一年 ┐
- 2nd year 第二年 │ secondary xylem 次生木質部 ┐ annual rings 年輪
- 3rd year 第三年 ┘ │
- secondary phloem 次生韌皮部
- cambium 形成層

three year old root
三齡根
- cork 木栓 ┐
- cork cambium 木栓形成層 ┘ periderm 周皮
- secondary phloem 次生韌皮部
- 1st year 第一年 ┐
- 2nd year 第二年 │ secondary xylem 次生木質部
- 3rd year 第三年 ┘

one year old root
一齡根
- cork 木栓 ┐
- cork cambium 木栓形成層 ┘ periderm 周皮
- secondary xylem 次生木質部
- secondary phloem 次生韌皮部

維管形成層的形成
formation of vascular cambium
- vascular cambium forming 維管形成層形成

僅有初生組織
primary tissues only
- epidermis 表皮
- cortex 皮層
- endodermis 內皮層
- pericycle 中柱鞘
- primary xylem 初生木質部
- primary phloem 初生韌皮部

fascicular (*adj*) of meristematic (p. 169) cambium (p. 86) between the xylem (p. 84) and phloem (p. 84) in vascular tissue (p. 83).

interfascicular (*adj*) of meristematic (p. 169) cambium (p. 86) that consists of a single layer of actively dividing cells between the phloem (p. 84) and xylem (p. 84) bundles in stems.

secondary xylem xylem (p. 84) that has been formed by the vascular cambium (p. 86) following the formation of the primary tissues (p. 83).

secondary phloem phloem (p. 84) that has been produced by the cambium (p. 86) following the formation of the primary tissues (p. 83).

annual rings the rings of lighter and darker wood that can be seen in a cross-section of the trunk of a tree living in temperate conditions. Each pair marks the annual increase in the girth of the tree as a result of activity of the cambium (p. 86). The lighter ring is large-celled xylem (p. 84) tissue (p. 83) produced in spring and the darker wood is smaller-celled summer wood.

phellogen (*n*) a layer of cells immediately beneath the epidermis (p. 131) of the stems undergoing secondary growth. It is a lateral meristem (p. 170) whose cells give rise to the phellem (↓) and phelloderm (↓). Also known as **cork cambium.**

bark (*n*) the protective outer layer of the stems of woody plants. It may consist of cork cells only or alternating layers of cork (↓) and dead phloem (p. 84).

cork (*n*) = phellem (↓).

phellem (*n*) an outer layer of dead, waterproof cells formed from the activity of the phellogen (↑) on the stems of woody plants.

phelloderm (*n*) the inner layer of the bark (↑) produced by the activity of the phellogen (↑).

suberin (*n*) a mixture of substances derived from fatty acids (p. 20) and present in the walls of phellem (↑) cells rendering these cells waterproof.

exogenous (*adj*) of branching generated on the outside of the plant.

endogenous (*adj*) of branching generated on the inside of the plant.

維管束中的(形)　指維管束組織(第83頁)木質部(第84頁)和韌皮部(第84頁)間的分生組織(第169頁)形成層(第86頁)而言。

維管束間的(形)　指分生組織(第169頁)形成層(第86頁)而言，它由莖內韌皮部(第84頁)和木質部(第84頁)束之間不斷進行分裂的單層細胞組成。

次生木質部　繼初生組織(第83頁)形成後，由維管形成層(第86頁)形成的木質部(第84頁)。

次生韌皮部　繼初生組織(第83頁)形成之後，由形成層(第86頁)產生的韌皮部(第84頁)。

年輪　在生長於溫帶氣候條件下的樹木莖幹橫切面上，可以看到的許多淡色和深色的木質輪紋。每一對輪紋表示一年中由於形成層(第86頁)活動結果而增加的樹幹圍長。淡色輪紋是在春季產生的細胞較大木質部(第84頁)組織(第83頁)；深色輪紋是細胞較小的晚材。

木栓生長帶(名)　緊靠莖表皮(第131頁)下方、進行次生生長的一層細胞。它是一種側生分生組織(第170頁)，其細胞可產生木栓(↓)和栓內層(↓)。亦稱是**木栓形成層**。

樹皮(名)　木本植物莖幹的保護外層。它可以僅僅由木栓細胞構成，或者由木栓(↓)的替代層和死的韌皮部(第84頁)構成。

木栓(名)　同木栓層(↓)。

木栓層(名)　由於木本植物莖幹上的木栓形成層(↑)活動而形成的一層處於體表的，死的防水細胞。

栓內層(名)　由木栓形成層(↑)活動產生的樹皮(↑)內層。

木栓質(名)　一種由脂肪酸(第20頁)衍生而來、由眾多物質組成的混合物。木栓質存在於木栓(↑)細胞壁中，可使木栓細胞防水。

外生的(形)　指在植物外面產生的分枝而言。

內生的(形)　指在植物內面產生的分枝而言。

annual rings 年輪

sapwood (still conducts water) 邊材(仍然能輸導水)

annual rings 年輪

heartwood (non-functional) 心材(無輸導功能)

bark 樹皮

reproduction (*n*) the means whereby organisms ensure the continued existence of the species (p. 40) beyond the life span of an individual by generating new individuals.

sexual reproduction the generation of new individuals of an organism to continue the life of the species (p. 40) by fusion of haploid (p. 36) nuclei (p. 13) or gametes (p. 175) to form a zygote (p. 166). In most animals a highly motile male spermatozoon (p. 188), generated by the testes (p. 187) and produced in large numbers, unites with a non-motile female ovum (p. 190) produced in small numbers in the ovary (p. 189).

asexual reproduction the generation of new individuals of an organism to continue the life of the species (p. 40) from a single parent by such means as budding (↓) or sporulation (↓). Multiplication is rapid and the offspring are genetically (p. 191) identical to one another and to the parent. For example, binary fission (p. 44) can occur very rapidly and is exponential so that one cell divides into two, two into four, four into eight, and so on, and all the cells are identical to the parent.

motile (*adj*) able to move. **motility** (*n*).

non-motile (*adj*) unable to move.

budding (*n*) asexual reproduction (↑), typical of corals (p. 61) and sponges, in which the parent produces an outgrowth or bud which develops into a new individual.

fragmentation (*n*) asexual reproduction (↑), occurring only in simple organisms such as sponges and green algae (p. 44), in which the parent fragments or breaks up and each fragment develops into a new individual.

sporulation (*n*) asexual reproduction (↑), typical of fungi (p. 46), in which the parent produces often large numbers of small, usually lightweight, single-celled structures, or spores (p. 178), which detach from the parent and may be widely distributed by wind or other mechanisms. Provided they fall into suitable conditions, each spore germinates (p. 168) to produce a new individual.

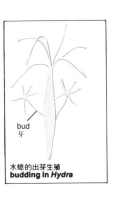

bud
芽

水螅的出芽生殖
budding in Hydra

生殖、繁殖（名）　生物在一個個體生命期限之外，藉產生新的個體確保物種（第40頁）延續的方式。

有性生殖　藉單倍體（第36頁）細胞核（第13頁）或配子（第175頁）融合，形成合子（第166頁），產生生物的新個體，以延續物種（第40頁）的生殖方式。大多數動物的精巢（第187頁）產生大量具高度游動性的雄精子（第188頁）與卵巢（第189頁）產生的少量非游動性的雌性卵（第190頁）相結合。

無性生殖　藉芽殖（↓）或孢子形成（↓）之類手段，由單個親體產生生物的新個體，以延續物種（第40頁）的生殖方式。這種方式增殖迅速，後代的遺傳性（第191頁）在相互之間以及與親體之間，均是相同的。例如，二分裂（第44頁）可以很快地進行，呈指數增殖，結果一個細胞分裂為兩個細胞，兩個細胞分裂為四個細胞，四個細胞則分裂為八個細胞，等等。所有的細胞都與親體相同。

游動的（形）　能夠運動的。（名詞形式為 motility）

非游動的（形）　不能運動的。

出芽生殖（名）　一種無性生殖（↑）方式。在這種生殖方式中，親體產生能發育成新的體的旁枝或芽。珊瑚（第61頁）和海綿的生殖是芽殖的代表性例子。

斷裂生殖（名）　一種無性生殖（↑）方式。在這種生殖方式中，親體分裂或斷裂，每一個裂片則發育成一個新個體。裂殖僅發生在諸如海綿和綠藻（第44頁）這些簡單生物中。

孢子形成　一種無性生殖（↑）方式。在這種生殖方式中，親體常產生大量小而輕單細胞結構或孢子（第178頁）；這些單細胞結構或孢子離開親體後，可以由風其他媒介廣泛散佈。如果它們落入適當的環境中，則每一個孢子即可萌發（第168頁）成為一個新個體。孢子形成的典型（例子是真菌（第46頁）的生殖。

vegetative propagation asexual reproduction (p. 173), occurring in plants, in which part of the plant, such as a leaf or even a specially produced shoot, detaches from the plant and develops into a new individual.

perennating organ any special structure, found in some biennial (p. 58) and perennial (p. 58) plants, which enables them to survive hostile conditions, such as drought. If more than one structure is produced, asexual reproduction (p. 173) is also achieved. When the conditions deteriorate, the plant dies back, leaving only the perennating organ which, with the onset of suitable conditions, develops into a new individual or individuals.

runner (n) an organ in the form of a stem that develops from an axillary (p. 83) bud, runs along the ground, and produces new individuals at its axillary buds or at the terminal bud only.

stolon (n) an organ which takes the form of a long, erect branch that eventually bends over, under its own weight, so that the tip touches the ground and roots. At the axillary (p. 83) bud a new shoot grows into a new individual.

rhizome (n) a perennating organ (↑) in which the stem of the plant remains below ground and continues to grow horizontally.

bulb (n) a perennating organ (↑) which takes the form of an underground condensed shoot with a short stem and fleshy leaves that are close together, overlap and form the food store for the plant. At each leaf base is a bud which can grow into a new bulb. In the growing season, the plant produces leaves and flowers, and exhausts the food store of the bulb. But the new plant makes more food material by photosynthesis (p. 93) and a new bulb (or bulbs) is formed at the leaf base bud.

corm (n) a perennating organ (↑) which takes the form of a special enlarged, fleshy, underground stem that acts as a food store.

tuber (n) a perennating organ (↑) similar to a rhizome (↑) in which food is stored in the underground stem or root as swellings which eventually detach from the parent plant.

營養繁殖　植物的一種無性生殖(第173頁)方式，在這種生殖方式中，植物的組成部份，例如一片葉子，甚至一根爲某種特定功能而生長出來的枝條，可以脫離植物而發育成一個新個體。

多年生器官　存在於某些二年生(第58頁)和多年生(第58頁)植物中的任何特殊結構。這種結構使它們能在乾旱的惡劣環境下生存下來。如果能產生一個以上的這種結構，則還可以實現無性生殖(第173頁)。當環境惡化時，植物枯萎，僅留下此多年生器官，在環境合適時，該器官即發育成一個新的個體或多個新個體。

長匍莖(名)　一種呈莖狀的器官。這種器官由腋(第83頁)芽發育而成，沿着地面匍匐生長，僅在其腋芽或頂芽處產生新個體。

匍匐莖(名)　一種呈直立長枝形態的器官。這種直立長枝受其自重的作用而彎下來，以致頂尖觸及地面和根。在腋(第83頁)芽處，新枝可長成新個體。

根莖(名)　植物的莖保留在地下，並能繼續水平生長的一種多年生器官(↑)。

鱗莖(名)　一種呈短縮地下莖形態的多年器官(↑)。其莖較短，着生有許多肉質葉，它們緊密重疊在一起，形成植物的養料貯存所。每一葉基都是一個可以長成新鱗莖的芽。在生長季節，植物長葉、開花，耗盡鱗莖中貯存的養料。但是，新的植物藉光合作用(第93頁)製造更多的養料，並在葉基芽處形成一個新的鱗莖(或多個新的鱗莖)。

球莖(名)　一種特別肥大的、肉質地下莖形態的多年生器官(↑)。這種地下莖起養料貯存所的作用。

塊莖(名)　一種似根莖的多年生器官(↑)。在塊莖中，養料貯存在最終可脫離親本植株的膨大地下莖或根內。

vegetative propagation 營養繁殖

bulb 鱗莖　new shoot arising from leaf axil in bulb 從鱗莖葉腋處長出的新枝

thick fleshy leaves for food storage 貯存養料的密集肉質葉

stem 莖

corm 球莖　new shoot, arising from leaf axil 從葉腋處長出的新枝

old leaf bases 老葉基

rhizome 根莖　aerial shoots 地上枝

bud 芽

rhizome 根狀莖　adventitious roots 不定根

direction of growth 生長方向

stolon 匍匐莖　flowering shoots 花枝

stolon 匍匐莖　adventitious roots 不定根

direction of growth 生長方向

runner 長匍莖

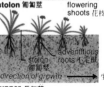

1 young plant grows at apex of runner
1. 幼苗在長匍莖的頂端

runner 長匍莖

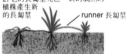

2 old runner dies, mature new plant produces new runner
2. 老的長匍莖死亡，新的成熟的植株產生新的長匍莖

runner 長匍莖

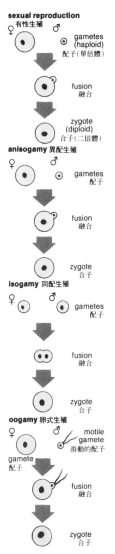

sexual reproduction
有性生殖

gametes (haploid)
配子（單倍體）

fusion
融合

zygote (diploid)
合子（二倍體）

anisogamy 異配生殖

gametes
配子

fusion
融合

zygote
合子

isogamy 同配生殖

gametes
配子

fusion
融合

zygote
合子

oogamy 卵式生殖

motile gamete
游動的配子

gamete
配子

fusion
融合

zygote
合子

apomixis (*n*) asexual reproduction (p. 173) which superficially resembles sexual reproduction (p. 173) although fertilization (↓) and meiosis (p. 38) do not occur. **apomictic** (*adj*).

fertilization (*n*) the process in sexual reproduction (p. 173) in which the nucleus (p. 13) of a haploid (p. 36) male gamete (↓) fuses with the nucleus of a haploid female gamete to form a diploid (p. 36) zygote (p. 166).

fertile (*adj*) of organisms able to produce young.

mature (*adj*) fully grown, fully developed.

immature (*adj*) not mature (↑).

gamete (*n*) a reproductive (p. 173) or sex cell. Each gamete is haploid (p. 36) and male gametes, or spermatozoons (p. 188) which are small and highly motile, fuse with larger female gametes or ova (p. 190) in the process of fertilization (↑) to form diploid (p. 36) zygotes (p. 166) which are capable of developing into new individuals.

gametangium (*n*) an organ which produces gametes (↑). **gametangia** (*pl.*).

syngamy (*n*) the actual fusion of two gametes (↑) which occurs during fertilization (↑).

isogametes (*n.pl.*) gametes (↑), produced by some organisms, e.g. some fungi (p. 46), which are not differentiated into male or female forms. All the gametes produced by the organism are similar.

anisogametes (*n.pl.*) gametes (↑) which are differentiated in some way either simply by size or by size and form.

heterogametes (*n.pl.*) anisogametes (↑) which are differentiated by size and form into small, highly motile male spermatozoons (p. 188), which are produced in large numbers, and large non-motile female ova (p. 190).

oogamy (*n*) fertilization (↑) which takes place by the union of heterogametes (↑).

dioecious (*adj*) of organisms in which the sex organs are borne on separate individuals which are themselves then described as either males or females.

monoecious (*adj*) of organisms in which the male and female sex organs are borne on the same individual.

hermaphrodite (*n, adj*) = monoecious (↑).

無融合生殖（名）　雖然不發生受精（↓）和減數分裂（第38頁），但表面上與有性生殖（第173頁）相似的一種無性生殖（第173頁）。（形容詞形爲 apomictic）

受精（作用）　指有性生殖（第173頁）的過程，在這種過程中，一個單倍體（第36頁）雄配子（↓）的核（第13頁）與一個單倍體雌配子的核融合，形成一個二倍體（第36頁）合子（第166頁）。

能育的（形）　指生物體能夠生育後代而言。

成熟的（形）　完全長成的、充分發育的。

未成熟的（形）　不成熟的（↑）。

配子（名）　一種生殖（第173頁）細胞或性細胞。每一個配子均是單倍體（第36頁）；小而游動性強的雄配子或精子（第188頁），在受精（↑）過程中，與大的雄配子或卵（第190頁）融合，形成能夠發育成新個體的二倍體（第36頁）合子（第166頁）。

配子囊（名）　可產生配子（↑）的一種器官。（複數形式爲 gametangia）

配子配合（名）　在受精（↑）過程中發生的兩個配子（↑）的有效融合。

同形配子（名、複）　某些生物如某些眞菌（第46頁）所產生的、不分化成雄性或雌性類型的配子（↑）。該生物產生的所有配子均是相同的。

異形配子（名、複）　僅僅在大小或大小和形的態某一方面可加以區分的配子（↑）。

異形配子（名、複）　專指可根據其大小和形態而區分爲小而游動性強的大量雄性精子（第188頁）以及大而非游動性的雌性卵（第190頁）的那些異形配子（↑）。

卵式生殖（名）　由異形配子（↑）結合進行的受精（↑）。

雌雄異體的（形）　指一些生物體而言，在這些生物體中，兩種性器官分別生長在各個獨立的個體上，這些獨立的個體或者被稱爲雄性個體或者被稱爲雌性個體。

雌雄同體的（形）　指一些生物體而言，在這些生物體中，雄性性器官和雌性性器官均生長在同一個體上。

兩性同體（名、形）　同雌雄同體（↑）。

parthenogenesis (*n*) reproduction (p. 173), occurring in some plants and animals such as the dandelion or aphids, in which the female gametes (p. 175) develop into new individuals without having been fertilized (p. 175). The offspring of parthenogenesis are always female and usually genetically (p. 196) identical with the parent and with one another. If the ovum (p. 178 and p. 190) has been produced by meiosis (p. 38), the offspring are haploid (p. 36) while, if it has been produced by mitosis (p. 37) the offspring are diploid (p. 36).

alternation of generations a life cycle of an organism in which reproduction (p. 173) alternates with each generation between sexual reproduction (p. 173) and asexual reproduction (p. 173). It is found, for example, among Coelenterata (p. 60) which have both a polyp (p. 61) and a medusa (p. 61) stage and among bryophytes (p. 52) in which haploid (p. 36) gametes (p. 175) from one stage – gametophyte (↓) – fuse to form a diploid (p. 36) zygote (p. 166) which germinates (p. 168) to form a sporophyte (↓) that, in turn, produces haploid spores (p. 178) by meiosis (p. 38). These develop into a haploid plant body (gametophyte). Each of the generations may be quite different in form.

generation (*n*) a set of individuals of the same stage of development or age, or the time taken for one individual to reproduce (p. 173) and for the progeny (p. 200) to develop to the same stage as the parent.

haplontic (*adj*) of a life cycle, found in some algae (p. 44) and fungi (p. 46), in which a haploid (p. 36) adult form occurs by meiosis (p. 38) of the diploid (p. 36) zygote (p. 166).

diplontic (*adj*) of a life cycle, found in all animals, as well as some algae (p. 44) and fungi (p. 46), in which haploid (p. 36) gametes (p. 175) are produced by meiosis (p. 38) from the diploid (p. 36) adults.

diplohaplontic (*adj*) of a life cycle, found in most plants, in which a diploid (p. 36) sporophyte (↓) generation alternates with a haploid (p. 36) gametophyte (↓) generation.

單性生殖、孤雌生殖(名) 蒲公英或蚜蟲之類的某些植物或動物所具的生殖(第173頁)方式。在這種生殖方式中,雌配子(第175頁)不經受精(第175頁)而發育成新的個體。單性生殖產生的後代總是雌性的,通常在遺傳性(第196頁)上與親體相同,相互之間也是相同的。如果卵(第178頁和第190頁)是減數分裂(第38頁)產生的,則後代為單倍體(第36頁);如果卵是有絲分裂(第37頁)產生的,則後代為二倍體(第36頁)。

世代交替 生物的一種生活史。在這種生活史中,每一世代中的生殖(第173頁)均在有性生殖(第173頁)和無性生殖(第173頁)之間交替進行。例如,在腔腸動物(第60頁)和苔蘚植物(第52頁)中,就出現這種世代交替的現象。前者具有水螅型(第61頁)和水母型(第61頁)兩個階段;在後者中,由配子體階段產生的單倍體(第36頁)配子(第175頁)融合形成二倍體(第36頁)合子(第166頁),這種合子萌發(第168頁)形成孢子體(↓),進而藉減數分裂(第38頁)又產生單倍體孢子(第178頁),這些孢子再發育成單倍體植物(配子體)。每個世代在形態上可完全不同。

世代、代(名) 指一罿處於同一發育階段或同一年齡的個體;也指一個個體用於生殖(第173頁)和其子代(第200頁)用於發育到與親代同一發育階段所需要的時間。

單倍性生物的(形) 指某些藻類(第44頁)和眞菌(第46頁)所具的一種生活史而言,在這種生活史中,單倍體(第36頁)成體型是經由二倍體(第36頁)合子(第166頁)減數分裂(第38頁)產生的。

二倍性生物的(形) 指所有動物以及某些藻類(第44頁)和眞菌(第46頁)所具的一種生活史而言,在這種生活史中,單倍體(第36頁)配子(第175頁)由二倍體(第36頁)成體減數分裂(第38頁)產生的。

雙單倍體的(形) 指大多數植物所具的一種生活史而言,在這種生活史中,二倍體(第36頁)孢子體(↓)世代與單倍體(第36頁)配子體(↓)世代交替出現。

**alternation of generations
and the major plant
divisions**
世代交替和主要的植物區分

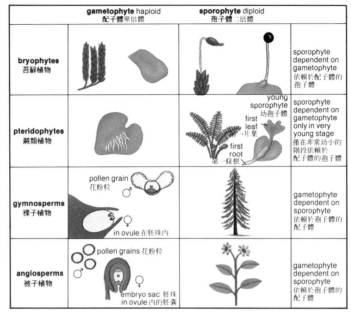

	gametophyte haploid 配子體單倍體		sporophyte diploid 孢子體二倍體		
bryophytes 苔蘚植物					sporophyte dependent on gametophyte 依賴於配子體的 孢子體
pteridophytes 蕨類植物			young sporophyte 幼孢子體 first leaf 第一片葉 first root 第一條根		sporophyte dependent on gametophyte only in very young stage 僅在非常幼小的 階段依賴於 配子體的孢子體
gymnosperms 裸子植物	pollen grain 花粉粒 ♂ ♀ in ovule 在胚珠內				gametophyte dependent on sporophyte 依賴於孢子體的 配子體
angiosperms 被子植物	pollen grains 花粉粒 ♂ ♀ embryo sac 胚珠 in ovule 內的胚囊				gametophyte dependent on sporophyte 依賴於孢子體的 配子體

sporophyte(*n*) the stage of an alternation
of generations (↑), found in most plants, in
which the diploid (p. 36) plant produces spores
(p. 178) by meiosis (p. 38) which then germinate
(p. 168) to produce the gametophyte (↓).

gametophyte (*n*) the stage of an alternation of
generations (↑), found in most plants, in which
the haploid (p. 36) plant produces gametes (p. 175)
by mitosis (p. 37) which fuse to form a zygote
(p. 166) that develops into the sporophyte (↑).

archegonium (*n*) the female sex organ of
Hepaticae (liverworts) (p. 52), Musci (mosses)
(p. 52), Filicales (ferns) (p. 56) and most
gymnosperms (p. 57). It is a multicellular (p. 9)
structure which is shaped like a flask with a
narrow neck and a swollen base which contains
the female gamete (p. 175).

oosphere (*n*) the large, unprotected, non-motile
female gamete (p.175) found in an archegonium (↑).

孢子體（名） 大多數植物中的一種世代交替（↑）階
段，在這種階段中，二倍體（第36頁）植物經由
減數分裂（第38頁）產生孢子（第178頁），然後
孢子萌發（第168頁）產生配子體（↓）。

配子體（名） 大多數植物中的一種世代交替（↑）階
段，在這種階段中，單倍體（第36頁）植物經由
有絲分裂（第37頁）產生配子（第175頁），配子
則融合而成合子（第166頁），合子再發育成孢
子體（↑）。

頸卵器（名） 苔綱（苔類植物）（第52頁）、蘚綱（蘚
類植物）（第52頁）、眞蕨綱（蕨類植物）（第56
頁）和大多數裸子植物（第57頁）的雌性性器
官。頸卵器是一種形狀類似於燒瓶的多細胞
（第9頁）結構，其頸部狹窄，底部膨大，內含
雌配子（第175頁）。

卵球（名） 頸卵器（↑）中所具的大而無保護的非游
動性雌配子。（第175頁）。

ovum[p] (*n*) the haploid (p.36) female gamete (p.175).

antheridium (*n*) the male sex organ of algae (p. 44), liverworts (p. 52), mosses (p. 52), ferns (p. 56) and fungi (p. 46). It may be unicellular (p. 9) or multicellular (p. 9) and produces small, motile gametes (p. 175) – the antherozoids (↓). **antheridia** (*pl.*).

antherozoid (*n*) the male gamete (p. 175) produced within the antheridium (↑).

spermatozoid (*n*) = antherozoid (↑).

sporogonium (*n*) the sporophyte (p. 177) generation of mosses (p. 52) and liverworts (p. 52) which produces the seed and parasitizes (p. 110) the gametophyte (p. 177) generation.

sporangium (*n*) (1) the organ in which, in the sporophyte (p.177) generation, the haploid (p. 36) spores (↓) are formed after meiotic (p. 38) division of the spore mother cells (↓): (2) in fungi (p. 46) a swelling occurring at the ends of certain hyphae (p. 46) in which protoplasm (p. 10) breaks up to form spores during asexual reproduction (p. 173). **sporangia** (*pl.*).

spore (*n*) a tiny, asexual, unicellular (p. 9) or multicellular (p. 9) reproductive (p. 173) body which is produced in vast numbers by fungi (p. 46) or the sporangia (↑) of plants.

spore mother cell a diploid (p. 36) cell that gives rise to four haploid (p. 36) cells by meiosis (p. 38). Also known as **sporocyte**.

microsporangium (*n*) a sporangium (↑) present in heterosporous (p. 54) plants, which produces and disperses the microspores (↓).

microspore (*n*) the smaller of the two different kinds of spore (↑) produced by ferns (p. 56) and spermatophytes (p. 57) and which gives rise to the male gametophyte (p. 177) generation.

microsporophyll (*n*) a modified leaf that bears the microsporangium (↑).

megasporangium (*n*) a sporangium (↑) present in heterosporous (p. 54) plants, which produces and disperses the megaspores (↓).

megaspore (*n*) the larger of the two different kinds of spore (↑) produced by ferns (p. 56) and spermatophytes (p. 57) and which gives rise to the female gametophyte (p. 177) generation.

卵[P]（名） 單倍體（第36頁）雌配子（第175頁）。

精子器、藏精器（名） 藻類（第44頁）、苔類植物（第52頁）、蘚類植物（第52頁）、蕨類植物（第56頁）和眞菌（第46頁）的雄性性器官。它可以是單細胞的（第9頁），也可以是多細胞的（第9頁）；可產生小而能游動的配子（第175頁），即游動精子（↓）。（複數形式爲 antherdia）

游動精子（名） 精子器（↑）內產生的雄配子（第175頁）。

游走精子（名） 同游動精子（↑）。

孢子體（名） 指蘚類植物（第52頁）和苔類植物（第52頁）的孢子體（第177頁）世代的孢子體。它可產生種子並寄生（第110頁）於配子體（第177頁）世代。

孢子囊（名） （1）在孢子體（第177頁）世代中，於孢子母細胞（↓）減數分裂（第38頁）以後，形成單倍體（第36頁）孢子（↓）的器官；（2）眞菌（第46頁）的某些菌絲（第46頁）端部上的一個膨大物，在無性生殖（第173頁）過程中，原生質（第10頁）在其內分解，形成孢子。（複數形式爲 sporangia）

孢子（名） 眞菌（第46頁）或植物孢子囊（↑）產生的大量微小的無性單細胞（第9頁）或多細胞（第9頁）繁殖（第173頁）體。

孢子母細胞 一種經減數分裂（第38頁）可產生四個單倍體（第36頁）細胞的二倍體（第36頁）細胞。也稱爲孢子、孢子被。

小孢子囊（名） 具異型孢子的（第54頁）植物中所具的一種孢子囊（↑）。它能產生和散播小孢子（↓）。

小孢子（名） 蕨類植物（第56頁）和種子植物（第57頁）產生的兩種不同類型孢子（↑）中較小的一種。它可產生雄配子體（第177頁）世代。

小孢子葉（名） 生有小孢子囊（↑）的一種變態葉。

大孢子囊（名） 具異形孢子的（第54頁）植物所具的一種孢子囊（↑）。它能產生和散播大孢子（↓）。

大孢子（名） 蕨類植物（第56頁）和種子植物（第57頁）產生的兩種不同類型孢子（↑）中較大的一種。它可產生雌配子體（第177頁）世代。

heterosporous plants 具異形孢子的植物
the production of megaspores 大孢子的產生
♀ gametophyte develops into old megaspore wall 配子體在老的大孢子壁內發育
megaspores 大孢子
megasporophyll 大孢子葉
megasporangium 大孢子囊
clubmoss 石松
megaspore develops into ♀ gametophyte with archegonia 大孢子發育成具有頸卵器的配子體
megasporangium 大孢子囊
megaspore 大孢子
megasporophyll 大孢子葉
conifer L.S. part of a ♀ cone 針葉樹球花的縱切面

megasporophyll (carpel) 大孢子葉（心皮）
ovule 胚珠
被子植物 angiosperm
megaspore develops into ♀ gametophyte (embryo sac) 大孢子發育成配子體（胚囊）
megaspore 大孢子

young ovule 幼胚珠

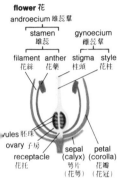

flower 花
androecium 雄蕊羣
stamen 雄蕊
gynoecium 雌蕊羣
filament 花絲　anther 花藥　stigma 柱頭　style 花柱
vules 胚珠
ovary 子房
receptacle 花托
sepal (calyx) 萼片（花萼）
petal (corolla) 花瓣（花冠）

perianth 花被

perianth 花被
corolla 花冠
etals 花瓣
epals 萼片
calyx 花萼
generalized flower 普通的花

gynoecium 雌蕊羣
stigma 柱頭
style 花柱
locule 室
ovary, containing ovules 藏有胚珠的子房

megasporophyll (*n*) a modified leaf that bears the megasporangium (↑). They can be grouped in a strobilus (p. 55).

flower (*n*) the structure concerned with sexual reproduction (p. 173) in angiosperms (p. 57). It is a modified vegetative shoot.

corolla (*n*) the part of the flower which is made up of all the petals (↓). It varies considerably in size, shape, form and colour and often attracts insects (p. 69) to visit the flower, pollinating (p. 183) the plant in the process.

petal (*n*) one of the often brightly coloured and scented individual elements which make up the corolla (↑). They are thought to be modified leaves. Those flowers which are pollinated (p. 183) by the wind have petals which are greatly reduced in size or absent.

calyx (*n*) the outermost part of a flower which comprises a number of sepals (↓) that protect the flower while it is developing in the bud stage.

sepal (*n*) the usually green, often hairy, leaf-like structures which make up the calyx (↑).

perianth (*n*) the part of the flower, comprising the corolla (↑) and calyx (↑), which surrounds the stamens (p. 181) and carpels (↓).

gynoecium (*n*) the female reproductive (p. 173) structure of a flower which is made up of the carpels (↓).

carpel (*n*) one of the single or more individual female reproductive (p. 173) structures of a plant which make up the gynoecium (↑). Each carpel is made up of an ovary (p. 180), a style (p. 181) and a stigma (p. 181). If there is more than one carpel, they may be fused together or separate.

大孢子葉（名）　生有大孢子囊（↑）的變態葉。它們可以形成孢子葉球（第55頁）。

花（名）　被子植物（第57頁）中與有性生殖（第173頁）相關的結構。它是一種變態的營養枝。

花冠（名）　花整體中由所有花瓣（↓）組成的部份。花冠在大小、形狀、結構和顏色等方面都有顯著的不同，常常吸引昆蟲（第69頁）來到花上，並在此過程中爲植物傳粉（第183頁）。

花瓣（名）　組成花冠（↑）的通常呈鮮艷色彩並富香氣的各個單體。花瓣可看成是變態葉。由風媒傳粉（第183頁）的那些花很小或者沒有花瓣。

花萼（名）　花的最外面部份，由若干枚萼片（↓）組成，萼片在花芽期發育，對花有保護作用。

萼片（名）　組成花萼（↑）的那些通常呈綠色而有茸毛的葉狀結構。

花被（名）　花整體中的花冠（↑）和花萼（↑）組成的那一部份。花被包圍着雄蕊（第181頁）和心皮（↓）。

雌蕊羣（名）　花的雌性生殖（第173頁）結構。它由心皮（↓）組成。

心皮（名）　構成植物雌蕊羣（↑）的單個或多個雌性生殖（第173頁）結構。每一心皮均由一個子房（第180頁）、一個花柱（第181頁）和一個柱頭（第181頁）構成。如果心皮多於一個，它們可融合在一起或者相互分開。

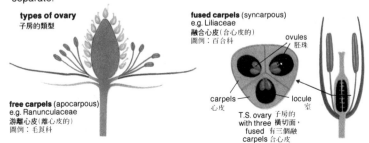

types of ovary 子房的類型

free carpels (apocarpous) e.g. Ranunculaceae 游離心皮（離心皮的）圖例：毛茛科

fused carpels (syncarpous) e.g. Liliaceae 融合心皮（合心皮的）圖例：百合科
ovules 胚珠
carpels 心皮
locule 室
T.S. ovary with three fused carpels 子房的橫切面，有三個融合心皮

ovary[p] (*n*) the part of the carpel (p. 179) which contains the ovules (↓).

ovule[p] (*n*) the structure containing the female gametes (p. 175) and which, after fertilization (p. 175), becomes the seed.

funicle (*n*) the stalk which attaches the base of the ovule (↑) to the wall of the carpel (p. 179).

placenta[p] (*n*) the part of the wall of the ovary (↑) to which the ovules (↑) are attached.

apocarpous (*adj*) of a gynoecium (p. 179) in which the carpels (p. 179) are not fused.

syncarpous (*adj*) of a gynoecium (p. 179) in which the carpels (p. 179) are fused.

placentation (*n*) the position and arrangement of the placentas (↑) in a syncarpous (↑) gynoecium (p. 179).

parietal (*adj*) of placentation (↑) in which the carpels (p. 179) are fused only by their margins with the placentas (↑) becoming ridges on the inner side of the wall of the ovary (↑).

axile (*adj*) of placentation (↑) in which the carpels (p. 179) fold inwards at their margins, fuse, and become a central placenta (↑).

free central of placentation (↑) in which the placenta (↑) grows upwards from the base of the ovary (↑).

nucellus (*n*) the central tissue (p. 83) of the ovule (↑) enclosing the megaspore (p. 178) or ovum (p. 178).

micropyle (*n*) a canal through the integument (↓) near the apex of the ovule (↑) which, in the seed, becomes a pore (p. 120) through which water may enter to enable germination (p. 168).

integument (*n*) the outermost layer of the ovule (↑) which forms the seed coat.

chalaza (*n*) the base of the ovule (↑) to which the funicle (↑) is attached. It is situated at the point where the nucellus (↑) and integuments (↑) merge.

embryo sac a large, oval-shaped cell, surrounded by a thin cell wall (p. 8), in the nucellus (↑) in which fertilization (p. 175) of the ovum (p. 178) takes place and the embryo (p. 166) develops.

polar nuclei a pair of haploid (p. 36) nuclei (p. 13) found towards the centre of the embryo sac (↑).

egg cell the female gamete (p. 175).

子房[p]（名） 心皮（第179頁）中藏有胚珠（↓）的部份。

胚珠[p]（名） 含有雌配子（第175頁）的結構，在受精（第175頁）以後，變成種子。

珠柄（名） 使胚珠（↑）的基部着生於心皮（第179頁）壁上的柄。

胎座[p]（名） 子房（↑）壁上着生胚珠（↑）的部份。

離心皮的（形） 指雌蕊羣（第179頁）而言，其內的心皮（第179頁）不融合。

合心皮的（形） 指雌蕊羣（第179頁）而言，其內的心皮（第179頁）融合。

胎座式（名） 胎座（↑）在合心皮（↑）雌蕊羣（第179頁）內的位置和排列方式。

側膜的（形） 指胎座式（↑）而言，處於這種胎座式時，心皮僅由其邊緣與胎座（↑）相融合，從而在子房壁的內側形成了一些脊狀隆起部份。

中軸的（形） 指胎座式（↑）而言，處於這種胎座式時，心皮（第179頁）的邊緣向內折疊，融合成一個中央胎座（↑）。

分離中央的（名） 指胎座式（↑）而言，處於這種胎座式時，胎座（↑）由子房（↑）的基部向上生長。

珠心（名） 包圍大孢子（第178頁）或卵（第178頁）的胚珠（↑）的中央組織（第83頁）。

珠孔（名） 靠近胚珠（↑）頂端、穿過珠被（↓）的一條通道。在種子中，它成為一個孔（第120頁），水份可通過該孔進入種子內，使其發芽（第168頁）。

珠被（名） 胚珠（↑）的最外層。它可形成種皮。

合點（名） 胚珠（↑）的基部，珠柄（↑）着生在該處。合點位於珠心（↑）和珠被（↑）的連合處。

胚囊 珠心（↑）內被一層薄的細胞壁（第8頁）包圍的一個卵形大細胞。卵（第178頁）的受精（第175頁）和胚（第166頁）的發育均在其內進行。

極核 存在於胚囊（↑）內，其位置朝向胚囊中心部位的一對單倍體（第36頁）核（第13頁）。

卵細胞 雌配子（第175頁）。

ovule structure 胚珠的結構

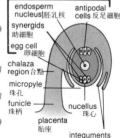

embryo sac (♀ gametophyte) 胚囊（配子體）
endosperm nucleus 胚乳核
antipodal cells 反足細胞
synergids 助細胞
egg cell 卵細胞
chalaza region 合點
micropyle 珠孔
funicle 珠柄
nucellus 珠心
placenta 胎座
integuments 珠被

placentation types 胎座式
ovaries cut through to show internal structure
子房內部結構剖面圖

axile 中軸胎座

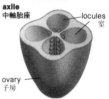

locules 室
ovary 子房

parietal 側膜胎座

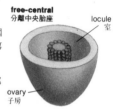

locule 室
ovary 子房

free-central 分離中央胎座

locule 室
ovary 子房

male floral parts
雄花的各組成部份

pollen sac　pollen grains
花粉囊　　　花粉粒

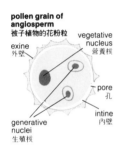

T.S.
anther
花藥的
橫切面

anther
花藥

filament
花絲

stamen
雄蕊

**pollen grain of
angiosperm**
被子植物的花粉粒

exine
外壁

vegetative
nucleus
營養核

pore
孔

intine
內壁

generative
nuclei
生殖核

actinomorphic flower
(radial symmetry)
輻射對稱的花

zygomorphic flower
(bilateral symmetry)
兩側對稱的花

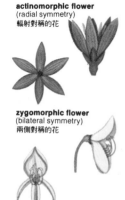

synergid (*n*) one of the two haploid (p. 36) cells which occur at the micropyle (↑) end of the embryo sac (↑) near the egg cell (↑).

antipodal cell one of the three haploid (p. 36) cells that move to the end of the embryo sac (↑) nearest the chalaza (↑). They do not take part in fertilization (p. 175).

style (*n*) the part of the carpel (p. 179) which joins the ovary (↑) and the stigma (↓).

stigma (*n*) the receptive tip of the carpel (p. 179) to which pollen (↓) becomes attached during pollination (p. 183).

androecium (*n*) the male reproductive (p. 173) structure of a flower which is made up of the stamens (↓).

stamen (*n*) one of the male reproductive (p. 173) structures which make up the androecium (↑). A stamen consists of an anther (↓) and a filament (↓).

anther (*n*) the tip of the stamen (↑) which produces the pollen (↓) contained in pollen sacs (↓).

filament (*n*) (1) the stalk of the stamen (↑) to which the anther (↑) is attached; (2) in plants, a chain of cells, e.g. some green algae (p. 44) are filamentous; (3) in animals, any fine threadlike structure.

pollen sac the chamber in which pollen is formed.

pollen (*n*) grain-like microspores (p. 178) produced in the pollen sac (↑) in huge numbers by meiotic (p. 38) division of their spore mother cells (p. 178). They contain the male gametes (p. 175).

tapetal cell one of the layer of cells which surrounds the spore mother cells (p. 178) and which provide nutrients for the spore mother cells and the developing spores (p. 178).

generative nucleus one of the two nuclei (p. 13) found in each grain of pollen (↑), both of which are transferred to the ovule (↑) via growth of the pollen tube (p. 184).

receptacle (*n*) the part of the stem of a flower which is often expanded and which bears the organs of the flower.

zygomorphic (*adj*) of a flower, such as a snapdragon, which is bilaterally symmetrical (p. 62).

actinomorphic (*adj*) of a flower, such as a buttercup, which is radially symmetrical (p. 60).

助細胞（名）　胚囊（↑）的珠孔（↑）端，鄰接卵細胞（↑）含的兩個單倍體（第36頁）細胞。

反足細胞　移向胚囊（↑）端部、與合點（↑）最爲鄰近的三個單倍體（第36頁）細胞。這些細胞不參於受精作用（第175頁）。

花柱（名）　將子房（↑）和柱頭（↓）連接起來的那一部份心皮（第179頁）。

柱頭（名）　在傳粉（第183頁）過程中，接受花粉（↓）、使之附着於其上的心皮（第179頁）頂端。

雄蕊羣　由雄蕊（↓）組成的花的雄性生殖（第173頁）結構。

雄蕊（名）　組成雄蕊羣（↑）的雄性生殖（第173頁）結構。雄蕊由花藥（↓）和花絲（↓）組成。

花藥（名）　產生花粉（↓）的雄蕊（↑）頂端，花粉貯存於花粉囊（↓）內。

花絲；絲狀體；絲（名）　（1）着生花藥（↑）的雄蕊（↑）的柄；（2）植物中，一些結合成鏈狀的細胞，例如某些綠藻（第44頁）是絲狀的；（3）動物中，任何細絲狀結構。

花粉囊　形成花粉的腔室。

花粉（名）　花粉囊（↑）內藉孢子母細胞（第178頁）減數分裂（第38頁）產生的大量顆粒狀小孢子（第178頁）。這些小孢子最終可形成雄配子（第175頁）。

絨氈層細胞　圍繞在孢子母細胞（第178頁）周圍的那一層細胞。這層細胞可爲孢子母細胞和發育中的孢子（第178頁）提供營養素。

生殖核　每個花粉（↑）粒中含有的兩個核（第13頁）。這兩個生殖核經由花粉管（第184頁）生長轉移到胚珠（↑）中。

花托（名）　通常爲花梗的膨大部份。花托上生有花的各個器官。

兩側對稱的（形）　指花而言，例如金魚草的花是兩側對稱的（第62頁）。

輻射對稱的（形）　指花而言，例如毛茛的花是輻射對稱的（第60頁）。

inflorescence 花序
e.g. capitate inflorescence of a composite
圖例：菊科植物的頭狀花序

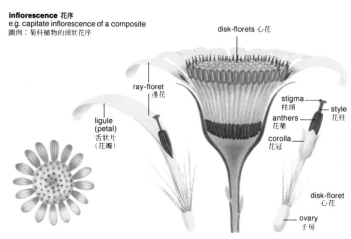

disk-florets 心花

ray-floret
漫花

ligule
(petal)
舌狀片
（花瓣）

stigma
柱頭

style
花柱

anthers
花藥

corolla
花冠

disk-floret
心花

ovary
子房

unisexual (*adj*) of a flower, which has the stamens (p. 181) and carpels (p. 179) on separate flowers. Unisexual flowers may be either monoecious (p. 175) or dioecious (p. 175).

nectary (*n*) a glandular (p. 87) swelling found on the receptacle (p. 181) or other parts of some flowers, which produces nectar (↓).

nectar (*n*) a sweet, sugary solution (p. 118) produced by the nectaries (↑). Many insects (p.69) visit flowers which produce nectar to feed on it and, in so-doing, pollinate (↓) the flower.

inflorescence (*n*) a group of flowers sharing the same stem.

單性的(形)　指花而言，在各朵單性花上都具有雄蕊(第181頁)和心皮(第179頁)。單性花可以是雌雄同株的(第175頁)，也可以是雌雄異株的(第175頁)。

蜜腺(名)　位於某些花的花托(第181頁)或其他部份的一種膨大腺體(第87頁)。這種腺體能分泌花蜜(↓)。

花蜜(名)　由蜜腺(↑)分泌的一種甜味的含糖溶液(第118頁)。許多昆蟲(第69頁)飛到分泌花蜜的花上探蜜，這樣即起到爲花傳粉(↓)的作用。

花序(名)　着生在同一花梗上的一簇花。

Inflorescence types 花序的類型

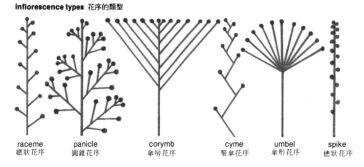

raceme	panicle	corymb	cyme	umbel	spike
總狀花序	圓錐花序	傘房花序	聚傘花序	傘形花序	穗狀花序

spikelet 小穗狀花序

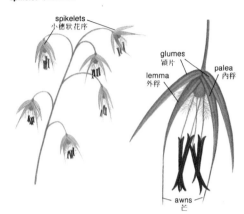

spikelets
小穗狀花序

glumes
穎片

palea
內稃

lemma
外稃

awns
芒

spathe 佛焰苞

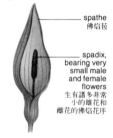

spathe
佛焰苞

spadix,
bearing very
small male
and female
flowers
生有諸多非常
小的雄花和
雌花的佛焰花序

cross-pollination 異花傳粉

self-pollination 自花傳粉

spikelet (*n*) the inflorescence (↑) of a grass.

spathe (*n*) a leaf-like structure or bract which encloses the spadix (↓) of certain monocotyledonous (p. 58) flowers.

spadix (*n*) the inflorescence (↑) of certain monocotyledonous (p. 58) flowers, bearing unisexual (↑) or hermaphrodite (p. 175) flowers.

floral formula a 'shorthand' way of describing the structure of a flower. It is given by a combination of capital letters and numbers as follows: K = calyx (p. 179); C = corolla (p. 179); A = androecium (p. 181); and G = gynoecium (p. 179). Thus, a flower with the floral formula K6 C6 G1 A5 would have six sepals, six petals, one carpel and five stamens.

pollination (*n*) the process in which pollen (p. 181) is transferred from the anther (p. 181) to the stigma (p. 181). **pollinate** (*v*).

self-pollination (*n*) pollination (↑) within the same flower or flowers from the same plant.

cross-pollination (*n*) pollination (↑) between flowers of different plants.

wind pollination pollination (↑) in which the pollen (p. 181) is carried from one flower to the next by the wind.

anemophily = wind pollination (↑).

小穗序花序（名） 禾草的花序（↑）。

佛焰苞（名） 包圍某些單子葉植物的（第58頁）花，其佛焰花序（↓）之葉狀結構或苞片。

佛焰花序（名） 生有單性（↑）花或兩性（第175頁）花的某些單子葉植物的（第58頁）花之花序（↑）。

花程式 描述花的結構的一種"速寫"方法。花程式可以用若干個大寫字母和數字的組合如下表示：K = 花萼（第179頁）；C = 花冠（第179頁）；A = 雄蕊羣（第181頁）；G = 雌蕊羣（第179頁）。例如，一朵花的花程式爲K6C6G1A5，即表示它含有6枚萼片、6枚花瓣、1個心皮和5個雄蕊。

傳粉（作用）（名） 花粉（第181頁）從雄蕊花藥（第181頁）傳送至雌蕊柱頭（第181頁）的過程。（動詞形式爲 pollinate）

自花傳粉（名） 在同一朵花內或同株植物的數朵花內進行的傳粉（↑）。

異花傳粉（名） 在不同植物的花之間進行的傳粉（↑）。

風媒傳粉（名） 藉風力將花粉（第181頁）從一朵花傳送到另一朵花所進行的傳粉（↑）。

風媒 同風媒傳粉（↑）。

insect pollination pollination (p. 183) in which the pollen (p. 181) is transferred from one flower to the next on the bodies of insects (p. 69) which are attracted to the flowers by the brightly coloured petals (p. 179), the scent, and the promise of nectar (p. 182).

entomophily = insect pollination (↑).

pollen tube a tubular outgrowth which forms when the pollen (p. 181) grain germinates (p. 168) and which is the means through which male gametes (p. 175) are carried to the egg.

double fertilization in flowering plants, the union of one generative nucleus (p. 181) with an ovum (p. 178) to form a zygote (p. 166) and the other with the two polar nuclei (p. 180) to form the primary endosperm (p. 168) nucleus (p. 13) which is triploid (p. 207). Subsequent division of the primary endosperm nucleus produces the endosperm.

seed (n) the structure which develops after the fertilization (p. 175) of the ovule (p. 180) and which is made up of the testa (p. 168) surrounding the embryo (p. 166). In suitable conditions, each seed may germinate (p. 168) and form a fully independent plant. Seeds in flowering plants may be contained within a fruit.

fruit (n) the ripened ovary (p. 180) wall of a flower which contains the seeds. Depending upon the method by which the seeds of the plant are distributed, the fruit may be fleshy (distributed by animals) or dry (distributed by wind or water).

pericarp (n) the outer wall of the ovary (p. 180) which develops into the fruit.

endocarp (n) the inner layer of the pericarp (↑) which develops into the stony covering of the seed of a drupe (↓), such as a cherry.

mesocarp (n) the middle layer of the pericarp (↑) which can form the fleshy part of a drupe (↓), such as a cherry, or the hard shell of a nut, like an almond.

exocarp (n) the tough outer 'skin' of a fruit.

epicarp (n) = exocarp (↑).

aleurone layer the outermost, protein-rich layer of the endosperm (p. 168) of the seeds of grasses.

蟲媒傳粉　由昆蟲(第69頁)身體將花粉(第181頁)從一朵花傳到另一朵花所進行的傳粉(第183頁)。昆蟲來到花上，是由於花的艷麗花瓣(第179頁)、香氣和花蜜(第182頁)的吸引所致。

蟲媒　同蟲媒傳粉(↑)。

花粉管　花粉(第181頁)粒萌發(第168頁)時形成的一種管狀突出物。它是將雄配子(第175頁)輸送到卵處的管道。

雙重受精作用　指有花植物形成三倍體(第207頁)的一種結合。其結合方式是：一個生殖核(第181頁)和一個卵(第178頁)結合形成合子(第166頁)，另一個生殖核和兩個極核(第180頁)結合形成初生胚乳(第168頁)核(第13頁)。其後，初生胚乳核分裂，形成胚乳。

種子(名)　胚珠(第180頁)受精(第175頁)後長成的結構。種子由包圍胚(第166頁)的種皮(第168頁)組成。在合適的環境條件下，每一粒種子都可發芽(第168頁)，長成一株完全獨立的植物。有花植物的種子可包含在果實內。

果實(名)　包含種子的，花的成熟的子房(第180頁)壁。根據植物種子的散播方式，果實可以分成肉質果(種子藉動物散播)或乾果(種子藉風或水散播)。

果皮(名)　可發育成果實的子房(第180頁)的外壁。

內果皮(名)　果皮(↑)的內層。它可發育成如櫻桃核果(↓)的種子的石質層。

中果皮(名)　果皮(↑)的中層。它可形成如櫻桃核果(↓)的肉質部份，或者像扁桃這樣的堅果的硬殼。

外果皮(名)　果實的堅韌外"皮"。
英文亦稱外果皮(↑)為 epicarp。

糊粉層(名)　禾本科植物種子胚乳(第168頁)的最外層，其內富含蛋白質。

fertilization in angiosperms
被子植物的受精

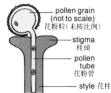

1 pollen grain lands on stigma, pollen tube grows through tissues of style carrying the male gametes 1. 花粉粒落在柱頭上，花粉管穿過花柱的組織而生長，從而可輸送雄配子

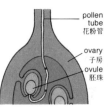

2 pollen tube grows through ovary wall and into micropyle of ovule 2. 花粉管穿過子房壁，伸入胚珠的珠孔而生長

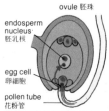

3 one male gamete fertilizes egg cell, the other fertilizes the endosperm nucleus forming endosperm mother cell 3. 一個雄配子使卵細胞受精，另一個雄配子使胚乳核受精，形成胚乳母

各類果實及其結構
fruits and fruit structure

berry e.g. tomato 圖例：番茄
漿果

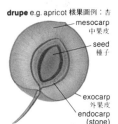

exocarp
外果皮

seeds
種子

mesocarp
中果皮

endocarp
內果皮

drupe e.g. apricot 核果圖例：杏

mesocarp
中果皮

seed
種子

exocarp
外果皮

endocarp
(stone)
內果皮(核)

legume e.g. pea
英果圖例：豌豆

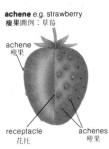

seeds
種子

pod
英

achene e.g. strawberry
瘦果圖例：草莓

achene
瘦果

receptacle
花托

achenes
瘦果

scutellum (n) the part of the embryo (p. 166) of a grass seed, which is situated next to the endosperm (p. 168).

coleoptile (n) the protective sheath with a hard pointed tip which protects the plumule (p. 168) in a germinating (p. 168) grass seedling.

dehiscence (n) the process in which the mature fruit wall opens, sometimes violently, to release the seeds. **dehisce** (v).

berry (n) a fruit, such as a blackberry, which, unlike a drupe (↓) does not have a stony endocarp (↑) so that the seeds are surrounded by a fleshy mesocarp (↑) and endocarp.

drupe (n) a fruit, such as a plum, formed from a single carpel (p. 179) which has a stony endocarp (↑) that surrounds the seed.

follicle[P] (n) a dry fruit, such as a delphinium, which has formed from a single carpel (p. 179) and in which, during dehiscence (↑), the fruit or pod splits along one line to release the seed.

legume (n) a dry fruit, such as a pea, which has formed from a single carpel (p. 179) and in which, during dehiscence (↑), the fruit or pod splits along two sides to release the seed.

siliqua (n) a dry, elongated fruit or special type of capsule (p. 53), such as that found in the cabbage family, formed from two carpels (p. 179) which are fused together but separated by a false septum or wall. During dehiscence (↑), the siliqua splits as the carpel walls separate leaving the seeds attached to the septum.

silicula (n) a type of siliqua (↑), found in plants such as the shepherd's purse, which is short and broad in shape.

achene (n) a dry fruit, such as that of the buttercup, which is formed from a single carpel (p. 179), contains only one seed, has a leathery pericarp (↑), and has no particular method of dehiscence (↑).

cypsela (n) a dry fruit, such as that of the dandelion, which is formed from two carpels (p. 179) of an inferior ovary (p. 180) which retains a plumed calyx (p. 179) to aid in wind dispersal (p. 186).

盾片（名）　禾本科植物種子的胚（第166頁）鄰接於胚乳（第168頁）的那一部份。

胚芽鞘（名）　在發芽（第168頁）的禾本科植物籽苗中，起保護胚芽（第168頁）作用的、具有堅硬尖端的保護鞘。

開裂（名）　成熟果實的壁張開，有時猛烈地張開，以釋出種子的過程。（動詞形式爲 dehisce）

漿果（名）　一種如黑莓的果實。它不同於核果（↓），沒有石質的內果皮（↑），所以，其種子是由肉質的中果皮（↑）和內果皮包裹的。

核果（名）　一種如李的果實。它由單心皮（第179頁）形成，有一層石質的、包圍種子的內果皮（↑）。

蓇葖[P]（名）　一種如飛燕草的乾果。由單心皮（第179頁）形成。在開裂（↑）過程中，其果實或莢沿着一條線裂開，釋出種子。

莢果（名）　一種如豌豆的乾果。它是由單心皮（第179頁）形成的。在開裂（↑）過程中，其果實或莢沿着兩側裂開，以釋出種子。

長角果（名）　一種長形的乾果或特殊類型的蒴果（第53頁），例如甘藍科中含有的那種果實。它由兩個融合在一起的心皮（第179頁）組成，但中間由一假隔膜或壁隔開。在開裂（↑）過程中，當心皮壁分開，使種子附着在假隔膜時，長角果即裂開。

短角果（名）　一類如薺屬植物中含有的長角果（↑）。其形狀短而寬。

瘦果（名）　一種由單心皮（第179頁）形成的乾果，例如毛茛屬植物的那種果實。它僅含一粒種子，有一層革質的果皮（↑），無特殊的開裂（↑）方式。

連萼瘦果（名）　一種乾果，例如蒲公英的果實。它由下位子房（第180頁）的兩個心皮（第179頁）形成。它保留一個羽狀花萼（第179頁），有助於風媒散播（第186頁）。

caryopsis (*n*) a dry fruit, such as that of grasses, which is similar to an achene (p. 185) except that the pericarp (p. 184) is united with the testa (p. 168).

nut (*n*) a dry fruit, such as that of the hazel, which is similar to an achene (p. 185), except that the pericarp (p. 184) is stony.

samara (*n*) a dry fruit, such as that of the elm, which is similar to an achene (p. 185), except that part of the pericarp (p. 184) forms wings that aid in wind dispersal (↓).

false fruit a fruit that includes other parts of the flower, such as the inflorescence (p. 182), as well as the ovary (p. 180).

pome (*n*) a fleshy false fruit (↑), such as that of the apple. The main flesh is made of the swollen receptacle (p. 181).

fruit dispersal the various methods by which a flower distributes its seeds and which may include the fruit or just the seeds alone.

mechanical dispersal fruit dispersal (↑) in which the fruit itself is responsible for distributing the seed by opening explosively when the seed is mature and scattering it widely.

wind dispersal fruit dispersal (↑) in which the seed is carried on the wind either because it is small and lightweight or by bearing wing-like structures which give them extra lift. In some cases, such as that of the poppy, the capsule (p. 53) itself sways in the wind to distribute the seed, like a censer.

animal dispersal fruit dispersal (↑) in which the seed is distributed by being transported by animals, including humans. The seed or fruit may have hooks or spines which stick to the animals' coats or their fruits may be palatable while the seeds may be indigestible so that the fruits are eaten and the seeds pass through undamaged. Indeed, some seeds can only germinate (p. 168) when they have passed through the digestive (p. 98) system of certain animals.

water dispersal fruit dispersal (↑) in which the fruit or seed is specially adapted to be carried in running water.

穎果（名）　一種如禾本科植物果實的乾果。這種乾果與瘦果（第185頁）類似，只是其果皮（第184頁）與外種皮（第168頁）相愈合。

堅果（名）　一種如榛屬植物果實的乾果。這種乾果與瘦果（第185頁）類似，只是其果皮（第184頁）是石質的。

翅果（名）　一種如榆樹果實的乾果。這種乾果與瘦果（第185頁）類似，只是其果皮（第184頁）的一部份延展成有助於果實風媒散播（↓）的翅。

假果　一種包括花的其他部份如花序（第182頁）及子房（第180頁）在內的果實。

梨果（名）　一種如蘋果果實的肉質假果（↑）。其果肉的主要部份由膨大的花托（第181頁）形成。

果實散播　花散播其種子的各種方式。果實散播可包括果實或僅僅是散播種子。

機械散播　一種果實敞佈（↑）方式。這種方式是由果實本身行使散播種子功能的。當種子成熟時，果實即爆裂開，將種子廣爲散播。

風媒散播　一種果實散播（↑）方式。種子或者由於體積小，重量輕，或者由於具有賦予其額外昇力的翅狀結構，而在風中得到傳送。在某些情況下，例如在罌粟屬植物中，蒴果（第53頁）本身可像一隻香爐一樣，在風中搖着散播種子。

動物散播　一種果實散播（↑）方式。由於動物包括人類携帶了種子，使種子散播開來。這種種子或果實可能長有可以附着在動物身上的鈎或刺；也可能果實很好吃，而種子則不能被消化，以致吃了果實，而種子卻完爲無損地被排泄出來。實際上，有些種子只有在通過某些動物的消化（第98頁）系統後才能發芽（第168頁）。

水媒散播　一種果實散播（↑）方式。利用這種方式，進行散播的果實或種子，特別適應於在流水中進行傳送。

caryopsis e.g. wheat
穎果圖例：小麥

nut e.g. hazelnut
堅果圖例：榛子

samara e.g. sycamore
翅果圖例：懸鈴木

seed
種子

wing
翅

pome e.g. apple
梨果圖例：蘋果

gonad (n) the male or female organ of reproduction (p. 173) in sexual animals that produce gametes (p. 175). In some cases, the gonads also produce hormones (p. 130).

生殖腺、性腺(名) 有性動物產生配子(第175頁)的雄性或雌性生殖(第173頁)器官。在某些情況下，生殖腺還分泌激素(第130頁)。

人的雄性生殖器官和其他結構 **human male reproductive organs and other structures**

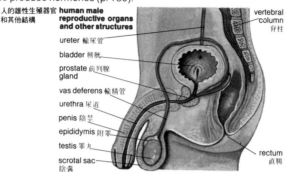

- ureter 輸尿管
- bladder 膀胱
- prostate 前列腺 gland
- vas deferens 輸精管
- urethra 尿道
- penis 陰莖
- epididymis 附睪
- testis 睪丸
- scrotal sac 陰囊
- vertebral column 脊柱
- rectum 直腸

spermatogenesis 精子發生

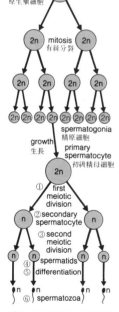

- primordial germ cell 原生殖細胞
- mitosis 有絲分裂
- spermatogonia 精原細胞
- growth 生長
- primary spermatocyte 初級精母細胞
- ① first meiotic division
- ② secondary spermatocyte
- ③ second meiotic division
- ④⑤ spermatids differentiation
- ⑥ spermatozoa

① 第一次減數分裂　④ 精子細胞
② 次數精母細胞　　⑤ 分化(作用)
③ 第二次減數分裂　⑥ 精子

testis (n) the male reproductive (p. 173) organ which produces spermatozoa (p. 188) by spermatogenesis (↓). In vertebrates (p. 74) there are two testes which usually lie in a sac of skin, the scrotum or scrotal sac (p. 188), outside the main body cavity and behind the penis (p. 189). In vertebrates, the testes also produce androgens (p. 195).

testicle (n) = testis (↑).

seminiferous tubule one of several hundred tiny coiled tubes which compose the testis (↑) and in which all stages of spermatogenesis (↓) take place.

Sertoli cell one of a number of large, specialized cells made of germinal epithelium (p. 87) and found in the testis (↑) which are thought to nourish the spermatids (p. 188) to which they are attached.

spermatogenesis (n) the process by which spermatozoa (p. 188) are produced in the testes (↑). A germ cell (p. 36) divides a number of times during a multiplication phase to produce spermatogonia (p. 188) each of which then grows into a primary spermatocyte (p. 188). In turn, the spermatocyte undergoes two phases of meiotic (p. 38) division to produce spermatids (p. 188) which then differentiate into the spermatozoa.

睪丸、精巢(名) 藉精子發生(↓)過程產生精子(第188頁)的雄性生殖(第173頁)器官。脊椎動物(第74頁)有兩個睪丸，通常存在於主體腔外和陰莖(第189頁)後面的皮囊——陰囊(第188頁)內。在脊椎動物中，睪丸還分泌雄激素(第195頁)。

睪丸(↑)的另一英文名稱為testicle。

精細管 構成睪丸(↑)的幾百根彎曲的細管。精子發生(↓)過程中的各個階段都是在精細管內進行的。

塞爾托利氏細胞、足細胞 在睪丸(↑)中、由生殖上皮(第87頁)組成的許多大型特化細胞。塞爾托利氏細胞對其所附着的精子細胞(第188頁)起滋養作用。

精子發生(名) 睪丸(↑)內產生精子(第188頁)的過程。生殖細胞(第36頁)在增殖期，分裂若干次，產生精原細胞(第188頁)，然後，每一個精原細胞發育成一個初級精母細胞(第188頁)。初級精母細胞又經歷兩次減數分裂(第38頁)產生精子細胞(第188頁)，再變態成精子。

spermatogonium (*n*) one of the large numbers of cells found in the testes (p. 187) and which grows into a primary spermatocyte (↓) during spermatogenesis (p. 187). **spermatogonia** (*pl.*).

spermatocyte (*n*) one of the large numbers of reproductive cells found in the seminiferous tubules (p. 187) and which is produced by the growth of a spermatogonium (↑).

spermatid (*n*) one of the large numbers of reproductive (p. 173) cells found in the seminiferous tubules (p. 187) during spermatogenesis (p. 187). It is produced as a result of two phases of meiosis (p. 38) of a spermatocyte (↑). Each spermatid, nourished by the Sertoli cells (p. 187), differentiates and matures into a spermatozoon (↓).

spermatozoon (*n*) the small, differentiated, highly motile mature male gamete (p. 175) or reproductive (p. 173) cell. Spermatozoa (*pl.*) are produced continuously in large numbers in the seminiferous tubules (p. 187). Locomotion (p. 143) takes place by movements of a flagellum (p. 12).

vas efferens one of the small channels through which spermatozoa (↑) are transported from the seminiferous tubules (p. 187) to the epididymis (↓).

epididymis (*n*) a muscular (p. 143), coiled tubule between the vas efferens (↑) and the vas deferens (↓) which functions as a temporary storage vessel for spermatozoa (↑) until they are released during mating.

vas deferens one of the pair of muscular (p. 143) tubules, with mucous (p. 99) glands (p. 87), which leads from the epididymis (↑) and through which spermatozoa (↑) are released into the urethra (↓) during mating.

urethra (*n*) a duct which leads from the bladder (p. 135) to the exterior and through which urine (p. 135) is excreted (p. 134). In males it also connects with the vas deferens (↑).

scrotal sac the external sac of skin which is divided into two, each carrying one testis (p. 187). Thus, the testes are maintained at a lower temperature than the rest of the body to ensure the best conditions for the development of spermatozoa (↑).

精原細胞（名） 睾丸（第187頁）中含有的大量細胞。這種細胞在精子發生（第187頁）過程中，發育成初級精母細胞（↓）。（複數形式爲spermatogonia）

精母細胞（名） 精細管（第187頁）中含有的大量生殖細胞。這種細胞是由精原細胞（↑）發育而成的。

精子細胞（名） 在精子發生（第187頁）過程中，存在於精子管（第187頁）中的大量生殖（第173頁）細胞。這種細胞是由精母細胞（↑）經兩次減數分裂（第38頁）產生的。由塞爾托利氏細胞（第187頁）滋養的每個精子細胞，經變態、成熟、形成精子（↓）。

精子（名） 經分化的、小而高度能動的成熟雄配子（第175頁）或生殖（第173頁）細胞。精子在精細管（第187頁）中不斷地大量產生。其運動（第143頁）藉鞭毛（第12頁）的運動而產生。

輸出管（名） 將精子（↑）從精細管（第187頁）輸送至附睾（↓）的諸多小管道。

附睾（名） 在輸出管（↑）和輸精管（↓）之間的一根肌肉質（第143頁）盤曲管。附睾起暫時貯存精子（↑）的作用，直至在交配時釋出精子爲止。

輸精管（名） 一對起始於附睾（↑），具有黏液（第99頁）腺（第87頁）的肌肉質（第143頁）管道。在交配時，精子（↑）通過輸精管釋入尿道（↓）中。

尿道（名） 一根從膀胱（第135頁）通向體外的管道。尿液（第135頁）通過尿道排泄（第134頁）。雄性的尿道還與輸精管（↑）相連接。

陰囊 指一種附於體表的皮膚囊。這種皮膚囊分成爲兩個部份，每一部份均含有一個睾丸（第187頁）。因此，睾丸就能保持比身體其他部份較低的溫度，從而可保證精子（↑）在最佳條件下發育。

spermatozoon 精子
e.g. human sperm 圖例：人的精子

acrosome (contains agent which dissolves egg membrane during fertilization)
頂體（含有受精時可溶解卵膜的因子）

head 頭

nucleus (rich in DNA)
核（富含脫氧核糖核酸）

neck 頸

centriole 中心粒

middle piece 中段

axial filament 軸絲

mitochondria 線粒體

centriole 中心粒

axial filament 軸絲

tail 尾

tail sheath 尾鞘

end piece 末段

prostate gland a gland (p. 87) surrounding the urethra (↑) which, under the control of androgens (p. 195), secretes (p. 106) alkaline substances that reduce the urine's (p. 135) acidity and aid in the motility of spermatozoa (↑).

seminal vesicle one of the two organs connected to the vas deferens (↑) in most male mammals (p. 80). It is under hormonal (p. 130) control and secretes (p. 106) fluid which makes up the bulk of the semen (p. 191) improving the motility of the spermatozoa (↑).

Cowper's gland one of two glands (p. 87) connected to the vas deferens (↑) which secretes (p. 106) fluid for the semen (p. 191).

penis (*n*) the organ through which the urethra (↑) connects with the exterior and which functions, during mating, to transport spermatozoa (↑) to the female reproductive (p. 173) organs. It contains spongy tissue (p. 83) which fills with blood (p. 90) during mating to become more rigid or erect.

ovary[a] (*n*) one of the pair of female reproductive (p. 173) organs in which ova (p. 190) are produced during oogenesis (↓). Female hormones (p. 130) are also produced in the ovaries.

oogenesis (*n*) the process by which ova (p. 190) are produced in the ovaries (↑). A germ cell (p. 36) divides by mitosis (p. 37) to form a number of oogonia (↓) each of which grows to give rise to a primary oocyte (↓). By two phases of meiotic (p. 38) division – with the second phase usually following fertilization (p. 175) – an ovum is produced together with additional polar bodies (↓).

oogonium (*n*) a specialized cell found within the ovary (↑) which is produced by mitotic (p. 37) division of the germ cell (p. 36) and which grows to give rise to a primary oocyte (↓) during oogenesis (↑). **oogonia** (*pl.*).

oocyte (*n*) a reproductive (p. 173) cell found within the ovary (↑) during oogenesis (↑). It results from the growth of an oogonium (↑).

polar body a tiny cell produced during oogenesis after the second meiotic (p. 38) division when the ovum (p. 190) is formed. The polar body contains a nucleus (p. 13) but virtually no cytoplasm (p. 10).

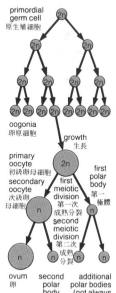

oogenesis 卵子發生

primordial germ cell 原生殖細胞

oogonia 卵原細胞

growth 生長

primary oocyte 初級卵母細胞

secondary oocyte 次級卵母細胞

first meiotic division 第一次成熟分裂

second meiotic division 第二次成熟分裂

first polar body 第一極體

ovum 卵

second polar body 第二極體

additional polar bodies (not always formed) 多餘的極體 （並非總會形成）

前列腺 在尿道(↑)周圍的一個腺體(第87頁)。在雄激素(第195頁)的控制下，該腺體可分泌(第106頁)一些降低尿液(第135頁)酸度，有助於提高精子(↑)游動性的鹼性物質。

貯精囊 大多數雄性哺乳動物(第80頁)體內，與輸精管(↑)相連的兩個器官。它們受激素(第130頁)控制，可分泌(第106頁)精液(第191頁)的大部份組成液體，以改善精子(↑)的游動性。

尿道球腺、高柏氏腺 與輸精管(↑)相連的兩個腺體(第87頁)。它們可分泌(第106頁)組成精液(第191頁)的液體。

陰莖(名) 尿道(↑)與體外相通所經由的器官。在交配時，陰莖的功能是將精子(↑)輸送至雌性生殖(第173頁)器官。陰莖內含有海綿組織(第83頁)，在交配時，海綿組織充血(第90頁)，從而使陰莖變得堅硬而直立。

卵巢[a](名) 一對雌性生殖(第173頁)器官。卵(第190頁)在卵子發生(↓)過程中，產生於卵巢內。雌性激素(第130頁)也產生於卵巢內。

卵子發生(名) 卵巢(↑)內產生卵(第190頁)所經歷的過程。一個生殖細胞(第36頁)經有絲分裂(第37頁)形成若干個卵原細胞(↓)，每一個卵原細胞長大成為一個初級卵母細胞(↓)。經過兩次成熟分裂(第38頁)——第二次成熟分裂通常發生於受精(第175頁)之後——即產生一個卵和一些多餘的極體(↓)。

卵原細胞(名) 卵巢(↑)內含有的一種特化細胞。這種細胞由生殖細胞(第36頁)有絲分裂(第37頁)產生；在卵子發生過程中，該細胞長大成為初級卵母細胞(↓)。(複數形式為 oogonia)

卵母細胞(名) 在卵子發生(↑)過程中，卵巢(↑)內含有的一種生殖(第173頁)細胞。它由卵原細胞(↑)生長而來。

極體 在卵子發生過程中，於第二次成熟分裂(第38頁)以後，形成卵(第190頁)時產生的一種極小的細胞。極體含有一個核(第13頁)，但實際上不含細胞(第10頁)。

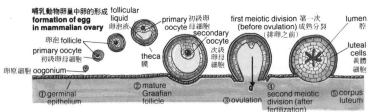

哺乳動物卵巢中卵的形成
formation of egg in mammalian ovary
卵泡液 follicular liquid
卵泡 follicle
primary oocyte 初級卵母細胞
primary oocyte
初級卵母細胞
卵原細胞 oogonium
theca 膜
secondary oocyte
次級卵母細胞
primary oocyte 初級卵母細胞
① 生殖上皮 germinal epithelium
② mature Graafian follicle 成熟的囊狀卵泡
first meiotic division 第一次 (before ovulation) 成熟分裂 (排卵之前)
③ ovulation 排卵
second meiotic division (after fertilization) 第二次成熟分裂 (受精之後)
lumen 腔
luteal cells 黃體細胞
⑤ corpus luteum 黃體

① 生殖上皮
② 成熟的囊狀卵泡
③ 排卵
④ 第二次成熟分裂 (受精之後)
⑤ 黃體

ovum[a] (*n*) the large, immotile female gamete (p. 175) produced in the ovary (p. 189) during oogenesis (p. 189). If it is fertilized (p. 175) by a spermatozoon (p. 188) it develops into a new individual. Fertilization may take place at the oocyte (p. 189) stage following the first meiotic (p. 38) division. **ova** (*pl.*).

Graafian follicle a fluid-filled, spherical mass of cells with a cavity that is found in the ovary (p. 189) and contains an oocyte (p. 189) attached to its wall. It is the site of development of the ovum (↑) and grows from one of the large number of follicles within the ovary.

corpus luteum a gland (p. 87) which forms temporarily in the Graafian follicle (↑) after rupture during ovulation (p. 194). It secretes (p. 106) the hormone (p. 130) progesterone (p. 195) which, if the ovum (↑) is fertilized (p. 175), continues to be released to prepare the female reproductive (p. 173) tract for pregnancy (p. 195). If fertilization does not take place, the corpus luteum degenerates. **corpora lutea** (*pl.*).

oviduct (*n*) a muscular (p. 143) tube lined with cilia (p. 12) by which ova (↑) are transported from the ovaries (p. 189) to the exterior.

uterus (*n*) a thick-walled organ in which the embryo (p. 166) develops. It is muscular (p. 143) and the smooth muscle increases in amount during pregnancy (p. 195) so that it is able to expel the young at birth. The size of the uterus as well as the thickness of its wall, which provides a point of attachment and nourishment for the developing embryo, varies cyclically and with sexual activity or inactivity under the influence of reproductive (p.173) hormones (p. 130). Also known as **womb**.

卵[a](名) 在卵子發生(第189頁)過程中，卵巢(第189頁)內產生，大而不活動的雌配子(第175頁)。如果它與精子(第188頁)結合(第175頁)，即可發育成一個新的個體。受精作用可在第一次成熟分裂(38頁)之後的卵母細胞(第189頁)階段進行。(複數形式為 ova)

囊狀卵泡、格拉夫氏泡 卵巢(第189頁)內含有的一種具有空腔的充液球形細胞團。這種細胞團含有一個附着於其壁上的卵母細胞(第189頁)。囊狀卵泡是卵(↑)發育之地，由卵巢內大量卵泡中的一個生長而成。

黃體 在排卵(第194頁)過程中，於囊狀卵泡(↑)破裂之後，在其內形成的一種暫時性腺體(第87頁)。這種腺體可分泌(第106頁)激素(第130頁)孕酮(第195頁)。如果卵(↑)受精(第175頁)，則孕酮繼續釋放，以使雌性生殖(第173頁)道好姙娠(第195頁)準備。如果並未受精，黃體即退化消失。(複數形式為 corpora lutea)

輸卵管(名) 一種內壁襯有纖毛(第12頁)的肌肉質(第143頁)管，卵(↑)經由該管從卵巢(第189頁)向外部輸送。

子宮(名) 一種厚壁器官，胚胎(第166頁)即於其內發育。子宮是肌肉質(第143頁)器官，在姙娠(第195頁)期間，其平滑肌增厚，因此，在分娩時能將幼體擠出。子宮爲發育中的胚胎提供附着的場所及營養。子宮的大小及壁厚，受各種生殖(第173頁)激素(第130頁)影響發生週期性變化，也隨有無性慾活動而變化。其另一英文名稱爲 womb。

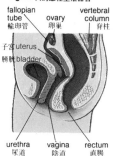

human female reproductive organs 人的雌性生殖器官
fallopian tube 輸卵管
ovary 卵巢
vertebral column 脊柱
子宮 uterus
膀胱 bladder
urethra 尿道
vagina 陰道
rectum 直腸

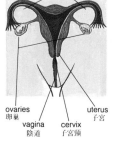

fallopian tubes 輸卵管
ovaries 卵巢
vagina 陰道
cervix 子宮頸
uterus 子宮

cervix (*n*) a ring of muscle (p. 143) between the uterus (↑) and the vagina (↓) which also contains mucous glands (p. 87).

vagina (*n*) the muscular (p. 143) duct which connects the uterus (↑) to the exterior and which receives the penis (p. 189) during mating.

copulation (*n*) the sexual union of male and female animals during mating in which, in mammals (p. 80), the penis (p. 189) is received by the vagina (↑) and ejaculation (↓) takes place. Also known as **coitus**.

semen (*n*) a fluid containing spermatozoa (p. 188) produced by the testes (p. 187) and other liquids produced by the prostate gland (p. 189). During copulation (↑) semen is passed from male to female.

ejaculation (*n*) the rhythmic and forcible discharge of semen (↑) from the penis (p. 189).

orgasm (*n*) the climax of sexual excitement which takes place during mating and involves a complex series of reactions of the reproductive (p. 173) organs and other parts of the body including the skin.

foetus in uterus 子宮內的胚兒

foetus
胎兒

umbilical
cord
臍帶

placenta
胎盤

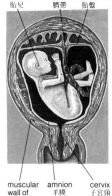

muscular
wall of
uterus
子宮的
肌肉壁

amnion
羊膜

cervix
子宮頸

implantation (*n*) following fertilization (p. 175), the process in which the developing zygote (p. 166) embeds itself in the wall of the uterus (↑).

foetus (*n*) an embryo (p. 166) with an umbilical cord (p. 192) which is sufficiently developed to show the main features that the mammal (p. 80) will possess after birth.

foetal membrane any one of those membranes (p. 14) or structures which are developed by the embryo (p. 166) for nourishment and protection but which do not form part of the embryo itself.

amnion (*n*) the fluid-filled sac in which the embryo (p. 166) develops in mammals (p. 80). The amnion offers the embryo protection from any pressure exerted on it by the organs of the mother and a liquid environment (p. 218) in which to develop (important for land animals). The sac wall consists of two layers of epithelium (p. 87) and sometimes only the inner layer is referred to as the amnion. **amniotic** (*adj*), **amniote** (*adj*).

amniotic cavity the amnion (↑) or the fluid-filled cavity within the amnion which contains the developing embryo (p. 166).

子宮頸（名） 子宮（↑）和陰道（↓）之間的一圈環狀肌肉（第143頁）。陰道含有黏液腺（第87頁）。

陰道（名） 子宮（↑）與外部連通的肌肉質（第143頁）管道。它在交配時接納陰莖（第189頁）。

交配（名） 雌雄動物配對時進行的性結合。哺乳動物（第80頁）交配時，陰莖（第189頁）爲陰道（↑）所接納，然後發生射精（↓）。也稱爲**性交**。

精液（名） 睾丸（第187頁）產生的一種含精子（第188頁）液體和前列腺（第189頁）產生的其他一些液體。在交配（↑）過程中，精液由雄性動物傳遞給雌性動物。

射精（名） 陰莖（189頁）進行的有節律的、強有力的精液（↑）釋出。

性慾高潮（名） 交配過程中發生的性興奮高潮。性慾高潮涉及各生殖（第173頁）器官和包括皮膚在內的身體其他部份的一系列複雜反應。

植入（名） 受精（第175頁）之後，發育中的受精卵（第166頁）將自身埋入子宮內（↑）壁內的過程。

胎兒（名） 帶有臍帶（第192頁）的胚胎（第166頁），經充份發育即顯示出哺乳動物（第80頁）在出生以後具有的各主要特徵。

胎膜 胚胎（第166頁）爲營養和保護需要而發育的那些膜（第14頁）或結構中的任何一個，但它不構成胚胎本身的組成部份。

羊膜（名） 哺乳動物（第80頁）的胚胎（第166頁）於其內進行發育的充液囊。羊膜可保護胚胎免受母體器官對其產生的任何壓力，並可爲胚胎發育提供一種液態環境（第218頁）（對陸生動物來說是重要的）。其囊壁由兩層上皮（第87頁）組成，有時僅將其內層稱之爲羊膜。（形容詞形式爲 amniotic 及 amniote）

羊膜腔 指羊膜（↑），或指羊膜內的充液腔，其中，含有發育中的胚胎（第166頁）。

allantois (n) a sac-like extension of the gut (p. 98) which is present in the embryos (p. 166) of reptiles (p. 78), birds and mammals (p. 80) and which grows out beyond the embryo itself. The connective tissue (p. 88) which covers it is liberally supplied with blood vessels (p. 127) and functions for gas exchange (p. 112) of the embryo as well as for storing the products of excretion (p. 134).

chorion (n) the outermost membrane (p. 14), the outer epithelium (p. 87) of the amnion (p. 191) wall which surrounds the embryo (p. 166) of mammals (p. 80) and which unites with the allantois (↑) to develop into the placenta (↓).

placentaᵃ (n) a disc-shaped organ which develops within the uterus (p. 190) during pregnancy (p. 195) and which is in close association with the embryo (p. 166) and with tissues (p. 83) of the mother. The placenta serves for attachment and nourishment over its large surface area.

umbilical cord a cord which connects the placenta (↑) to the navel of the foetus (p. 191) allowing interchange of materials via two arteries (p. 127) and a vein (p. 127).

viviparity (n) the condition in which embryos (p. 166) develop within a uterus (p. 190), are attached to a placenta (↑), and are born alive. **viviparous** (adj).

gestation period the time which elapses between fertilization (p. 175) of the ovum (p. 190) and the birth of the young in viviparous (↑) animals. It varies from species (p. 40) to species.

parturition (n) the process of giving birth to live young in viviparous (↑) animals by rhythmic contractions stimulated by the secretion (p. 106) of certain hormones (p. 130).

lactation (n) the production of milk in the mammary glands (p. 87) to nourish the young in mammals (p. 80).

puberty (n) the sexual maturity of a mammal (p. 80).

menopause (n) the period in females during which the menstrual cycle (p. 194) becomes irregular with increasing age of the individual before ceasing totally.

尿囊（名） 爬蟲動物（第78頁）、鳥類和哺乳動物（第80頁）的胚胎（第166頁）具有的一種消化道（第98頁）的囊狀突出物，長出於胚胎本體之外。覆蓋尿囊的結締組織（第88頁）有豐富的血管（第127頁），因而可起爲胚胎進行氣體交換（第112頁）和貯存排泄（第134頁）物的作用。

絨毛膜（名） 哺乳動物（第80頁）胚胎（第166頁）周圍的羊膜（第191頁）壁的外上皮（第87頁），即最外層的膜（第14頁）。它與尿囊（↑）結合發育成胎盤（↓）。

胎盤ᵃ（名） 一種於姙娠（第195頁）期間在子宮（第190頁）內發育的盤狀器官。它與胚胎（第166頁）、母體組織（第83頁）均有密切關係。胎盤的表面積巨大，可起附着母體和爲胎兒提供營養的作用。

臍帶 一根將胎盤（↑）與胎兒（第191頁）臍部相連的索帶。它可借助於兩支動脈（第127頁）和一支靜脈（第127頁）進行物質交換。

胎生（名） 指胚胎（第166頁）在子宮（第190頁）內發育，附着於胎盤（↑）之上，並活着出生的情況。（形容詞形式爲 viviparous）

姙娠期 胎生（↑）動物從卵（第190頁）受精（第175頁）到幼體出生所經歷的一段時間。姙娠期因種（第40頁）而異。

分娩、生產（名） 胎生（↑）動物藉某些激素（第130頁）的分泌（第106頁），刺激一些節律性收縮，使活的幼體產出的過程。

泌乳（名） 哺乳動物（第80頁）爲哺育幼體而在其乳腺（第87頁）中分泌出乳汁的情況。

青春期（名） 哺乳動物（第80頁）的性成熟期。

絕經期（名） 女性個體隨着年齡增長，在月經完全停止之前，月經周期（第194頁）變得不規則的時期。

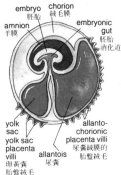

embryonic membranes of a mammal 哺乳動物的胎膜

embryo 胚胎
chorion 絨毛膜
amnion 羊膜
embryonic gut 胚胎消化道
yolk sac 卵黃囊
yolk sac placenta villi 卵黃囊胎盤絨毛
allantois 尿囊
allanto-chorionic placenta villi 尿囊絨膜的胎盤絨毛

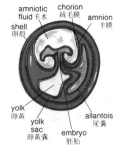

embryonic membranes in a reptilian egg 爬蟲動物卵的胎膜

amniotic fluid 羊水
chorion 絨毛膜
amnion 羊膜
shell 卵殼
yolk 卵黃
yolk sac 卵黃囊
allantois 尿囊
embryo 胚胎

sexual cycle the sequence of events which occurs in the females of animals that reproduce (p. 173) sexually and which, in humans, takes place on a monthly pattern with menstruation (p. 194) alternating with ovulation (p. 194).

oestrus cycle the rhythmic sexual cycle (↑) which occurs in mature females of most mammals (p. 80) assuming that the female does not become pregnant (p. 195). There are four main events in the oestrus cycle of which the most important is oestrus (p. 194) itself. In the *follicular phase*, there is growth of the Graafian follicles (p. 190), a thickening of the lining of the uterus (p. 190) and an increase in the production of oestrogen (p. 194). This is followed by *oestrus*. Then comes the *luteal phase* during which a corpus luteum (p. 190) grows from the Graafian follicle which secretes (p. 106) progesterone (p. 195) with a reduction in the secretion of oestrogen. If fertilization (p. 175) and pregnancy occur then the cycle is interrupted and the fourth phase does not follow. If fertilization does not occur, then the corpus luteum diminishes, hormone (p. 130) levels fall, and a new Graafian follicle begins to grow.

性周期 指以性方式進行生殖(第173頁)的各類雌性動物，其體內所發生事件之先後次序，在人體內，這些事件之先後次序是按月來月經(第194頁)和排卵(第194頁)兩者交替進行的方式發生的。

動情周期 大多數性成熟的雌性哺乳動物(第80頁)，在未懷孕(第195頁)情況下出現的有規律的性周期(↑)。動情周期有四個主要階段，其中最重要的是動情期(第194頁)本身。在卵泡期，囊狀卵泡(第190頁)生長，子宮(第190頁)壁增厚，雌激素(第194頁)分泌量增加；隨後是動情期；再後是黃體期，在此期間，由囊狀卵泡長出黃體，黃體分泌(第106頁)孕酮(第195頁)，同時，減少雌激素分泌量。如果發生受精(第175頁)和出現姙娠，動情周期即中止，這四個時期便停止循環。如果未發生受精，則黃體縮小，激素(第130頁)水平降低，新的囊狀卵泡又開始生長。

relationships between hormones secreted by pituitary, the oestrus cycle and pregnancy in a human
人的垂體分泌的激素、動情周期和姙娠之間的關係

① 促卵泡激素 (FSH)　　⑨ 月經
② 促黃體激素 (LH)　　　⑩ 月經
③ 動情期　　　　　　　⑪ 受精卵
④ 動情期　　　　　　　⑫ 修復
⑤ 囊狀卵泡　　　　　　⑬ 增生
⑥ 黃體　　　　　　　　⑭ 未受精
⑦ 排卵　　　　　　　　⑮ 植入
⑧ 排卵　　　　　　　　⑯ 姙娠

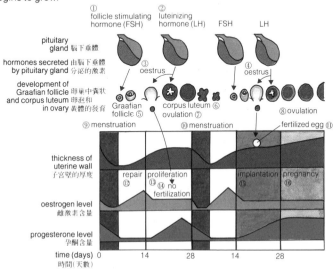

menstrual cycle in humans and some other primates a modified version of the oestrus cycle (p. 193) in which the oestrus (↓) is not obvious so that the female is continuously attractive and receptive to males. There is a regular discharge of blood (p. 90) and the lining of the uterus (p. 190) (menstruation) which occurs, following ovulation (↓) when fertilization (p. 175) does not occur.

ovulation (*n*) the release from the Graafian follicle (p. 190) of an immature ovum (p. 190) or oocyte (p. 189). It takes place, under the influence of a hormone (p. 130) released by the pituitary gland (p. 157) at regular intervals (approximately every 28 days in humans) and in the presence of oestrogen (↓). **ovulate** (*v*).

oestrus (*n*) a short period during the sexual cycle (p. 193) of animals in which the female ovulates (↑) and is also sexually attractive to males so that copulation (p. 191) takes place.

follicle-stimulating hormone FSH. A hormone (p. 130) produced by the pituitary gland (p. 157), following the completion of the oestrus cycle (p. 193) or pregnancy (↓) which stimulates the growth of the Graafian follicles (p. 190) and the ova (p. 190) in females and spermatogenesis (p. 187) in males.

oestrogen (*n*) a female sex hormone (p. 130), produced in the Graafian follicle (p. 190), which stimulates the production of a suitable environment (p. 218) for fertilization (p. 175) and then growth of the embryo (p. 166) by repairing the walls of the uterus (p. 190) following menstruation (↑). During the first part of the oestrus cycle (p. 191), it builds up until it stimulates the production of luteinizing hormone (↓) by the pituitary gland (p. 157). It is also involved in the development of other female organs associated with the sexual cycle.

luteinizing hormone LH. A hormone (p. 130) secreted (p. 106) by the pituitary gland (p. 157) under the influence of oestrogen (↑). It stimulates ovulation (↑) and development of a Graafian follicle (p. 190) into a corpus luteum (p. 190) which produces progesterone (p. 195).

月經周期　指人和其他一些靈長類動物所具有的變型動情周期(第193頁)。在這種變型動情周期中，動情期(↓)不顯著，因此，雌性對雄性能夠持續不斷地保持吸引力和接納能力。此外，還存在着一種有規則地排出血液(第90頁)和子宮(第190頁)內膜剝落的現象(月經)，這種現象在排卵(↓)之後而未受精(第175頁)時發生。

排卵(名)　未成熟的卵(第190頁)或卵母細胞(第189頁)從囊狀卵泡(第190頁)釋出的過程。在腦下垂體(第157頁)釋出一種激素(第130頁)的影響下，在雌激素(↓)參與下，排卵即以一定的間隔時間(人每隔28天左右)發生。(動詞形式爲 ovulate)

動情期(名)　動物性周期(第193頁)中的一個短暫時期。在此期間，雌性動物排卵(↑)，而對雄性動物也具有性吸引力，所以便發生交配(第191頁)。

促卵泡激素　即 FSH。動情周期(第193頁)或姙娠(↓)結束之後，由腦下垂體(第157頁)分泌的一種激素(第130頁)。這種激素可促進雌性動物的囊狀卵泡(第190頁)和卵(第190頁)生長，同時可促進雄性動物的精子發生(第187頁)。

雌激素(名)　囊狀卵泡(第190頁)中產生的一種雌性性激素(第130頁)。這種激素可借助於月經(↑)之後對子宮(第190頁)壁的修復，而爲受精(第175頁)創造一個適合的環境(第218頁)，進而促進胚胎(第166頁)的發育。在動情周期(第191頁)的第一階段，這種激素開始累積，直至它刺激腦下垂體(第157頁)分泌促黃體激素(↓)時爲止。該激素也參與其他與性周期有關的一些雌性器官的發育。

促黃體〔生成〕激素、黃體生成素　即 LH。在雌激素(↑)影響下，腦下垂體(第157頁)分泌(第106頁)的一種激素(第130頁)。這種激素可刺激排卵(↑)，促進囊狀卵泡(第190頁)發育成可分泌孕酮(第195頁)的黃體(第190頁)。

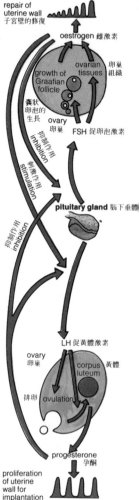

Interaction of hormones in the female sexual cycle
各種激素在雌性性週期中的相互作用

repair of uterine wall 子宮壁的修復

oestrogen 雌激素

growth of Graafian follicle 囊狀卵泡的生長

ovarian tissues 卵巢組織

ovary 卵巢

inhibition 抑制作用

stimulation 激發作用

inhibition 抑制作用

FSH 促卵泡激素

pituitary gland 腦下垂體

LH 促黃體激素

ovary 卵巢

corpus luteum 黃體

ovulation 排卵

progesterone 孕酮

proliferation of uterine wall for implantation 受精卵植入而產生的子宮壁增生

progesterone (*n*) a hormone (p. 130) secreted (p. 106) by the corpus luteum (p. 190) which stops further Graafian follicles (p. 190) from developing by preventing the secretion of follicle-stimulating hormone (↑). It also prepares the uterus (p. 190) for implantation (p. 191) of the ova (p. 190), and assists in the development of the placenta (p. 192) and mammary glands (p. 87).

pregnancy (*n*) the condition which occurs in a female following successful fertilization (p. 175) and implantation (p. 191). The oestrus cycle (p. 193) is suspended in the luteal phase. The production of hormones (p. 130) is altered so that they are produced by the placenta (p. 192) as well as the pituitary gland (p. 157) to ensure that parturition (p. 192) and lactation (p. 192) take place properly. **pregnant** (*n, adj*).

oxytocin (*n*) a hormone (p. 130) produced by the pituitary gland (p. 157) at the end of pregnancy (↑) which stimulates the contraction of uterine (p. 190) muscles (p. 143) during labour and prepares the mammary glands (p. 87) for the production of milk during lactation (p. 192).

prolactin (*n*) a hormone (p. 130) produced by the pituitary gland (p. 157) which stimulates and controls the production of milk during lactation (p. 192).

breeding season in animals in which the oestrus cycle (p. 193) does not occur continuously throughout the year, the time during which it does take place and which is usually under the influence of climate or other environmental (p. 218) factors.

androgens (*n.pl.*) the male sexual hormones (p. 130), such as testosterone (↓), produced essentially by the testes (p. 187), and which stimulate and control spermatogenesis (p. 187) as well as other male characteristics, such as the growth of facial hair.

testosterone (*n*) an androgen (↑) produced by male vertebrates (p. 74).

interstitial cell-stimulating hormone a luteinizing hormone (↑) which stimulates the secretion (p. 106) of androgens (↑) by the testes (p. 187) in males.

孕酮(名) 黃體(第190頁)分泌(第106頁)一種激素（第130頁）。這種激素可阻止分泌促卵泡激素（↑），從而阻碍囊狀卵泡(第190頁)的發育；並可使子宮(第190頁)作好植入(第191頁)卵(第190頁)的準備；還有助胎盤(第192頁)和乳腺(第87頁)的發育。

姙娠、受孕(名) 雌性動物成功受精(第175頁)和植入(第191頁)卵之後出現的狀況。動情周期(第193頁)中止於黃體期。各種激素(第130頁)的分泌也隨之發生變化，改由胎盤(第192頁)及腦下垂體(第157頁)分泌激素，以確保正常分娩(第192頁)和泌乳(第192頁)(名詞及形容詞形式爲 pregnant)

催產素(名) 在姙娠(↑)末期，腦下垂體(第157頁)分泌的一種激素(第130頁)。這種激素在分娩期間，可刺激子宮(第190頁)肌肉(第143頁)收縮；在泌乳(第192頁)期間，則使乳腺(第87頁)作好分泌乳汁的準備。

促乳素(名) 腦下垂體(第157頁)分泌的一種激素(第130頁)。這種激素在分泌乳(第192頁)期間，可刺激和控制分泌乳汁。

生育季節 指並非全年連續發生動情周期(第193頁)的動物處於發生動情周期時的那一段時間。這段時間通常受氣候和其他環境(第218頁)因素的影響。

雄激素(名、複) 主要由睾丸(第187頁)分泌的雄性性激素(第130頁)，例如睾酮(↓)。雄激素可刺激和控制精子發生(第187頁)以及其他一些雄性特徵如面毛的生長。

睾酮(名) 雄性脊椎動物(第74頁)分泌的一種雄激素(↑)。

促間質細胞激素(名) 一種促黃體素(↑)。這種激素可刺激雄性動物的睾丸(第187頁)分泌(第106頁)雄激素(↑)。

genetics (*n*) the study or science of inheritance concerning the variations between organisms and how these are affected by the interaction of environment (p. 218) and genes (↓).

inherit (*v*) to receive genetic (↓) material from one's parents or ancestors. **inheritance** (*n*).

genotype (*n*) the actual genetic (↓) make-up of an organism which may, for example, define the limits of its growth that are then affected by the environment (p. 218).

phenotype (*n*) the total characteristics and appearance of an organism. Organisms may have the same genotype (↑) while the phenotypes may be different because of the effects of the environment (p. 218).

genome (*n*) the genetic (↓) material.

gene (*n*) the smallest known unit of inheritance that controls a particular characteristic of an organism, such as eye colour. A gene may be considered to be a complex set of chemical compounds sited on a chromosome (p. 13). A gene may replicate to produce accurate copies of itself or mutate (p. 206) to give rise to new forms. **genetic** (*adj*).

Mendelian genetics the system of genetics (↑) developed by the Austrian monk, Gregor Mendel (1822–84), in which he studied inheritance by a series of controlled breeding experiments with the garden pea. He studied simple characteristics, controlled by a single gene (↑), and, using statistics, analyzed the results of cross breeding. In this way he showed that phenotypes (↑) did not result from a blending of genotypes (↑) but that the phenotypes were passed on in different ratios.

first filial (F₁) generation the first generation of offspring resulting from cross breeding pure lines (↓) or parentals (↓) of a single species (p. 40).

second filial (F₂) generation the generation of offspring resulting from the cross breeding between individuals of the first filial generation (↑).

pure line the succession of generations which results from the breeding of a homozygous (↓) organism so that they breed true and produce genetically (↑) identical offspring.

遺傳學（名）　研究遺傳的科學。其內容涉及生物體之間的種種變異及其如何受到環境（第218頁）和基因（↓）之間相互作用的影響。

遺傳（動）　指生物體從其親代或祖先獲得遺傳（↓）物質。（名詞形式爲 inheritance）。

基因型、遺傳型（名）　一個生物體的實際遺傳（↓）組成。例如，基因型可以確定生物體生長方面的種種限度，而這種限度以後受環境（第218頁）影響的。

表現型、表型（名）　一個生物體的性狀之總和。生物體可以具有相同的基因型（↑），但由於受環境（第218頁）的影響，它們的表現型則是不相同的。

基因組（名）　指遺傳（↓）物質。

基因（名）　已知的最小遺傳單位。它控制一個生物體的某種特定性狀，例如眼睛的顏色。一個基因可以被認爲是位於一個染色體（第13頁）上的一組複雜的化合物。基因可以複制，產生與自己一模一樣的複制品，或者發生突變（第206頁），變成新的型式。（形容詞形式爲 genetic）

孟德爾遺傳學　奧地利修道士格列高里·孟德爾（1822—84）創建的遺傳學（↑）體系。在創建該體系的過程中，他利用豌豆在控制的條件下進行一系列繁殖試驗，研究遺傳現象。他研究由單個基因（↑）控制的簡單性狀，並且利用統計方法分析雜交繁育的結果。從而證明表現型（↑）並不是由基因型（↑）融合產生，而是以不同的比例傳遞的。

第一子代、子一代 (F₁) 代　單一物種（第40頁）雜交繁育產生純系（↓）後代之第一代或單一物種親代（↓）產生後代之第一代。

第二子代、子二代 (F₂) 代　子一代（↑）個體間雜交繁育產生的子代。

純系　純合（↓）體繁育產生的連續世代。這些連續世代都是進行眞實繁育的，從而產生遺傳上（↑）相同的後代。

phenotype the actual appearance
表現型實際外觀

genotype the actual genetic makeup as determined by the chromosomes
基因型由染色體所決定的實際遺傳組成

chromosome
染色體

parental (*n*) the succession of generations which leads to filial generations (↑).

monohybrid inheritance the result of cross breeding from pure lines (↑) with one pair of contrasting characteristics to give offspring with one of the characteristics, such as Mendel's cross of tall and dwarf garden peas to give a tall monohybrid.

dominant (*adj*) of (1) a gene (↑) which gives rise to a characteristic that always appears in either a homozygous (↓) or a heterozygous (p. 198) condition e.g. in Mendel's cross of tall and dwarf garden peas, all the F_1 generation (↑) were tall while, in the F_2 generation (↑), tall individuals were in a ratio of 3:1 to dwarf individuals. Thus, the dominant gene was for tallness. (2) a plant species (p. 40) which, in any particular community (p. 217) of plants, is the most common and characteristic species of that community in its numbers and growth. The dominant species has a direct effect on the other plants in the community.

recessive (*adj*) of a gene (↑) which gives rise to a characteristic that can only appear in a homozygous (↓) condition and is suppressed by the dominant (↑) gene in the heterozygous (p. 198) condition. For example, in Mendel's cross of tall and dwarf garden peas, the recessive gene was for dwarfness.

allele (*n*) one of the alternative forms of a gene (↑) e.g. from the pair of genes designated BB giving rise to brown eyes and the pair of genes designated bb giving rise to blue eyes, the genes B and b are said to be alleles of the same gene and B is dominant (↑) while b is recessive (↑).

homozygous (*adj*) of an organism which has the same two alleles (↑) for a particular characteristic, such as eye colour. If a homozygote is crossed with a similar homozygote, it breeds true for that characteristic. If an organism is homozygous for every characteristic and breeds with a genetically (↑) identical organism, the offspring will be identical to the parents. This gradually occurs with constant inbreeding so that, while the organisms may well be adapted to their particular environment (p. 218), if it should change, they would be slow to respond.

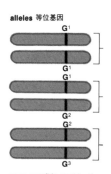

alleles 等位基因

G¹
G¹
G¹
G²
G²
G³

some possible combinations of 3 alleles on a chromosome pair 一對染色體上3個等位基因的幾種可能的組合

親代（名） 在連續不斷的世代中，每一代對其子代（↑）而言均屬親代。

單基因雜種遺傳 指具有一對對比性狀的純系（↑）進行雜交繁育的結果。這種結果使後代具有一種孟德爾式的性狀，例如孟德爾將高莖豌豆與矮莖豌豆進行雜交，產生了一種高莖豌豆的單基因雜種。

顯性的、優勢的（形） 1.指基因（↑）而言，這種基因產生的性狀總是或者表現在純合（↓）狀態或者表現在雜合（第198頁）狀態之中。例如，孟德爾對高莖豌豆與矮莖豌豆雜交的結果，所有的子一代（↑）都是高莖的，而在子二代（↑）中，高莖個體與矮莖個體之比爲3：1。由此可見，顯性基因是控制高度性狀的。(2)指某個植物種（第40頁）而言，在任何特定植物羣落（第217頁）中，該植物種在數量和生長方面都是那個植物羣落中最常見和最具特徵的種。優勢種對其羣落中的其他植物具有直接影響。

隱性的（形） 指一種基因（↑）而言，這種基因產生的性狀僅能表現在純合（↓）狀態之中，而在雜合（第198頁）狀態中，則被顯性（↑）基因所控制。例如，在孟德爾所進行的高莖豌豆與矮莖豌豆的雜交中，隱性基因是控制矮莖性狀的。

等位基因（名） 一個基因（↑）的許多可能狀態之一，例如從名爲 BB 的這對基因可產生出褐色的眼睛；而名爲 bb 的這對基因可產生出藍色的眼睛；基因 B 和 b 即被認爲是同一基因的等位基因，而 B 是顯性的（↑），b 則是隱性（↑）的。

純合的（形） 指某種生物體而言，該生物體對於如眼睛顏色這一特定性狀來說，具有相同的兩個等位基因（↑）。如果一個純合體與另一個類似的純合體進行雜交，則對那種性狀來說，純合體是能夠真實遺傳的。如果一個生物體對於每一種性狀來說，都是純合的，並與一個在遺傳上（↑）完全相同的生物體進行繁育，則其後代將與親代完全相同。上述這種情況是隨着不斷地進行近交而逐漸發生的，結果，雖然這些生物體可以充分適應其所處的特定環境（第218頁），但倘若環境發生變化，它們就難以很快地作出反應。

heterozygous (*adj*) of an organism which has two different alleles (p. 197) for a particular characteristic, such as eye colour, so that the dominant (p. 197) allele is expressed in the phenotype (p. 196). If a heterozygote breeds with a genetically (p. 196) identical heterozygote, some recessive (p. 197) characteristics will appear in some of the offspring. Heterozygous organisms are more adaptable to changing conditions than homozygous (p. 197) ones.

law of segregation Mendel's first law. One of the two laws formulated by the Gregor Mendel, to explain the way in which inheritance occurred. It states that in two alleles (p. 197) on a gene (p. 196) for a pair of characters, only one can be carried in a single gamete (p. 175).

雜合的(形)　指某種生物體而言，該生物體對於如眼睛顏色這一特定性狀來說，具有兩個不同的等位基因(第197頁)，所以，顯性(第197頁)等位基因便在表現型(第196頁)中表現出來。如果一個雜合體與另一個在遺傳上(第196頁)完全相同的雜合體進行繁育，則在一些後代中會出現某些隱性(第197頁)性狀。雜合體比純合(第197頁)更能適應環境的變化。

分離定律　即孟德爾第一定律。格列高里·孟德爾為解釋遺傳發生方式而提出的兩個定律之一。該定律認為，決定一對性狀的一個基因(第196頁)的兩個等位基因(第197頁)中，只有一個可以爲單配子(第175頁)攜帶。

homozygosity and heterozygosity 純合性和雜合性

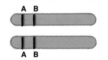

chromosome pair 染色體對
homozygous for gene **A** and gene **B**
基因 A 和基因 B 均是純合的

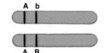

chromosome pair 染色體對
homozygous for gene **A**
heterozygous for gene **B**
基因 A 是純合的 基因 B 是雜合的

**Mendel's laws
1** segregation
孟德爾第一定律
分離定律

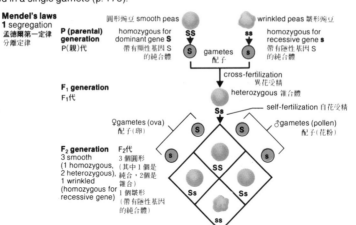

P (parental) generation
P(親)代

圓形豌豆 smooth peas
SS
homozygous for dominant gene **S**
帶有顯性基因 S 的純合體

wrinkled peas 皺形豌豆
ss
homozygous for recessive gene **s**
帶有隱性基因 S 的純合體

gametes 配子
S　**s**

cross-fertilization 異花受精

F₁ generation
F₁代
heterozygous 雜合體
Ss

self-fertilization 自花受精

♀gametes (ova) 配子(卵)
S　**s**

♂gametes (pollen) 配子(花粉)

F₂ generation F₂代
3 smooth (1 homozygous, 2 heterozygous), 1 wrinkled (homozygous for recessive gene)
3 個圓形 (其中 1 個是純合，2個是雜合) 1 個皺形 (帶有隱性基因的純合體)

SS　**Ss**　**Ss**　**ss**

test cross a test to show whether or not an organism, which shows a characteristic associated with a dominant (p. 197) gene (p. 196), is heterozygous (↑) or homozygous (p. 197) for that characteristic, by crossing it with a double recessive (↓) for the characteristic. If the organism under test is homozygous, all the offspring will show the characteristic of the dominant gene while, if it is heterozygous, half show the dominant character and half show the recessive (p. 197).

測交　一種測試方法，用於對表現出一種與顯性(第197頁)基因(第196頁)有關的性狀的一個生物體進行測試，以確定就這種性狀來說，該生物體到底是雜合的(↑)還是純合的(第197頁)。其具體步驟是，將該生物體與一個有關該性狀的雙隱性個體(↓)進行雜交。如果試驗的生物體是純合的，則其所有的後代都將表現出該顯性基因所決定的性狀；如果該生物體是雜合的，則其後代中的一半將表現出顯性性狀，另一半則表現出隱性(第197頁)性狀。

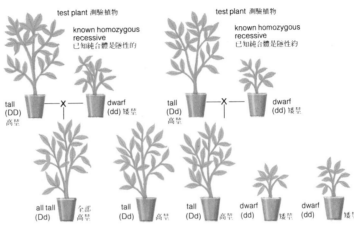

example of a test cross to
see whether a tall individual
is heterozygous or
homozygous. If
homozygous the progeny
are all tall; if heterozygous
half are tall and half dwarf

測定一個高莖個體是雜合體還是純合體的一個測定例子。如果該
個體是純合體，則其後代全部是高莖的；如果該個體是雜合體，
則其後代的一半是高莖的，一半是矮莖的。

test plant 測驗植物

known homozygous
recessive
已知純合體是隱性的

tall
(DD)
高莖

X

dwarf
(dd) 矮莖

test plant 測驗植物

known homozygous
recessive
已知純合體是隱性約

tall
(Dd)
高莖

X

dwarf
(dd) 矮莖

all tall
(Dd)

全部
高莖

tall
(Dd)

高莖

tall
(Dd)

高莖

dwarf
(dd)

矮莖

dwarf
(dd)

矮莖

double recessive an individual in which the
alleles (p. 197) of a particular gene (p. 196) are
identical for a recessive (p. 197) characteristic
so that the recessive characteristic is
expressed in the phenotype (p. 196).

carrier (*n*) an organism which may carry a
recessive (p. 197) gene (p. 196) for a
characteristic which may be harmful and which
is not expressed in the carrier because it is
masked by the dominant (p. 197) gene for that
characteristic.

dihybrid inheritance the result of cross breeding
from pure lines (p. 196) of homozygous (p. 197)
organisms with two different alleles (p. 197) for
different characteristics, such as Mendel's cross
of yellow round and wrinkled green garden
peas to give a yellow round dihybrid in which
the genes (p. 196) for yellow and round are
dominant (p. 197) and suppress the genes for
wrinkled and green which are recessive (p. 197).

dihybrid cross the result of dihybrid inheritance
(1). If the offspring of dihybrid inheritance are
self crossed, the characteristics are expressed
in the ratio 9:3:3:1, in other words, in the Mendel
cross, nine plants are yellow and round, three
are yellow and wrinkled, three are green and
round, and one is green and wrinkled.

雙隱性個體　指一種個體，在這種個體內，一個特
定基因(第196頁)的各等位基因(第197頁)，對
於表現一種隱性(第197頁)性狀來說，都是相
同的，所以，該隱性性狀便在其表現型(第196
頁)中表現出來。

攜帶者(名)　帶有一個決定某一性狀的隱性(第197
頁)基因(第196頁)的生物體。該隱性基因可能
是有害的，但由於它被決定那一性狀的顯性
(第197頁)基因所抑制，而未在攜帶者中表現
出來。

雙因子雜種遺傳　指由純合(第197頁)體構成的純
系(第196頁)進行雜交繁育的結果，其純合體
具有決定不同性狀的兩個不同的等位基因(第
197頁)。例如，孟德爾將黃色和圓形的豌豆與
綠色和皺形的豌豆進行雜交，產生了一種黃
色、圓形的雙因子雜種豌豆。在這種豌豆中，
決定黃色和圓形的各基因(第196頁)都是顯性
(第197頁)基因，抑制了決定綠色和皺形的各
隱性(第197頁)基因。

雙因子雜種雜交　指雙因子雜種遺傳(↑)的結果。
如果雙因子雜種遺傳的後代進行自交，則各性
狀以 9：3：3：1的比例表現出來，換言
之，在孟德爾的雜交中，有九株植物是黃色和
圓形的，三株是黃色和皺形的，三株是綠色和
圓形的。一株是綠色和皺形的。

law of independent assortment Mendel's second law. One of the two laws of inheritance formulated by the Gregor Mendel, which states that each member of one pair of alleles (p. 197) is as likely to be combined with one member of another pair of alleles as with any other member because they associate randomly (and independently).

獨立分配定律、自由組合定律　即孟德爾第二定律。格列高里·孟德爾提出兩個遺傳定律之一。該定律認為，一對等位基因(第197頁)中的每一個等位基因可能與另一對等位基因中的一個等位基因相結合，也可能與其他任何基因結合，因為它們是隨機地(而且是獨立地)結合的。

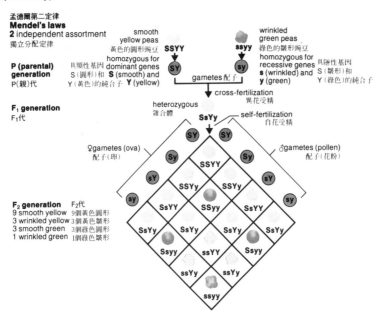

孟德爾第二定律
Mendel's laws
2 independent assortment
獨立分配定律

P (parental) generation
P(親)代

smooth yellow peas
黃色的圓形豌豆
homozygous for dominant genes
具顯性基因 **S**(圓形)和 **Y**(黃色)的純合子
SSYY
S (smooth) and **Y** (yellow)

wrinkled green peas
綠色的皺形豌豆
homozygous for recessive genes
具隱性基因 **s**(皺形)和 **y**(綠色)的純合子
ssyy
s (wrinkled) and **y** (green)

gametes 配子

cross-fertilization
異花受精

F₁ generation
F₁代

heterozygous
雜合體
SsYy

self-fertilization
自花受精

♀gametes (ova)
配子(卵)

♂gametes (pollen)
配子(花粉)

F₂ generation　F₂代
9 smooth yellow　9個黃色圓形
3 wrinkled yellow　3個黃色皺形
3 smooth green　3個綠色圓形
1 wrinkled green　1個綠色皺形

SSYY
SSYy SSYy
SsYY SSYy SsYY
SsYy SsYy SsYy SsYy
Ssyy ssYy Ssyy
ssYy ssYy
ssyy

progeny (*n.pl.*) the offspring which result from reproduction (p. 173).

後代(名、複)　繁殖(第173頁)產生的子孫。

linkage (*n*) the situation in which genes (p. 196) on the same chromosome (p. 13) are said to be linked so that they are unable to assort according to the Law of Independent Assortment (↑) and are inherited together.

連鎖(名)　同一染色體(第13頁)上有多個基因(第196頁)相環連的情況，結果這些基因不能按照獨立分配定律(↑)進行分配，而是一起遺傳給後代。

linkage group a group of linked (↑) genes (p. 196) on the same chromosome (p. 13) which are inherited together.

連鎖羣　位於同一染色體(第13頁)上、能一起遺傳給後代的一羣連鎖(↑)基因(第196頁)。

sex chromosomes 性染色體

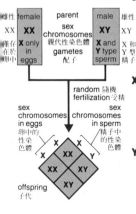

sex chromosomes the chromosomes (p. 13) which control whether or not a given individual of most animals should be male or female. There is a homologous (p. 39) pair of chromosomes in the nucleus (p. 13) of one sex, usually the female, and an unlike or single chromosome in the nucleus of the other, usually the male.

X chromosomes the sex chromosomes (↑) which occur as a like pair XX in the nuclei (p. 13) of the homogametic (↓) sex and usually are responsible for the female sex in most animals. All the gametes (p. 175) of the homogametic sex will contain one X chromosome.

Y chromosomes the sex chromosomes (↑) which occur either as an unlike pair with an X chromosome (↑) or unpaired in the nuclei (p. 13) of the heterogametic (↓) sex and are usually responsible for the male sex in most animals. The gametes (p. 175) of the heterogametic sex are of two kinds, with or without an X chromosome, which are equal in number.

hetersomes (*n.pl.*) homologous chromosomes (p. 39), such as the sex chromosomes (↑) which are not normally identical in appearance.

autosomes (*n.pl.*) homologous chromosomes (p. 39) which are not sex chromosomes (↑) and which are normally identical in appearance.

homogametic sex the sex, usually the female, which contains sex chromosomes (↑) that occur as a like pair of XX chromosomes (p. 13) in the nuclei (p. 13) of an organism.

heterogametic sex the sex, usually the male, which contains sex chromosomes (↑) that occur as an unlike pair of XY chromosomes (p. 13) or unpaired in the nuclei (p. 13) of an organism.

sex-linked (*adj*) of certain characteristics associated with recessive (p. 197) genes (p. 196) which are linked to the sex of the individual because they are attached to the X chromosome (↑).

colour blindness a sex-linked (↑) characteristic in which there is an inability to distinguish between pairs of colours, usually red/green, although the ability to distinguish shade and form is unaffected.

inheritance of colour blindness 色盲的遺傳

X = normal sex chromosome
X＝正常的性染色體

X^c = sex chromosome with gene for colour blindness
X^c＝具有色盲基因的性染色體

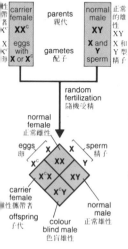

性染色體 決定大多數動物某一個體的性別爲雄性或雌性的染色體（第13頁）。在通常是雌性別的細胞核（第13頁）中，有一對同源（第39頁）染色體；而在通常是雄性的另一性別的細胞核中，則一對之中的兩個染色體彼此是不相同的，或只有單個染色體。

X 染色體 在同配（↓）性別的細胞核（第13頁）內，以一對相同的染色體 XX 形式出現的性染色體（↑）。X染色體通常決定大多數動物的雌性性別。同配性別的所有配子（第175頁）都含有一個 X 染色體。

Y 染色體 在異配（↓）性別的細胞核（第13頁）內，以其中具有一個 X 染色體（↑）的一對不相同的染色體 XY 形式出現，或者以單個而不成對的形式出現的性染色體（↑）。Y 染色體通常決定着大多數動物的雄性性別。異配性別的配子（第175頁）有兩種，一種具有一個 X 染色體，另一種沒有 X 染色體，這些配子在數量上是相同的。

異源染色體（名、複） 外表上通常是不相同的同源染色體（第39頁），例如性染色體（↑）。

常染色體（名、複） 在外表上通常是相同的那些非性染色體（↑）的同源染色體（第39頁）。

同配性別 通常爲雌性的性別，在這種性別的生物體的細胞核（第13頁）內，含有以一對相同的染色體（第13頁）XX 形式出現的性染色體（↑）。

異配性別 通常爲雄性的性別，在這種性別的生物體的細胞核（第13頁）內，含有以一對不相同的染色體（第13頁）XY 形式出現，或者以單個而不成對形式出現的性染色體（↑）。

伴性的、性連鎖的（形） 指與隱性（第197頁）基因（第196頁）有關的某些性狀而言，這些隱性基因與個體的性別是相聯係的，因爲它們附屬於 X 染色體（↑）。

色盲 一種伴性（↑）性狀。具有這種性狀時，缺乏辨別成對的一些不同顏色的能力，通常的一對爲紅色／綠色，但並不影響對色彩的濃淡和由色彩構成的形態加以區分的能力。

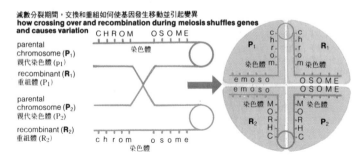

減數分裂期間，交換和重組如何使基因發生移動並引起變異
how crossing over and recombination during meiosis shuffles genes and causes variation

parental chromosome (**P₁**)
親代染色體 (p₁)

recombinant (**R₁**)
重組體 (P₁)

parental chromosome (**P₂**)
親代染色體 (P₂)

recombinant (**R₂**)
重組體 (R₂)

haemophilia (*n*) a sex-linked (p. 201) characteristic or disease, known only in males, in which the blood (p. 90) is unable to clot (p. 129) properly after wounding.

crossing over the exchange of genetic material (p. 203) during meiosis (p. 39), between male and female parentals (p. 197) in which the chromatids (p. 35) of homologous chromosomes (p. 39) break at the chiasmata (p. 39) and rejoin to allow the assortment of linked (p. 201) genes. Crossing over leads to increased variation (p. 213).

recombinants (*n.pl.*) the gametes (p. 175) which result from crossing over (↑) so that an exchange of genetic material (p. 203) between the parentals (p. 197) gives rise to some characteristics which are not present in either parental. This leads to increased variability in the offspring and greater change of adaptation to changing conditions. **recombine** (*v*).

crossover frequency the number of recombinants (↑) that are likely to occur as a result of the crossing over (↑) between two genes (p. 196) on different parts of the same chromosome (p. 13). It is usually expressed as a percentage of the number of recombinants compared with the total number of offspring produced. The crossover frequency is lower the closer the genes occur on chromosomes.

chromosome map a diagram of the order and distance between the genes (p. 196) on a chromosome (p. 13) worked out by experiments and an analysis of the crossover frequency (↑).

血友病（名）　一種伴性（第201頁）性狀或疾病。僅存在於男性中。患此病的病人，在受傷後，其血液（第90頁）不能正常凝固（第129頁）。

交換、互換　在減數分裂（第39頁）過程中，雄性與雌性親代（第197頁）之間進行的遺傳物質（第203頁）交換，進行交換時，同源染色體（第39頁）的染色單體（第35頁）在交叉（第39頁）處斷裂，爾後再接合，使連鎖（第201頁）基因得以分配。交換導致增加變異（第213頁）。

重組體（名、複）　由交換（↑）產生的配子（第175頁）。正是由於雙親（第197頁）之間遺傳物質（第203頁）的交換，才產生在雙親中都不存在的某些性狀。這就使後代中產生的變異性增加，也使其因適應環境變化而產生的變異加大。（動詞形式爲 recombine）

交換率、互換率　由同一染色體（第13頁）不同部份的兩個基因（第196頁）之間交換（↑）而可能產生的重組體（↑）數目。交換率通常以重組體數目與所產生後代的總數之百分比來表示。交換率越低，染色體上各基因之間的距離就越近。

染色體圖　根據實驗和交換率（↑）分析作出的染色體（第13頁）上基因（第196頁）之間的順序和距離的示意圖。

crossing-over
during first meiotic division
第一次減數分裂期間的交換

non-sister chromatids
非姊妹染色單體

sister chromatids
姊妹染色單體

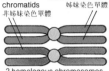

2 homologous chromosomes
2個同源染色體

chiasma
交叉

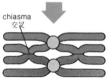

chiasmata formed
形成交叉

bivalent or tetrad
二價染色體或四聯體

genetic material exchanged, chromosomes separate
交換遺傳物質後，染色體分離

locus 座位(位點)
homologous chromosomes
同源染色體

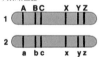

alleles **A, B, C, X, Y, Z**, occupy same loci (positions) on chromosome 1 as alleles **a, b, c, x, y, z**, on chromosome 2
等位基因 A, B, C, X, Y, Z 在第1個染色體上所佔據的座位(位置)與等位基因 a, b, c, x, y, z 在第2個染色體上所佔據的座位相同。

gene locus the precise position of a gene (p. 196) on the chromosome (p. 13). Alleles (p. 197) of the same gene occupy the same loci (*pl.*) on homologous chromosomes (p. 39).

multiple alleles a series of three or more alleles (p. 197) on the same gene (p. 196) which give rise to a particular characteristic. Only two of these alleles in various combinations can occupy the same gene locus (↑) on a pair of homologous chromosomes (p. 39) at the same time.

lethal alleles alleles (p. 197) which will kill the individual if they are dominant (p. 197) in a heterozygous (p. 198) individual or if they are recessive (p.197) in a homozygous (p.197) one.

partial dominance the situation which occurs between dominant (p. 197) alleles (p. 1.97) in which one may be slightly more dominant than the other. For example, if the allele for red is dominant in individuals of the same flower while the allele for white is dominant in other individuals, breeding them may produce pink flowers which will be reddish pink if the allele for red is more dominant than the allele for white. Also known as **co-dominance**.

epistasis (*n*) the interaction of non-allelic (p. 197) genes (p. 196) in which one gene suppresses the characteristics which would normally be expressed by another gene. It is similar to recessiveness (p. 197) and dominance (p. 197) between alleles.

genetic material the organic compounds (p. 15) which carry the genetic (p. 196) information from one generation to the next and from cell to cell. Chromosomes (p. 13) are composed of proteins (p. 21) and DNA (p. 24) which carries the genetic information.

genetic code the sequence of the four bases (p.22) adenine (p. 22), guanine (p. 22), cytosine (p.22), and thymine (p. 22) on a strand of DNA (p. 24) represents a code that controls the construction of proteins (p. 21) and enzymes (p. 28) which make up the cytoplasm (p. 10) of an organism and directs its functioning. Triplets of these bases code for the twenty different amino acids (p. 21), and groups of these triplets, code for whole proteins. More than one triplet can code for an amino acid.

基因座位、基因位點 基因(第196頁)在染色體(第13頁)上的準確位置。同一基因的各等位基因(第197頁)在同源染色體(第39頁)上佔據相同的座位。

複等位基因 指產生一特定性狀的、同一基因(第196頁)的三個或三個以上的一系列等位基因(第197頁)。在各種組合中,這些等位基因中只有兩個可以同時在一對同源染色體(第39頁)上佔據同一基因座位(↑)。

致死等位基因 會使個體致死的等位基因(第197頁)。如果這些等位基因在雜合(第198頁)個體中是顯性的(第197頁),或者在純合(第197頁)個體中是隱性的(第197頁),那麼,它們即可產生上述的致死效應。

不完全顯性 顯性(第197頁)等位基因(第197頁)之間存在的狀況,其中,一種顯性等位基因可能略比另一種顯性等位基因更呈顯性。例如,如果在同一種花中,決定紅花性狀的等位基因在開紅花的一些個體中是顯性的,而決定白花性狀的等位基因在其他一些個體中是顯性的,且決定紅花性狀的等位基因比決定白花性狀的等位基因更呈顯性,倘若將它們進行雜交,即可產生粉紅色的花,這種粉紅色則是偏紅的。它也稱爲**等顯性**。

上位(名) 非等位(第197頁)基因(第196頁)之間的相互作用,在這種相互作用中,一個基因抑制通常應該由另一個基因表現的性狀。這種作用與等位基因之間表現隱性(第197頁)和顯性(第197頁)的情況相似。

遺傳物質 將遺傳(第196頁)信息從一代傳遞給下一代,從一個細胞傳遞到另一個細胞的有機化合物(第15頁)。染色體(第13頁)由蛋白質(第21頁)和載有遺傳信息的 DNA(第24頁)組成。

遺傳密碼 一條 DNA(第24頁)鏈上的四個鹼基(第22頁),即腺嘌呤(第22頁)、鳥嘌呤(第22頁)、胞嘧啶(第22頁)和胸腺嘧啶(第22頁)的排列順序代表一種密碼,這種密碼控制着組成生物體細胞質(第10頁)的蛋白質(第21頁)和酶(第28頁)的合成,也操縱着它們所行使的功能。由這些鹼基組成的三聯體,爲20種不同的氨基酸(第21頁)編碼;而這些三聯體的各種組合,則可爲全部蛋白質編碼。能爲一種氨基酸編碼的三聯體不止一種。

amino acids and the genetic code
(see previous page) 氨基酸和遺傳密碼(見前頁)

amino acid general formula
氨基酸的通式

$$NH_3^+-\overset{COO^-}{\underset{R}{\overset{|}{C}}}-H$$

R = side group R＝側基

密碼子　　氨基酸

codon	amino acid	side group (R) 側基 (R)	side group (R) 側基 (R)	amino acid 氨基酸	codon 密碼子
AAA AAG	lysine 賴氨酸	$-CH_2CH_2CH_2CH_2NH_3^+$	$-H$	glycine 甘氨酸	GGU GGC GGA GGG
AAU AAC	asparagine 天冬醯胺	$-CH_2CONH_2$	$-CH_2COO^-$	aspartic acid 天冬氨酸	GAU GAC
ACU ACC ACA ACG	threonine 蘇氨酸	$-CHOHCH_3$	$-CH_2CH_2COO^-$	glutamic acid 谷氨酸	GAA GAG
AGU AGC	serine 絲氨酸	$-CH_2OH$	$-CH_3$	alanine 丙氨酸	GCU GCC GCA GCG
AGA AGG	arginine 精氨酸	$-CH_2CH_2CH_2NHC\overset{NH_2}{\underset{N^+H_2}{}}$	CH_3CHCH_3	valine 纈氨酸	GUU GUC GUA GUG
AUU AUC AUA	isoleucine 異亮氨酸	$CH_3CH_2CHCH_3$			
AUG	methionine 甲硫氨酸	$-CH_2CH_2SCH_3$	phenylalanine benzene ring	phenylalanine 苯丙氨酸	UUU UUC
CCU CCC CCA CCG	proline 脯氨酸	proline ring structure	$-CH_2-CH\overset{CH_3}{\underset{CH_3}{}}$	leucine 亮氨酸	UUA UUG
CAU CAC	histidine 組氨酸	histidine ring structure	tyrosine ring $-COH$	tyrosine 酪氨酸	UAU UAC
				NONSENSE 無意義	UAA UAG
CAA CAG	glutamine 谷氨醯胺	$-CH_2CH_2CONH_2$	$-CH_2SH$	cysteine 半胱氨酸	UGU UGC
CGU CGC CGA CGG	arginine 精氨酸	$-CH_2CH_2CH_2NHC\overset{N^+H_2}{\underset{NH_2}{}}$	tryptophan ring structure	tryptophan 色氨酸	UGG
CUU CUC CUA CUG	leucine 亮氨酸	$-CH_2CH\overset{CH_3}{\underset{CH_3}{}}$		NONSENSE 無意義	UGA
			$-CH_2OH$	serine 絲氨酸	UCU UCC UCA UCG

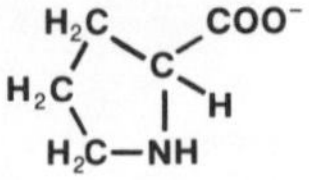

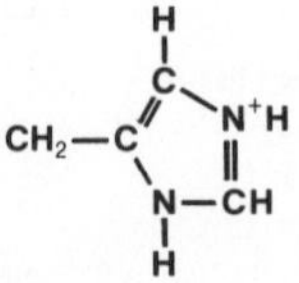

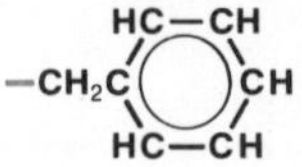

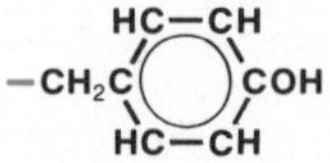

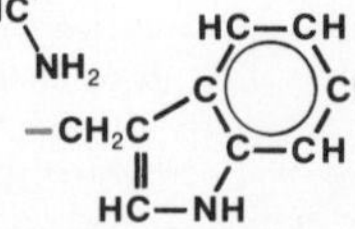

transcription (*n*) the process in which the genetic code (p. 203) is, in the first place, copied from the DNA (p. 24) on to a single strand of RNA (p. 24) in the nuclei (p. 13) of cells.

translation (*n*) the process in which the messenger RNA (p. 24) from the transcription (↑) then leaves the nucleus (p. 13) and passes into the ribosomes (p. 10) in the cytoplasm (p. 10) to function as a pattern from which amino acids (p. 21) are built into proteins (p. 21).

轉錄（作用）（名） 在細胞核（第13頁）內，首先將遺傳密碼（第203頁）從 DNA（第24頁）抄錄到單鏈 RNA（第24頁）上的過程。

轉譯、翻譯（名） 經過轉錄（↑）形成的信使 RNA（第24頁）離開細胞核（第13頁）進入細胞質（第10頁）中的核糖體（第10頁），從而起模板作用的過程。各種氨基酸（第21頁）根據這種模板合成種種蛋白質（第21頁）。

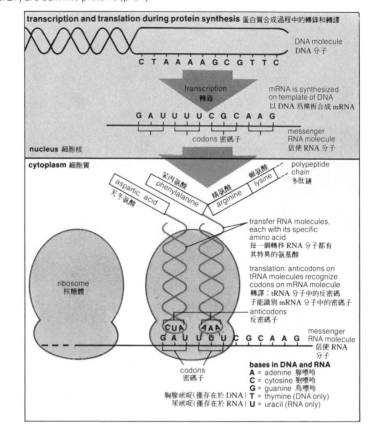

transcription and translation during protein synthesis 蛋白質合成過程中的轉錄和轉譯

mutation (*n*) a change in the structure of the genetic material (p. 203) of an organism which will be inherited if it occurs in the cells which produce the gametes (p. 175). It can occur as a result of changes in genes (p. 196) or of changes in the structure or number of chromosomes (p. 13). Most mutations are harmful but some allow the organism to adapt to changing circumstances and as a source of increased variation (p. 213) are the very material of evolution (p. 208). Mutations can be stimulated by increases in certain chemicals or ionizing radiation.

mutant (*n*) the result of a mutation (↑) which is usually recessive (p. 197) in the most common types of mutation.

mutagenic agent some stimulus, such as certain chemicals or ionizing radiation, which is likely to cause a mutation (↑).

chromosome mutation a change or mutation (↑) of the number or arrangement of the chromosomes (p. 13).

deletion (*n*) a chromosome (p. 13) mutation (↑) which occurs if a segment of a chromosome breaks away and is lost during nuclear division (p. 35) with a resulting loss of genetic material (p. 203).

inversion (*n*) (1) a chromosome (p. 13) mutation (↑) which occurs if a segment of a chromosome breaks away during nuclear division (p. 35) and rejoins the chromosome the wrong way round to reverse the sequence of genes (p. 196); (2) a gene mutation (↑) in which the order of the bases (p. 22) in a strand of DNA (p. 24) is changed.

translocation[2] (*n*) a chromosome (p. 13) mutation (↑) which occurs if a segment of the chromosome breaks away during nuclear division (p. 35) and rejoins the original chromosome in a different place or joins another chromosome.

duplication (*n*) a chromosome (p. 13) mutation (↑) in which a segment of the chromosome is duplicated either on the same or on another chromosome.

突變（名） 一個生物體的遺傳物質（第203頁）在結構上發生的變化。這種變化如果發生在在產生配子（第175頁）的細胞中，便可以遺傳。基因（第196頁）的變化或者染色體（第13頁）結構和數目的變化，均可產生突變。大多數突變是有害的，但是，有些突變使生物體能夠適應環境變化，而作爲增加變異（第213頁）根源的這些突變，正是進化（第208頁）的物質基礎。某些化學物質或電離輻射的增加均能促使發生突變。

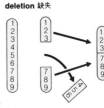
deletion 缺失

突變型、突變體（名） 在大多數普通突變類型中，通常爲隱性的（第197頁）突變（↑）的結果。

誘變劑、誘變因素 有可能引起突變（↑）的一些刺激物，例如某些化學物質或電離輻射。

染色體突變 染色體（第13頁）數目或結構的變化或突變（↑）。

缺失（名） 一種染色體（第13頁）突變（↑）。如果在核分裂（第35頁）過程中，一個染色體的一個片段斷裂、丟失，導致喪失遺傳物質（第203頁），即會發生這種突變。

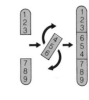
inversion 倒位

倒位（名） （1）一種染色體（第13頁）突變（↑）。如果在核分裂（第35頁）過程中，一個染色體的一個片段斷裂下來，倒轉了方向，錯誤地使染色體再接合，導致基因（第196頁）的順序顛倒，即會發生這種突變；（2）一種基因突變（↓）。發生這種基因突變時，DNA（第24頁）鏈上的鹼基（第22頁）的排列順序發生變化。

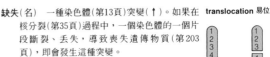
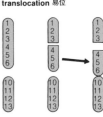
translocation 易位

易位（名） 一種染色體（第13頁）突變（↑）。如在核分裂（第35頁）過程中，一個染色體的一個片段斷裂後在一個不同位置與原來的染色體再結合，或者該斷裂片段與另一個染色體相連接，即會發生這種突變。

duplication 重複

複製（名） 一種染色體（第13頁）突變（↑）。在發生這種染色體突變的情況下，染色體的一個片段在同一染色體上或者在另一個染色體上出現重複現象。

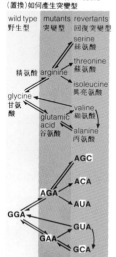

how changes of single
nucleotides in a triplet
(substitution) can cause
mutants to occur
三聯體中的單個核苷酸的變化
（置換）如何產生突變型

gene mutation a mutation (↑) in which the sequence of bases (p. 22) is not copied precisely in replicating a strand of DNA (p. 24) resulting in a change in the formation of the proteins (p. 21). Once it has occurred, it is replicated in the formation of further strands of DNA.

substitution (*n*) a gene mutation (↑) in which one DNA (p. 24) base (p. 22) is replaced by another.

insertion (*n*) a gene mutation (↑) in which another base (p. 22) is inserted in the existing sequence of bases in the strand DNA (p. 24).

sickle-cell anaemia a disease which is inherited and exhibits partial dominance (p. 203). A sickle cell contains a mutant (p. 196) gene (p. 196) which crystallizes haemoglobin (p. 126) in the erythrocytes (p. 91) of human blood (p. 90) and distorts them causing the blood vessels (p. 127) to clog. It is found usually among negroid people and is thought to offer some resistance to malaria.

polyploidy (*n*) the condition in which the cells of an organism contains at least three times the normal haploid (p. 36) number of chromosomes (p. 13). **polyploid** (*adj*).

triploid (*adj*) of a polyploid (↑) cell in which there is three times the normal haploid (p. 36) number of chromosomes (p. 13). It results from the fusion of a haploid and a diploid (p. 36) gamete (p. 175).

tetraploid (*adj*) of a polyploid (↑) cell in which there is four times the normal haploid (p. 36) number of chromosomes (p. 13). It occurs as a result of the fusion of two diploid (p. 36) cells.

aneuploidy (*n*) the condition in which chromosome (p. 13) mutation (↑) results in the gain or loss of chromosomes from a set.

euploidy (*n*) the condition in which chromosome (p. 13) mutation (↑) results in the gain of a whole set of chromosomes.

autopolyploid (*adj*) of the condition of polyploidy (↑) which results from euploidy (↑) in which the cell has multiple sets of its chromosomes (p. 13).

allopolyploid (*adj*) of the condition of polyploidy (↑) which results from euploidy (↑) in which the cell contains two different sets of chromosomes (p. 13) from the hybridization (p. 216) of two closely related organisms especially plants.

基因突變 在複製 DNA（第24頁）鏈時，未精確複製鹼基（第22頁）的順序，導致蛋白質（第21頁）形成發生變化的一種突變（↑）。這種突變一旦發生，就會在其他一些 DNA 鏈的形成中重現。

置換（名） DNA（第24頁）的一個鹼基（第22頁）被另一個鹼基取代所發生的一種基因突變（↑）。

插入（名） 在 DNA（第24頁）鏈的原有鹼基序列中，插入另一個鹼基（第22頁）時所發生的一種基因突變（↑）。

鐮形細胞貧血症 一種表現不完全顯性（第203頁）的遺傳病。鐮形細胞含有一個突變（↑）基因（第196頁），這個突變基因能使人血液（第90頁）紅細胞（第91頁）內的血紅蛋白（第126頁）結晶，並能使紅細胞變形，導致血管（第127頁）阻塞。該病通常見於黑種人，它對瘧疾有一些抗病作用。

多倍性（名） 指一個生物體的細胞內所含的染色體（第13頁）數目至少達三倍於正常的單倍（第36頁）數的狀態。（形容詞形式為 polyploid）

三倍體的（形） 指一種多倍體（↑）細胞而言，在該種細胞內，染色體（第13頁）的數目達三倍於正常的單倍（第36頁）數。它是由一個單倍體配子（第175頁）和一個二倍體（第36頁）配子融合而成的。

四倍體的（形） 指一種多倍體（↑）細胞而言，在該種細胞內，染色體（第13頁）的數目達四倍於正常的單倍（第26頁）數。它是由兩個二倍體（第36頁）細胞融合而成的。

非整倍性（名） 指染色體（第13頁）突變（↑）導致染色體組中的染色體增加或減少的狀態。

整倍性（名） 指染色體（第13頁）突變（↑）導致完整的染色體組增加的狀態。

同源多倍體的（形） 指由整倍性（↑）產生多倍性（↑）的狀態而言，處於這種狀態時，細胞具有多組染色體（第13頁）。

異源染色體的（形） 指由整倍性（↑）產生多倍性（↑）的狀態而言，處於這種狀態時，細胞含有兩組不同的染色體（第13頁），這些染色體是由兩種關係密切的生物，尤其是植物，雜交（第216頁）而獲得的。

evolution (*n*) the process whereby all organisms descend from the common ancestors which emerged on the Earth. Over successive generations throughout geological time, populations (p. 214) are modified in response to changes in environment (p. 218) by such processes as natural selection (↓) so that new species (p. 40) are formed which are all related, however distantly, by common descent.

Darwinism (*n*) the mechanism first put forward by the British naturalist, Charles Darwin (1809-82), following careful observation of animals and plants all over the world, e.g. Darwin's finches, to explain how organisms changed slowly, over millions of years, to evolve new forms. He suggested that in any given population (p. 214) of an organism there was considerable variation (p. 213) between individuals. Some would exhibit different characteristics which would be better fitted to their circumstances and environment (p. 218) than others. Thus, these individuals would be more likely to survive to maturity and breed so that their offspring would also exhibit these characteristics. Those individuals that were less well suited to their conditions would have less chance of breeding success so that eventually the population would contain more and more of the individuals that were better suited to their environment and the character of the species (p. 40) would change as a whole and result in a new species. Darwin was unable to explain, however, how the variations were produced in the first place. He called this process the theory of natural selection (↓).

natural selection one of the central deductions of Darwinism (↑). If the variations (p. 213), which occur among individuals within a population (p. 214) of animals, gives to those individuals a better chance of surviving, they are more likely to reach sexual maturity and breed so that their offspring will also inherit those advantageous characteristics. Eventually, the inheritance of variations in a particular direction over generations will lead to a new species (p. 40).

進化、演化（名）　指所有各種生物從地球上最初出現的那些共同祖先轉變而來的過程。在經歷整個地質時期的連續世代中，各種羣（第214頁）經歷了自然選擇（↓）的過程，隨着環境（第218頁）的變化而發生各種各樣的變異，從而由共同的祖先產生許多雖然彼此都有親緣關係但關係却很疏遠的新物種（第40頁）。

達爾文學說（名）　英國博物學家查理·達爾文（1809—82）經過對世界各地的動植物，例如達爾文雀，進行仔細的考察之後首先提出的理論，用以解釋各種生物在數百萬年時間中是如何緩慢地發生變化而進化成爲新物種的。他認爲，在任何一個特定的生物種羣（第214頁）中，個體之間存在着相當大的差異（第213頁）。某些個體會顯示出一些與衆不同的特性，這是指比其它一些個體更能適應其周圍的情況和環境（第218頁）。因而，這些個體更有可能存活下來，直至成熟，並進行繁育，從而使其後代也顯示出這些特性。而那些不太適應其環境的個體，能成功繁育後代的機會就比較少。因此，最終使該種羣中所包含的更能適應其環境的個體就越來越多，物種（第40頁）的性狀也就會在整體上發生變化，從而產生出新的物種。然而，達爾文却首先就無法解釋各種變異究竟是如何產生的。他把這一進化過程稱爲自然選擇（↓）學說。

自然選擇、天擇　達爾文學說（↑）的中心推論之一。如果在一個動物種羣（第214頁）的各個體中發生的變異（第213頁）給予那些個體以更多的生存機會，那麼，它們就更有可能達到性成熟和進行繁育，從而使它們的後代也能繼承那些有益的性狀。這些變異按照一個特定的方向，經過各世代的遺傳，最終就會導致產生新物種（第40頁）。

Darwin's finches on the Galapagos, Darwin observed that there were many species of finches which he thought had evolved from one species after it had arrived on the islands
達爾文雀
在加拉帕戈斯羣島，達爾文觀察到，羣島上存在着許多雀種，他認爲這些雀種都是由一個雀種在其到達該羣島之後才進化而來的

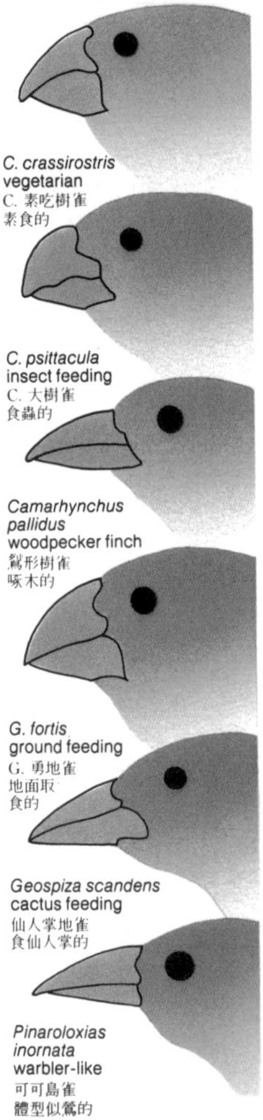
C. crassirostris
vegetarian
C. 素吃樹雀
素食的
C. psittacula
insect feeding
C. 大樹雀
食蟲的
Camarhynchus pallidus
woodpecker finch
鴷形樹雀
啄木的
G. fortis
ground feeding
G. 勇地雀
地面取食的
Geospiza scandens
cactus feeding
仙人掌地雀
食仙人掌的
Pinaroloxias inornata
warbler-like
可可島雀
體型似鶯的

survival of the fittest the idea behind natural selection (↑), which suggests that only the animals which are best fitted to their circumstances will survive in the struggle for existence while those that are less well fitted will tend to perish.

Neodarwinism (*n*) the modern, modified version of Darwinism (↑) which, with the aid of the theories (p. 235) of genetics (p. 196) based on the work of Gregor Mendel (p. 196), seeks to explain the mechanisms for the existence of advantageous variations (p. 213) which may occur naturally in a population (p. 214) of organisms which, because of lack of knowledge at the time, Darwin was unable to account for.

origin of species the theory (p. 235) leading from Darwinism (↑), which was developed by Charles Darwin in a paper published by him in 1859 and entitled *On the Origin of Species by Means of Natural Selection and the Preservation of Favoured Races in the Struggle for Life*. The theory suggests that within a population (p. 214) of one species (p. 40), various factors exist, such as geographical barriers (rivers, oceans, mountains, etc) or specific differences in behaviour (p. 164) which can isolate (p. 214) breeding groups within the population. This tends to maintain the integrity of the genes (p. 196) which carry the variations (p. 213) within the breeding group that are advantageous for the local environment (p. 218). In this way, the genetic differences between one group and another can build up, and over many generations lead to the development of new species, each fitted to its own conditions. This is referred to as speciation (p. 213). *See also* natural selection (↑).

Lamarckism (*n*) a theory (p. 235) which stemmed from the observations of the French biologist, Jean de Lamarck (1744-1829), who noticed that particular organs of an animal could fall into use or disuse if they were needed or not. From this he suggested that these acquired characteristics could be inherited. Modern genetic (p. 196) studies, however, have been unable to discover any mechanism whereby characters developed during an individual's lifetime could be passed on to its offspring so that the theory has fallen into disuse.

適者生存 自然選擇(↑)學說依據的基本觀念。該觀念認爲，僅有那些最能適應環境的動物才能在生存斗爭中存活下來，而那些不太適應環境的動物則將趨於滅亡。

新達爾文學說(名) 一種經過修正的近代達爾文學說(↑)。這種學說借助以格列高里‧孟德爾(196頁)的工作爲基礎的遺傳學(第196頁)理論(第235頁)，力圖解釋種種有益的變異(第213頁)——這在一個生物種羣(第214頁)內是會自然產生的——之所以存在的機理。關於這一點，當時由於知識的缺乏，達爾文未能作出解釋。

物種起源 達爾文學說(↑)中最主要的理論(第235頁)，這一理論由查理‧達爾文在他1859年發表的論著中提出，論著題目爲《物種起源於自然選擇，生存競爭中保存適宜物種》。該理論認爲，一個物種(第40頁)的種羣(第214頁)內，存在着可使該種羣內各個進行繁育的集羣隔離(第214頁)開來的種種因素，例如地理障礙(河流、海洋、山脈等)及行爲(第164頁)上的一些特殊差異。這就能保持帶有種種變異(第213頁)信息基因(第196頁)的完整性，而這些產生於進行繁育的集羣內的種種變異，是有利於適應當地環境(第218頁)的。這樣，一個集羣與另一個集羣之間就能建立起一些遺傳差異，經過許多世代之後，便導致產生一些新物種，而這些新物種中的每一種便都能適應其自身的環境。以上這一過程稱爲物種形成(第213頁)。參見"自然選擇"(↑)。

拉馬克學說(名) 法國生物學家讓‧拉馬克(1744—1829)經過種種觀察創立的一種學說(第235頁)。他注意到，動物的一些特殊器官的使用與否是根據對它們的是否需要來決定的，因而，他認爲，這些獲得性狀是能夠遺傳的。然而，現代遺傳學(第196頁)的研究，還未發現一個個體在其生存期內所獲得的各種性狀能夠遺傳給其後代的任何機理，因此，該學說現已廢棄不用。

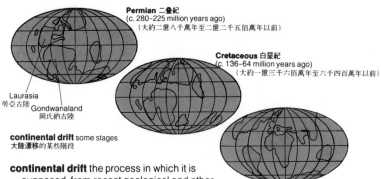

Permian 二叠紀
(*c.* 280-225 million years ago)
（大約二億八千萬年至二億二千五佰萬年以前）

Cretaceous 白堊紀
(*c.* 136-64 million years ago)
（大約一億三千六佰萬年至六千四百萬年以前）

Laurasia
勞亞古陸
Gondwanaland
岡氏納古陸

continental drift some stages
大陸漂移的某些階段

Miocene 中新紀
(*c.* 26-7 million years ago)（大約二千六百萬年至七百萬年以前）

continental drift the process in which it is
supposed, from recent geological and other
evidence, that the continental land masses
have not always occupied their present position
on the globe and that, powered by processes
within the Earth itself, they are slowly and
continuously on the move. In this way, new
land masses are created and destroyed, split
apart and joined, over geological time. This is
one of the processes that can lead to the
geographical isolation (p. 214) of a breeding
population (p. 214) and so to speciation (p. 213).

Pangea (*n*) the single landmass or
'supercontinent' which itself was formed during
Devonian times, some 395 to 345 million years
ago, by collision of the two original continents,
known as Gondwanaland and Laurasia. The
present continents evolved from Pangea by the
process of continental drift (↑).

plate tectonics the theory (p. 235) which has
been developed recently to provide a
mechanism for continental drift (↑). It supposes
that the surface layers of the Earth fit together,
rather like a spherical jigsaw puzzle, and that
the individual pieces are on the move in
relation to one another. In this way, pieces may
slide past one another, collide with one piece
being forced beneath the other, or separate
with new crust being formed as they move
apart. It is at the boundaries between the
individual plates that the majority of volcanic
eruptions and earthquakes take place.

大陸漂移 指根據最近地質學上和其他方面獲得的
證據而假設的一種過程。在這種過程中，地球
各大陸地塊並非始終是處於它們目前所處的位
置上，而且，由於受地球內部的一些變化過程
的驅使，各大陸地塊便處於緩慢地和連續不斷
地移動之中。因此，在整個地質時期中，有一
些新的陸塊產生和消失，分離和接合。這是導
致一個繁殖種羣（第214頁）的地理隔離（第214
頁），從而導致物種形成（第213頁）的過程之
一。

泛古陸（名） 指大約在三億九千五百萬年至三億四
千五百萬年以前的泥盆紀時期，由稱爲岡氏納
古陸和勞亞古陸的兩塊原始大陸碰撞而形成的
單個陸塊或"超大陸"。現有的各塊大陸都是經
由大陸漂移（↑）過程從泛古陸演變而來的。

板塊構造 最近提出的一種支持大陸漂移（↑）過程
的學說（第235頁）。該學說認爲，地球的各塊
表層是鑲嵌在一起的，頗象一個球形的拚板玩
具；而且，各個板塊相互之間都處於相對移動
之中。因此，各個板塊彼此之間可能相互滑
過；也可能發生碰撞，迫使一個板塊擠壓於另
一個板塊的下面；而在各個板塊相互移開時，
也可能發生分離，形成新的表層。大多數火山
噴發和地震就是發生在各個板塊之間的邊界地
帶的。

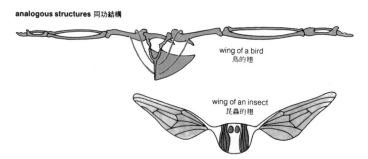

analogous structures 同功結構

wing of a bird
鳥的翅

wing of an insect
昆蟲的翅

homologous structures
同源結構

flipper of a turtle
海龜的鰭狀肢

arm of
a man
人的臂

wing of
a bird
鳥的翅

wing of
a bat
蝙蝠的翅

analogous (*adj*) of structures or organs which occur in different species (p. 40) of organisms and that have similar functions but a different evolutionary (p. 208) and embryological (p. 166) origin so that their structure is also different. For example, the wings of birds and those of the insects both enable the animals to fly but their origins and form are quite different.

homologous (*adj*) of structures or organs which occur in different species (p. 40) of organisms but which have similar evolutionary (p. 208) and embryological (p. 166) origins even though their functions may have been modified. For example, the limbs of all tetrapod (p. 77) vertebrates (p. 74) are based on the pattern of five digits. This suggests evolutionary relationships between different species.

divergent (*adj*) of evolution (p. 208) in which homologous (↑) structures have become adapted to perform different functions. For example, the flippers of sea mammals (p. 80), such as seals, are homologous with the limbs of land-based vertebrates (p. 74) but, are used in a different way as the seals have become better adapted to their marine environment (p. 218).

convergent (*adj*) of evolution (p. 208) in which analogous (↑) structures have become adapted to perform the same function. For example, the eye of a cephalopod (p. 72) performs the same function as that of a vertebrate (p. 74) but has a quite different origin and structure.

同功的(形) 指不同種(第40頁)生物體內包含的一些結構或器官而言，這些結構或器官具有相似的功能，但在進化(第208頁)起源和胚胎(第166頁)起源方面則不相同，因此，它們的構造也是不相同的。例如，鳥類的翅和昆蟲的翅均能使這些動物飛翔，但它們的起源和形態則完全不相同。

同源的(形) 指不同種(第40頁)生物體內包含的一些結構或器官而言，雖然這些結構或器官的功能可能已經起了變化，但它們卻具有相似的進化(第208頁)起源和胚胎(第166頁)起源。例如，所有四足(第77頁)脊椎動物(第74頁)的肢都是基於五足(趾)模式的。這種情況暗示不同物種之間的進化關係。

趨異的(形) 指一些同源(↑)結構已適應於執行不同功能的進化(第208頁)而言，例如，海洋哺乳動物(第80頁)如海豹的鰭狀肢與陸生脊椎動物(第74頁)的肢是同源的，但是使用方式不同，因爲海豹已經變得更爲適應其海洋環境(第218頁)了。

趨同的(形) 指一些同功(↑)結構已適應於執行同樣功能的進化(第208頁)而言，頭足綱(第72頁)動物的眼睛與脊椎動物(第74頁)的眼睛執行同樣的功能，但在起源和結構方面則完全不同。

vestigial (*adj*) of a structure or organ which originally performed a useful function but, through evolution (p.208), has become reduced to a remnant of its former self and no longer functions. For example, the appendix (p. 102) in humans.

primitive (*adj*) of a structure or organism that is at an early stage in evolution (p. 208), or is like an organism at an early stage.

phylogenetic (*adj*) of a classification (p. 40) which is based on the apparent evolutionary (p. 208) relationships between organisms.

palaeontology (*n*) the science or study of ancient life forms through their remains as fossils (↓).

fossil (*n*) any remains or trace of a once-living organism that has been preserved in some way such as in the rocks or in ice.

退化的(形) 指一種結構或器官而言，這種結構或器官原來執行過一種有用的功能，但經過進化（第208頁），已經縮小成爲其前身的殘留物，而不再起作用了。例如，人的闌尾（第102頁）便是這類殘留物。

原始的(形) 指處於進化（第208頁）早期階段的一種結構或生物而言，也可指某種生物與一種處於早期階段的生物相似。

系統發育的(形) 指一種分類（第40頁）而言，這種分類是以生物體之間顯著的進化（第208頁）關係爲基礎劃分的。

古生物學(名) 用殘遺化石（↓）研究古代生物類型的科學。

化石(名) 指以某種方式例如在岩石或冰塊內保存下來的一種曾經生存過的生物的任何殘遺物或遺迹。

fossil e.g.
Archaeopteryx
化石圖例：始祖鳥

fossil record the continuing record of the origins, development and existence of life on Earth as expressed through the finds of fossils (↑) preserved in the rocks from the origins of the planet to the present day.

geological column a tabular time scale which has been worked out by geologists on the basis of the fossil record (↑) and other evidence, such as radiometric dating, in which the history of the Earth is broken down into eras, periods, and epochs.

化石記錄 以從地球的起源至現今保存於岩石中的許多化石（↑）發現物，表示地球上生命的起源、生存和發展的連續紀錄。

地質柱狀圖 地質學根據化石記錄（↑）和其他數據如放射性年代測定法所取得的數據制定出的一種地質年代表，在該表中，將地球的歷史劃分成代、紀和世。

variation (*n*) the differences in form and structure which occur naturally among individuals within the same species (p. 40) and which may result from genetic (p. 196) changes, such as mutations (p. 206), or from differences in such factors as nutrition (p. 92) or the density of the population (p. 214).

diversity (*n*) the state of things being different from each other.

isolating mechanisms factors, such as the existence of geographical barriers, behaviour (p.164), or the timing of the breeding season (p.195) which tend to separate groups of individuals into reproductive (p. 173) communities (p. 217).

gene pool the total number and type of genes (p. 196) that exist at any given time within a breeding population (p. 214) that has been separated by various isolating (p. 214) mechanisms. The genes within a given gene pool may then be intermixed randomly by interbreeding within the group.

speciation (*n*) the process by which two or more new species (p. 40) evolve (p. 208) from one original species as breeding groups become separated by isolating (p. 214) mechanisms and develop a range of distinctive characters, as a result of natural selection (p. 208), to the extent that the isolated populations (p. 214) are no longer able to breed with one another.

變異（名） 同一物種（第40頁）內個體之間自然發生的形態和結構方面的種種差異，這類差異可由遺傳（第196頁）變化如突變（第206頁）引起，或者由諸如種羣（第214頁）的營養（第92頁）和密度這樣一些因素的差別引起。

多樣性（名） 指事物相互之間存在差異的狀態。

隔離機制 指能夠將一些個體分隔成爲生殖（第173頁）羣落（第217頁）的那些因素，例如地理障礙，行爲（第164頁）和生育季節（第195頁）的時機選擇。

基因庫 指在一個已被各種隔離（第214頁）機制所分隔的繁殖種羣（第214頁）中，在任何給定時間內所存在的基因（第196頁）總數和類型。一個特定基因庫內的各種基因經過在該種羣內進行雜交，即可隨機混合。

物種形成（名） 由一個原始物種演變（第208頁）成兩個或兩個以上的新物種（第40頁）所經歷的過程。由於各種隔離（第214頁）機制將一個原始物種分隔成多個進行繁殖的集羣，通過自然選擇（第208頁），它們便會顯現出一系列特殊的性狀，結果，這些被隔離的種羣（第214頁）相互之間就不再能進行繁殖了。

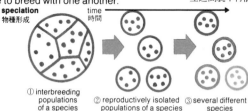

speciation 物種形成 / time 時間

① interbreeding populations of a species
② reproductively isolated populations of a species
③ several different species

①一個物種內的雜交種羣
② 一個物種內的生殖隔離種羣
③ 幾個不同的物種

differential mortality the basis of natural selection (p. 208) during periods of increasing population (p. 214) when those individuals of the overpopulated community which are best fitted to their environment (p. 218) survive to breed while those that are less well fitted die so that evolution (p. 208) takes place by natural selection.

差別死亡率 自然選擇（第208頁）的基礎。在種羣（第214頁）增大的各個時期中，當過密羣落中的那些最能適應其環境（第218頁）的個體幸存下來，進行繁殖，而那些不太適應其環境的個體死亡時，進化（第208頁）便通過自然選擇進行。

melanism (*n*) the condition in which such structures as hair, skin and eyes are coloured by the dark-brown pigment (p. 126) melanin. Melanistic skin protects the individual from the harmful effects of prolonged exposure to sunlight. Consequently, humans who have evolved (p. 208) in areas of high sunlight intensity have, by natural selection (p. 208), evolved darker skin colour.

gene frequency the occurrence of one particular gene (p. 196) in a given population (↓) in relation to all its other alleles (p. 197).

Hardy-Weinberg principle a law formulated in 1908 from which the effects of natural selection (p. 208) can be better understood. It suggests that in any population (↓), in which mating takes place at random, the proportion of dominant (p. 197) to recessive (p. 197) genes (p. 196) in the population remains unchanged from one generation to the next. Until the principle was worked out, it was thought, quite reasonably, that the numbers of recessive genes would decline while the dominant genes would increase.

gene flow the process by which genes (p. 196) move within a population (↓) by mating and the exchange of genes.

genetic drift the process by which the genetic (p.196) structure of a small population (↓) of organisms changes by chance rather than by natural selection (p. 208). In a small population the Hardy-Weinberg principle (↑) may not be maintained because the number of pairings will not be random.

isolation (*n*) the process by which two populations (↓) become separated by geographical, ecological (p. 217), behavioural (p. 164), reproductive (p. 173), or genetic (p. 196) factors. After two populations have become genetically or reproductively separated, they will not revert to the same species (p. 40) even if they come together geographically again.

population (*n*) a group of organisms of the same species (p. 40) which occupies a particular space over a given period of time. The actual numbers of individuals within a population may rise and fall as a result in changes of the birth and death rate and such factors as climate, food supply, and disease.

黑化(名) 指諸如毛髮、皮膚和眼睛這些結構爲深褐色的色素(第126頁)即黑色素所着色的狀態。黑化的皮膚可保護個體免受長期暴露陽光所致的有害作用。因此，其進化(第208頁)過程是在一些日照強度高的地區進行的那些人，經過自然選擇(第208頁)，就進化成爲膚色較黑的人。

基因頻率 在某一種羣(↓)中，一個特定基因(第196頁)相對於其所有其他等位基因(第197頁)的出現率。

哈迪──溫伯格定律 在1908年正式提出的一條定律，根據這一定律，可以更好理解自然選擇(第208頁)的作用。該定律認爲，在進行隨機交配的任何種羣(↓)中，其顯性(第197頁)基因(第196頁)與隱性(第197頁)基因的比例將一代一代地保持不變。在這個定律制訂出之前，人們頗有道理地認爲，隱性基因的數目會下降，而顯性基因的數目會增加。

基因流動 基因(第196頁)經由交配和基因的交換而在一個種羣(↓)內移動的過程。

遺傳漂變 指生物中一個小的種羣(↓)的遺傳(第196頁)結構的變化過程。這種變化過程是隨機的，而不是經由自然選擇(第208頁)的。對一個小的種羣而言，因爲配對數不是隨機的，所以就可能出現不遵循哈迪──溫伯格定律(↑)的情况。

隔離(名) 由地理、生態(第217頁)、行爲(第164頁)、生殖(第173頁)或遺傳(第196頁)因素等造成兩個種羣(↓)分隔所經歷的過程。在兩個種羣形成遺傳隔離或生殖隔離之後，即使它們在地理上重新歸於一起，也不會回復到同一物種(第40頁)。

種羣〔量〕、羣體(名) 在特定時間內，佔據特定空間的一羣同種(第40頁)生物。一個種羣內的個體的實際數目，可因出生率和死亡率的變化以及諸如氣候、食物供應和疾病這類因素的變化而增加或減少。

allopatric (*adj*) of two or more populations (↑) of the same or related species (p. 40) which could interbreed if they were not geographically isolated (↑) from one another.

sympatric (*adj*) of two or more related species (p. 40) which are not geographically isolated (↑) from one another and which could interbreed apart from differences in behaviour (p. 164) or the timing of the breeding season (p. 195) etc.

ecological isolation isolation (↑) which occurs within populations (↑) as a result of the different ways in which they relate to their environment (p. 218).

reproductive isolation isolation (↑) which occurs within populations (↑) as a result of differences in their breeding behaviour (p. 164) or timing of their breeding season (p. 195).

異地的、分佈區不重疊的（形）　指同一物種（第40頁）或相關物種的兩個或兩個以上的種羣（↑）而言，如果這些種羣相互之間在地理上未形成隔（↑）離，它們就可以雜交。

同地的、分佈區重疊的（形）　指兩個或兩個以上的相關物種（第40頁）而言，這些物種相互之間在地理上未形成隔離（↑），除行為（第164頁）或生育季節（第195頁）的時間選擇等方面存在一些差異之外，它們是可以雜交的。

生態隔離　指一些種羣（↑）由於與其環境（第218頁）相關聯的方式各異，而於種羣之內發生的隔離（↑）現象。

生殖隔離　指一些種羣（↑）由於其生育行為（第164頁）或生育季節（第195頁）的時機選擇方面的差異，而於種羣之內發生的隔離（↑）現象。

sympatric
e.g. two species occurring in the same place
同地的
圖例：存在於同一地區的兩個物種

allopatric
e.g. two species occurring in different places
異地的
圖例：存在於不同地區的兩個物種

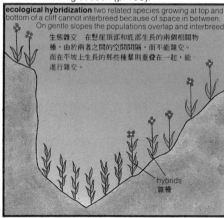

ecological hybridization two related species growing at top and bottom of a cliff cannot interbreed because of space in between. On gentle slopes the populations overlap and interbreed

生態雜交　在懸崖頂部和底部生長的兩個相關物種，由於兩者之間的空間間隔，而不能雜交。而在平坡上生長的那些種羣則重疊在一起，能進行雜交。

hybrids
雜種

genetic isolation isolation (↑) which occurs within populations (↑) as a result of their genetic (p. 196) incompatibility so that they are unable to produce fertile (p. 175) offspring.

artificial selection the process by which humans make use of the principles of genetics (p. 196) and evolution (p. 208) to create breeds or hybrids (p. 216) which would not be expected to occur as a result of natural selection (p. 208).

遺傳隔離　指一些種羣（↑）由於其遺傳（第196頁）不親和性，致使它們不能繁殖能育的（第175頁）後代，而於種羣之內發生的隔離（↑）現象。

人工選擇　人類利用遺傳學（第196頁）和進化（第208頁）原理培育一些品種和雜種（第216頁）所經歷的過程，這些品種和雜種不能指望經由自然選擇（第208頁）產生。

inbreeding (*n*) breeding by the mating of closely related individuals, including self-fertilization (p. 175) in plants. It tends to reduce the genetic (p. 196) variability of the population (p. 214) and leads to a greater frequency of expression of recessive (p. 197) characteristics. Humans make use of inbreeding during artificial selection (p. 215) to develop characteristics which are seen as useful.

outbreeding (*n*) breeding by the mating of individuals which are not closely related. The most extreme form of outbreeding is between organisms of different species (p. 40) which leads to the production of non-fertile (p. 175) offspring. Outbreeding normally gives rise to greater genetic (p. 196) variability and vigour and there may be various mechanisms within organisms to encourage it.

hybrid vigour an increase in the vigour of such factors as growth or fertility (p. 175) in the offspring as compared with the parents which results from the cross-breeding of individuals from lines which are genetically (p. 196) different leading to greater heterozygosity (p. 198) and an increased expression of dominant (p. 197) genes.

hybrid (*n*) the offspring of parents from genetically (p. 196) different lines. **hybridization** (*n*).

spontaneous generation the idea, disproved by the French bacteriologist, Louis Pasteur (1822–95) and others, that, in suitable conditions, organisms, especially microorganisms, could be generated from inorganic compounds (p. 15).

special creation a hypothesis (p. 235) which suggests that every form of life that exists or has ever existed was created separately by a deity or other supernatural force. Palaeontological (p. 212) and genetic (p. 196) evidence suggests that this is unlikely and few scientists take the hypothesis seriously today.

steady state a hypothesis (p. 235) which suggests that all organisms were created at some time in the past and have remained unchanged ever since with each generation being identical to its predecessor. Palaeontological (p. 212) evidence suggests that this cannot be the case.

近交、近親繁殖(名)　由親緣關係密切的一些個體交配所進行的繁殖,包括植物中的自株傳粉(第175頁)。這種繁殖往往會減少種羣(第214頁)的遺傳(第196頁)變異性,並使一些隱性(第197頁)性狀的表現頻度較高。人類在人工選擇(第215頁)的過程中,利用近交來發育那些被認爲是有用的性狀。

遠交、遠系繁殖　由親緣關係疏遠的一些個體交配所進行的繁殖。遠交的最極端形式是在不同種(第40頁)的生物體之間進行交配,這可導致產生不能育的(第175頁)後代。遠交通常產生較大的遺傳(第196頁)變異性和活力;生物體內可能存在各種促進遠交的機制。

雜種優勢　指由不同遺傳(第196頁)系的個體的雜交產生的後代,與其親代比較,在諸如生長和能育性(第175頁)這類因素的優勢上有所增加。這種雜交子代具有高度的雜合性(第198頁),也使顯性(第197頁)基因的表現增強。

雜種(名)　由不同遺傳(第196頁)系的親代交配所產生的後代。(名詞形式 hybridization 意爲雜交)。

無生源論、自然發生說　被法國細菌學家路易斯·巴斯德(1822—95)和其他一些人駁斥的一種概念。這種概念認爲,在合適條件下,可以從一些無機化合物(第15頁)產生出生物,尤其是微生物。

特創論　一種假說(第235頁)。這種假說認爲,現存的和曾經生存過的所有形式的生命都是由上帝或其他超自然力量各別地創造出來的。古生物學(第212頁)和遺傳學(第196頁)的證據表明,這是不可能的;在今天,幾乎沒有科學家認眞對待這種假說。

穩定狀態　一種假說(第235頁)。這種假說認爲,所有的生物都是在過去的某一時間創造出來的,並且從此保持不變,每一世代與其祖先是完全一樣的。古生物學(第212頁)的證據表明,情況並非如此。

ecology (*n*) the science or study of organisms in relation to one another and the environment (p. 218).

biosphere (*n*) the part of the Earth which includes all of the living organisms on the planet and their environment (p. 218).

biome (*n*) a part of the biosphere (↑) which might be a large, regional community (↓) of interrelated organisms and their environment (p. 218) and would include such habitats (↓) and communities as a tropical rainforest or grassland.

ecosystem (*n*) a self-contained and perhaps small unit or area, such as a woodland, which would include all the living and non-living parts of that unit.

生態學（名） 研究生物之間及生物與環境（第218頁）之間相互關係的科學。

生物圈（名） 包括地球上所有生存生物及其環境（第218頁）在內的地球的一部份。

生物羣落（名） 生物圈（↑）的一部份，它可以是由一些相互有關的生物及其環境（第218頁）構成的一個大區域羣落（↓），也包含諸如熱帶雨林和草地這類生境（↓）和羣落。

生態系統（名） 一個獨立的、多半是小的單位或區域，例如森林，它包括該單位中所有的生物及非生物部份。

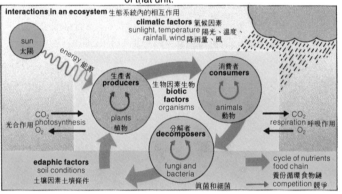

interactions in an ecosystem 生態系統內的相互作用

community (*n*) a localized group of a number of populations (p. 214) of different species (p. 40) living and interacting with one another within an ecosystem (↑). Communities can be described as open (niches (p. 219) unstable or 'empty' allowing new species into the community) or closed (niches stable and full).

habitat (*n*) a part of an ecosystem (↑), such as a desert, in which particular organisms live because the environmental (p. 218) conditions within the habitat are essentially uniform even though they may vary with the season or between, say, ground level and the tops of the trees.

microhabitat (*n*) a small area within a habitat (↑) such as the underside of a stone.

羣落（名） 指在一個生態系統（↑）內生活和相互作用的不同物種（第40頁）的若干種羣（第214頁）之局部組合。羣落可以被說成是開放的（生態位（第219頁）不穩定的或"未佔用的"，允許新的物種進入羣落）或封閉的（生態位是穩定的和已佔滿的）。

生境、棲息地（名） 生態系統（↑）的一部份，例如沙漠。其中，生活着一些特定的生物，因爲該生境內的環境（第218頁）條件基本上是均一的，但是這些環境條件可隨着季節而變化，或者說在地面和樹頂之間也可能是不同的。

小生境、小棲息地（名） 生境（↑）內的一小塊區域，例如一塊石頭的下面。

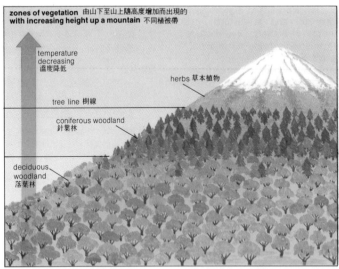

zones of vegetation 由山下至山上隨高度增加而出現的 **with increasing height up a mountain** 不同植被帶

temperature decreasing 溫度降低

herbs 草本植物

tree line 樹線

coniferous woodland 針葉林

deciduous woodland 落葉林

zone (*n*) a part of a biome (p. 217) which is characterized by one particular group of organisms depending upon the environmental (↓) conditions present in that area.

climate (*n*) the sum total of all the interrelating weather conditions, such as temperature, pressure, rainfall, sunshine, etc., that exists in a particular region throughout the year and averaged over a number of years.

microclimate (*n*) the climate (↑) which occurs in a small region, such as a town or woodland, which differs in some way from the overall climate of the region due to the effects of other factors within the area. For example, the temperature in a large city may be significantly higher than that of its rural surroundings because of the heat that is trapped by the buildings and re-released.

environment (*n*) the sum total of all the external conditions within which an organism lives.

territory (*n*) any area which is occupied and defended by an animal for purposes of breeding, feeding etc.

帶(名) 生物羣落(第217頁)的一部份,其特徵在於有一個特定的生物類羣,該生物類羣隨那個區域存在的環境(↓)條件而定。

氣候(名) 存在於一個特定地區全年的或若干年平均的、所有相互關聯的天氣條件之總和,諸如氣溫、氣壓、降雨量、日照等的總和。

小氣候(名) 局限於一個小地區例如一個城鎮或者一片森林的氣候(↑)。由於該地區內的其他一些因素的影響,這種氣候在某一方面不同於該地區總的氣候。例如,大城市的氣溫可大大高於其農村環境的氣溫,因爲大城市的建築物截獲了熱量並將它重新釋放出來。

環境(名) 與生物生存有關的所有外部條件之總和。

領域、勢力圈、活動區(名) 動物爲繁育、進食等目的而佔據和防衞的任何地區。

niche (*n*) the local physical and biological conditions which an organism fills in an ecosystem (p. 217). If, at some stage, more than one species (p. 40) of organism attempt to occupy the same niche, then they compete with one another until one is eliminated. On the other hand, it is possible for different species to occupy the same niche in geographically separated regions or for one species to evolve (p. 208), by natural selection (p. 208), to occupy different niches.

abiotic (*adj*) of the physical environment (↑) to which organisms are subjected, such as temperature, light intensity, availability of water, etc.

aquatic (*adj*) of a watery environment (↑) or a species (p. 40) which lives primarily in water.

freshwater (*adj*) of an aquatic (↑) environment (↑), such as a river, which does not contain salt and is, therefore, not marine (↓). Also, of a species (p. 40) which lives primarily in fresh water.

marine (*adj*) of an aquatic (↑) environment (↑), such as the ocean, which contains salt. Also describes a species (p. 40) which lives primarily in a marine environment.

littoral (*n*) the zone (↑) of a freshwater (↑) environment (↑) between the water's edge and a depth of about six metres or the zone of a marine (↑) environment between the high and low water marks. A littoral species (p. 40) is one which lives primarily in the littoral zone.

amphibious (*adj*) of an organism which is capable of or spends part of its time living in water and part on land.

terrestrial (*adj*) of those organisms that spend most or all of their lives on land.

subterranean (*adj*) of those organisms that spend most or all of their lives underground, in caves, for example.

arboreal (*adj*) of those organisms that spend most or all of their lives living among the branches of trees.

aerial (*adj*) of those organisms or parts of organisms that spend part or all of their lives in the air. The roots of certain trees grow in the air and are referred to as aerial.

生態位、生態龕(名)　一種生物在某一生態系統(第217頁)內的局部物理環境和生物環境中所佔的地位。如果在某一階段，生物的種(第40頁)在一個以上，而它們都試圖佔據同一生態位，那麼，相互之間就會發生競爭，直至失敗的一方被淘汰為止。另一方面，生物的不同種在地理上隔開一些區域，有可能佔據同一生態位，或者，生物的一個種，在進化(第208頁)過程中，通過自然選擇(第208頁)，也有可能佔據不同的生態位。

非生物的、無生命的(形)　指對生物施加影響的物理環境(↑)，諸如溫度、光強度、水的可獲量。

水的；水生的(形)　指一種充滿水的環境(↑)，也指主要生活在水中的一個物種(第40頁)。

淡水的；淡水生的(形)　指一種如河流的水生(↑)環境(↑)而言，這種水生環境不含鹽份，因此不是海洋(↓)環境。也可指主要生活於淡水中的一個物種(第40頁)而言。

海水的、海水生的(形)　指一種如海洋的水生(↑)環境(↑)而言，這種水生環境含有鹽份。也可指主要生活於海洋環境中的一個物種(第40頁)而言。

濱海帶、沿岸帶、潮汐帶(名)　位於淡水的邊緣至深度大約為六米的淡水之間的一個淡水(↑)環境(↑)區域(↑)；也指在高潮標記與低潮標記之間的一個海洋(↑)環境區域而言。沿岸物種(第40頁)是指主要生活在沿岸帶的物種。

兩棲的(形)　指一種生物而言，這種生物部份時間能夠生活在水中，部份時間能夠生活在陸上。

陸生的(形)　指大半生或一生都生活在陸地上的那些生物而言。

地下的(形)　指大半生或一生都生活在地下如洞穴中的那些生物而言。

樹棲的(形)　指大半生或一生都生活在樹枝上的那些生物而言。

氣生的(形)　指部份時間或一生都生活在空氣中的那些生物或那些生物的某些組成部份而言。某些樹的根生長在空氣中，稱為氣生根。

climatic factors those aspects of the environment (p. 218), grouped together as the climate (p. 218), including temperature, rainfall, etc., which affect the distribution of organisms.

edaphic factors those aspects of the environment (p. 218) concerned with the soil and including moisture content, pH (p. 15), etc., which affect the distribution of organisms.

biotic (*adj*) of those biological parts of the environment (p. 218) other than the abiotic (p. 219) factors to which organisms are subjected, and include their relationships with other organisms such as competition (↓) for habitat (p. 217) etc.

predation (*n*) the process by which certain animals gain nutrition (p. 92) by killing and feeding upon other animals. A predator is a secondary consumer (p. 223) and predators do not include parasites (p. 110).

氣候因素　指環境(第218頁)中可以歸結爲氣候(第218頁)的那些方面,其中包括溫度、降雨量等。這些方面都會影響生物的分佈。

土壤因素　指環境(第218頁)中與土壤有關的那些方面,其中包括含水量、pH值(第15頁)等。這些方面都會影響生物的分佈。

生物的、生命的（形）　指環境（第218頁）中除對生物施加影響的一些非生物（第219頁）因素之外的那些生物因素而言。這些生物因素包含一些生物與其他一些生物之間的種種關係,例如爲生境（第217頁）而進行的競爭（↓）。

捕食性(名)　某些動物將其他一些動物殺死,並以此爲食,借以獲得營養(第92頁)的過程。捕食者是次級消費者(第223頁),它不包括寄生物(第110頁)。

competition in a plant community 一個植物羣落內的競爭

leaves compete 各片葉子爭奪陽光、for light, CO₂, space 二氧化碳和空間

roots compete for nutrients and water 各條根爭奪養份和水

mimicry 擬態

hoverfly 食蚜蠅

harmless, hoverfly mimics unpleasant wasp 無害的食蚜蠅與討厭的黃蜂極爲相似

wasp 黃蜂

competition (*n*) the process in which more than one species (p. 40) or individuals of the same species attempt to make use of the same resources in the environment (p. 218) because there are not enough resources to satisfy the needs of all the organisms. Competition often leads to differential mortality (p. 213).

intraspecific (*adj*) of an action, for example, competition (↑), which takes place between individuals of the same species (p. 40).

interspecific (*adj*) of an action, for example, competition (↑), which takes place between different species (p. 40).

mimicry (*n*) the process in which one organism resembles another and thereby gains some advantage, e.g. a defenceless hoverfly closely resembles the form and colour of a wasp and may, therefore, be avoided by predators (↑).

競爭(名)　指生物經歷的一種特定過程,在這種過程中,一個以上的物種(第40頁)或同一物種內的各個個體試圖利用環境(第218頁)中的同類資源,因爲那裏沒有足夠的資源以滿足所有這些生物的需要。競爭常導致出現差別死亡率(第213頁)。

種內的(形)　指一種行動例如競爭(↑)在同一物種(第40頁)內的各個個體之間發生。

種間的(形)　指一種行動例如競爭(↑)在不同物種(第40頁)之間發生。

擬態(名)　指生物經歷的一種專門過程,在這種過程中,一種生物長得與另一種生物相似,從而獲得了某種益處。例如,無防禦能力的食蚜蠅,其形態和體色與黃蜂很相似,因而可以躲過捕食者(↑)。

synecology (*n*) the study or science of all communities (p. 217) and ecosystems (p. 217) within an environment (p. 218) and their relationships to one another.

autecology (*n*) the study or science of individuals of one species (p. 40) in relation to one another and to their environment (p. 218).

succession (*n*) a progressive sequence of changes which takes place, after the first colonization (↓) of a particular environment (p. 218), in the organisms which occupy that environment until a stable position is reached where no further changes can take place unless the abiotic (p. 219), edaphic (↑), or climatic factors (↑) themselves are altered. The process takes place rapidly at first and then slows down as stability is approached.

羣落生態學（名） 研究環境（第218頁）中所有羣落（第217頁）和生態系統（第217頁）及其相互關係的科學。

個體生態學（名） 研究一個物種（第40頁）內各個體相互關係及它們與環境（第218頁）關係的科學。

演替（名） 指一羣生物在首次移殖（↓）到一個特定環境（第218頁）之後，佔據該環境，直至取得穩定地位所發生的一系列漸進變化的過程。在那裏，不會再發生進一步的變化，除非非生物（第219頁）因素、土壤因素（↑）和氣候因素（↑）發生了變更。這一過程在開始階段進展迅速，以後隨着趨於穩定而減緩下來。

succession
演替

a pioneer species colonizes 一個先蜂物種移殖於一個生境
a habitat

pioneer plants grow and reproduce 一些先蜂植物的生長和繁殖

growth of plants alters edaphic 一些植物的生長改變了
and biotic factors and more 土壤因素和生物因素，
species colonize 使更多的物種得以移生

climax community with many 具有多個植物種的
plant species. Conditions no 頂極羣落。環境對
longer suitable for pioneer 先蜂植物不再適合

colonization (*n*) the arrival and growth to reproductive (p. 173) age of an organism in an area, i.e. the spread of species (p. 40) to places where they have not lived before. **colonize** (*v*), **colony** (*n*).

pioneer (*n*) a plant species (p. 40) that is found in the early stages of succession (↑).

climax (*adj*) of a community (p. 217) which, following succession (↑), has reached stability.

sere (*n*) a succession (↑) of plant communities (p. 217) which themselves affect the environment (p. 218) leading to the next community and resulting ultimately in the climax (↑) community.

集羣現象（名） 指一種生物到達一個地區並生長發育到生殖（第173頁）年齡的現象，也就是指物種（第40頁）散佈到前未曾生活過的地方的現象。（動詞形式爲 colonize、名詞形式爲 colony）

先鋒植物 演替（↑）的一些早期階段包含的一個植物種（第40頁）。

頂極的（形） 指一個羣落（第217頁）而言，這種羣落在演替（↑）之後達到穩定狀態。

演替系列（名） 指植物羣落本身都會影響環境（第218頁），爲下一個羣落的出現準備條件，並且，最終都會出現演替頂極（↑）羣落。

soil (*n*) the material which forms a surface covering over large areas of the Earth and in which organisms gain support, protection, and nutrients (p. 92). It results from the weathering and breakdown of rocks into inorganic (p. 15) mineral particles which are then further acted upon by climatic (p. 220) and biotic (p. 220) factors. The composition depends upon the composition of the original rock.

inorganic component the part of the soil which results from the action of weather on the parent rocks, breaking it down into mineral particles of varying size and composition depending upon the composition of the original rock.

organic component the part of the soil which is derived from the existence and activity of the large numbers of living organisms in the soil.

sand (*n*) the inorganic component (↑) in which the particles range in size from 0.02-2.0 millimetres and are angular. A soil with a high sand content tends to be dry, because of the ease with which water drains away, acidic (p. 15), and low in nutrient (p. 92) content.

clay (*n*) the inorganic component (↑) in which the particles are less than 0.02 millimetres in size and are relatively smooth and rounded. A soil with a high clay content tends to be easily waterlogged, can become compacted, and will harden on drying. It is usually rich in nutrient (p. 92) content, however.

humus (*n*) the organic component (↑) of soil which results from the activity and decomposition (↓) of the living organisms within a soil and which is a mixture of fibrous (p. 143) and colloidal materials made up essentially of carbon, nitrogen, phosphorus and sulphur. Humus improves the structure and texture of a soil, helps it to retain water and nutrients (p. 92), and raises the soil's temperature by absorbing more of the sun's energy because of its dark colour.

erosion (*n*) the process by which the products of weathering of a rock or a soil are worn away by the action of wind, running water, or moving ice, etc.

土壤（名） 構成覆蓋地球廣大區域表面的物質。許多生物可從其中獲得支托、保護和營養素（第92頁）。土壤是由岩石風化和分解，變成無機（第15頁）礦物粒子，然後進一步受氣候因素（第220頁）和生物（第220頁）因素作用而形成的。土壤的組成取決於原始岩石的組成。

無機成份 土壤中由天氣對母岩作用而產生的那一部份。天氣的作用可將母岩分解成大小不同的礦物粒子，其組成取決於原始岩石的組成。

有機成份 土壤中由於生存大量生物和及其活動而得到的那一部份。

砂土、砂（名） 粒子大小為0.02-2.0毫米的角狀無機成份（↑）。含砂量高的土壤往往是乾燥的，因為水份很容易從中流失；這種土壤呈酸性（第15頁），養份（第92頁）含量也低。

黏土（名） 粒子大小低於0.02毫米、比較光滑而呈圓形的無機成份（↑）。黏土含量高的土壤往往容易積水，變成堅實，且在乾燥時會硬化。然而，黏土通常含豐富的養份。

腐殖土（名） 土壤的有機成份（↑）。它是由土壤中各種生物活動和分解作用（↓）而產生的。腐殖土是一種主要由碳、氮、磷和硫組成的纖維（第143頁）狀和膠體物質的混合物。腐殖土可改良土壤的結構和質地，有助於土壤保持水份和養份（第92頁），由於其顏色較深，故還可吸收更多的太陽能，提高土壤的溫度。

侵蝕（名） 岩石或土壤的風化產物在風、流水和移動冰塊等的作用下被磨耗的過程。

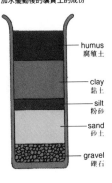

the constituents of loam
soil shaken up with water
加水擺動後的壤質土的成份

- humus 腐殖土
- clay 黏土
- silt 粉砂
- sand 砂土
- gravel 礫石

soil profile the series of distinct layers that can be observed in a vertical section through soil from the parent rock, through weathered rock and *subsoil* to *topsoil*.

土壤剖面 指從母岩經風化岩和底土至頂土的垂直剖面中，可以觀察到的一系列不同的土壤層。

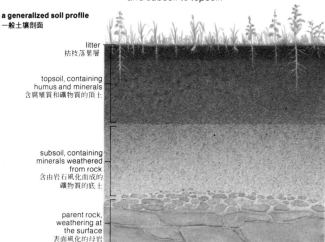

a generalized soil profile
一般土壤剖面

litter
枯枝落葉層

topsoil, containing humus and minerals
含腐殖質和礦物質的頂土

subsoil, containing minerals weathered from rock
含由岩石風化而成的礦物質的底土

parent rock, weathering at the surface
表面風化的母岩

pyramid of available energy at the trophic levels of a food web
在食物網的各營養級上可獲得的能量的金字塔

trophic level
營養級

higher order consumer (large carnivore) 高級
消費者(大型食肉動物)

4

secondary consumer (carnivore) 次級消費者(食肉動物)

3

primary 初級消費者
consumer (食草
herbivore) 動物)

2

producer
生產者

1

energy lost through respiration, heat radiation and other metabolic processes
經由呼吸、熱輻射和其他代謝過程失去的能量

energy available as food
以食物形式獲得的能量

producers (*n.pl.*) the organisms, especially green plants and some bacteria (p. 42), which are able to manufacture nutrients (p. 92) from inorganic (p. 15) sources by such processes as photosynthesis (p. 93).

consumers (*n.pl.*) the heterotrophic (p. 92) organisms which obtain their nourishment by consuming the producers (↑) or other consumers.

decomposers (*n.pl.*) the organisms which obtain their nutrients (p. 92) by feeding upon dead organisms, breaking them down into simpler substances and, in so doing, making other nutrients available for the producers (↑).

decomposition (*n*).

trophic level the particular position which an organism occupies in an ecosystem (p. 217) in respect of the number of steps away from plants at which the organism obtains its food. The producers (↑) are at the lowest trophic level while the predators (p. 220) at the highest trophic levels.

生產者(名、複) 一些生物，特別是各種綠色植物和某些細菌(第42頁)，它們能夠藉助光合作用(第93頁)之類過程，由無機(第15頁)物製造養份(第92頁)。

消費者(名、複) 一些異養(第92頁)生物，它們通過將生產者(↑)或其他一些消費者消耗掉的方式，來獲取營養。

分解者(名、複) 一些生物，它們以生物屍體為食，將其分解成為一些較簡單的物質，以獲取養份(第92頁)，與此同時，也製造了其他一些對生產者(↑)有用的養份。(名詞形式為 decomposition)

營養級 一種生物在一個生態系統(第217頁)中所佔的特定位置，而這種特定位置是與該生物藉以為食的那些植物所處的階梯相隔開的階梯數目而言的。生產者(↑)處於最低的營養級，而捕食者(第220頁)處於最高的營養級。

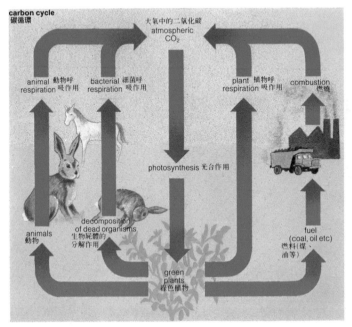

carbon cycle
碳循環

大氣中的二氧化碳
atmospheric
CO_2

animal 動物呼
respiration 吸作用

bacterial 細菌呼
respiration 吸作用

plant 植物呼
respiration 吸作用

combustion
燃燒

photosynthesis 光合作用

decomposition
of dead organisms
生物屍體的
分解作用

animals
動物

fuel
(coal, oil etc)
燃料(煤、
油等)

green
plants
綠色植物

carbon cycle the chain or cycle of events by which carbon is circulated through the environment (p. 218) and living organisms. Plants take in carbon dioxide from the atmosphere and turn it into carbohydrates (p. 17), proteins (p. 21) and fats. Some of the carbon dioxide is returned to the atmosphere during the plants' respiration (p. 112). The plants are eaten by herbivores (p. 105) which, in turn, are eaten by carnivores (p. 105). When the herbivores and carnivores die, they are fed upon by saprophytes (p. 92) and decomposers (p. 223) so that carbon is returned to the soil or to the atmosphere as a product of respiration of bacteria (p. 42) and fungi (p. 46).

oxygen cycle the chain or cycle of events by which oxygen is circulated through the environment (p. 218) and living organisms.

碳循環　指碳藉以在環境(第218頁)中和各種生物體內進行循環的那些鏈式活動過程或循環活動過程。植物從大氣中吸收二氧化碳，將其轉化爲碳水化合物(第17頁)、蛋白質(第21頁)和脂肪。一些二氧化碳在植物的呼吸作用(第112頁)過程中又回到大氣中。植物爲食草動物(第105頁)所食，食草動物進而又被食肉動物(第105頁)所吃。當食草動物和食肉動物死亡時，它們則被腐生生物(第92頁)和分解者(第223頁)所食，因此，碳作爲細菌(第42頁)和眞菌(第46頁)呼吸作用的產物又回到土壤和大氣中。

氧循環　指氧藉以在環境(第218頁)中和各種生物體內進行循環的那些鏈式活動過程或循環活動過程。

nitrogen cycle the chain or cycle of events by which nitrogen is circulated through the environment (p. 218) and living organisms. Some bacteria (p. 42) and algae (p. 44) can make use of nitrogen directly, and lightning, acting upon atmospheric nitrogen and oxygen, causes it to combine into nitrous and nitric oxide which dissolve in falling rain to enter the soil and form nitrates and nitrites. Most plants make use of nitrogen as nitrates and use them in the manufacture of proteins (p. 21). The plants are fed upon by herbivores (p. 105) which, in turn, are eaten by carnivores (p. 105) which also make use of the nitrogen in the manufacture of animal proteins. When animals and plants die, the nitrogen is returned to the soil by nitrifying bacteria as nitrites, ammonia, and ammonium compounds.

氮循環 指氮藉以在環境（第218頁）中和各種生物體內進行循環的那些鏈式活動過程或循環活動過程。某些細菌（第42頁）和藻類（第44頁）能直接利用氮。閃電作用於大氣中的氮和氧，使它們化合成爲氧化亞氮和一氧化氮，這些氮的氧化物溶解於降落的雨水中，而進入土壤，生成硝酸鹽和亞硝酸鹽。大多數植物以硝酸鹽形式利用氮，它們將硝酸鹽用於製造蛋白質（第21頁）。植物可爲食草動物（第105頁）所食，食草動物進而又被食肉動物（第105頁）所吃，食肉動物在製造動物蛋白質時也利用氮。當動物和植物死亡時，藉硝化細菌的作用，氮又以亞硝酸鹽、氨和銨的化合物形式回到土壤中。

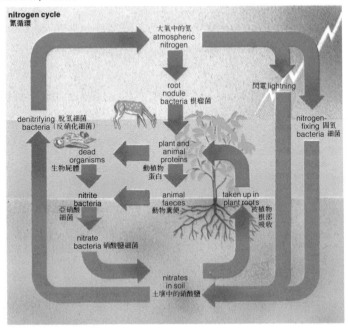

nitrogen cycle
氮循環

大氣中的氮
atmospheric nitrogen

root nodule bacteria 根瘤菌

閃電 lightning

denitrifying 脫氮細菌 bacteria（反硝化細菌）

nitrogen-fixing 固氮 bacteria 細菌

dead organisms
生物屍體

plant and animal proteins
動植物蛋白

nitrite bacteria
亞硝酸細菌

animal faeces
動物糞便

taken up in plant roots
被植物根部吸收

nitrate bacteria 硝酸鹽細菌

nitrates in soil
土壤中的硝酸鹽

water cycle the chain or cycle of events by which water, essential for life, is circulated through the environment (p. 218) and living organisms.

food chain the sequence of organisms from producers (p. 223) to consumers (p. 223) which feed at different trophic levels (p. 223). A simple food chain: grass grows; a cow eats the grass; a human eats the cow or drinks its milk.

food web an interconnected group of food chains (↑). There are few systems as simple as a food chain and many chains may interlink to form a complex web.

水循環　生活所必需的水在環境(第218頁)中和各種生物體內進入循環的鏈式活動過程或循環活動過程。

食物鏈　指從生產者(第223頁)到消費者(第223頁)的各種生物所形成的次序，這些生物以不同的營養級(第223頁)攝取食物。一個簡單的食物鏈是：草生長；牛吃草；人吃牛肉和喝牛奶。

食物網　一組相互連結的食物鏈(↑)。像一個食物鏈這樣簡單的系統很少；許多食物鏈可能相互連結起來，形成一張複雜的網。

a food chain 食物鏈

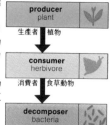

a food web 食物網

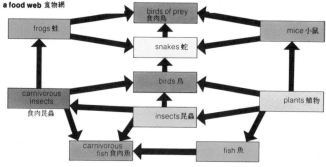

biomass (*n*) the total mass or volume of all the living organisms within a particular area, community (p. 217), or the Earth itself.

pyramid of biomass a diagrammatic representation, which forms the shape of a gently sloping pyramid, to show the biomass (↑) at every trophic level (p. 223).

standing crop the total amount of nutritional (p. 92) living material in the biomass (↑) of a given area at a particular time.

diurnal rhythm the rhythmic sequence of metabolic (p. 26) events, such as the motion of leaves in plants, that take place over a roughly twenty-four hour pattern, and which can be shown to occur in all living organisms even if they are isolated from their normal external environment (p. 218).

circadian rhythm = diurnal rhythm (↑).

生物量(名)　一個特定的區域或羣落(第217頁)之內包含的全部生物之總量或總體積；也指地球本身所擁有的全部生物總量或總體積。

生物量金字塔　用以表現每一營養級(第223頁)的生物量(↑)的一種圖形表示法。這種圖形呈坡度平緩的金字塔形。

現存量、現有量　在一特定時間、特定區域的生物量(↑)中所含的營養性(第92頁)生活物質總量。

日間節律　指如植物中葉子運動之類具有節律性，成序列的代謝(第26頁)活動。這些代謝活動大致以二十四小時爲節律周期的模式進行。而上述這種情況在所有的生物中均可證明其存在；即使在將這些生物與其正常的外部環境(第218頁)隔絕時，情況也仍然如此。

畫夜節律　同日間節律(↑)。

annual rhythm the rhythmic sequence of metabolic (p. 26) events, such as germination (p. 168), flowering and fruiting in plants, that take place over a roughly yearly pattern even if they are isolated from their normal external environment (p. 218).

年節律　如植物發芽(第168頁)、開花和結果之類具有節律性生成序列的代謝(第26頁)活動。這些代謝活動大致以年爲節律周期的模式進行。即使在將這些植物與其正常的外部環境(第218頁)隔絕時，情況也仍然如此。

zooplankton 浮游動物

phytoplankton 浮游植物

plankton (*n*) any of the various, usually tiny or microscopic (p. 9), organisms that float freely in an aquatic environment (p. 218) that have no visible means of locomotion (p. 143) and depend on the currents in the water for distribution. They are not attached to any other organism or substrate.

浮游生物(名)　通常爲細小或顯微鏡(第9頁)下可見的任何一種生物，這些生物自由漂浮在水生環境(第218頁)中。它們並不具備有形的運動(第143頁)器，而是依賴水流動進行散佈。它們也不附着在任何其他生物或基質之上。

phytoplankton (*n*) the plant plankton (↑), especially the diatoms, which are an important source of food for other organisms such as many species (p. 40) of whales.

浮游植物(名)　植物類型的浮游生物(↑)，尤其是矽藻。矽藻是其他一些生物如許多種(第40頁)鯨的重要食物來源。

zooplankton (*n*) the animal plankton (↑) including the larvae (p. 165) of many species (p. 40) of fish.

浮游動物(名)　動物類型的浮游生物(↑)，包括許多種(第40頁)魚的幼體(第165頁)。

pelagic (*adj*) of the upper waters of an aquatic, especially marine environment (p. 218), as opposed to the bed of the ocean or lake, and the organisms which inhabit them.

海面的、浮游的(形)　指水生環境，尤其是海洋環境(第218頁)中與海洋或湖泊的水底相對應的上層水域，以及棲息於其中的那些生物。

benthic (*adj*) of the bed of an aquatic, especially marine environment (p. 218), and the organisms which live on or in it.

海底的、底棲的(形)　指水生環境，尤其是海洋環境(第218頁)中的水底地層，以及棲息於水底地層之上或水底地層之內的那些生物。

association (*n*) any relationship which exists between organisms to the benefit of one or all of them. In plants, a climax (p. 221) community (p. 217) dominated by one or a small number of species (p. 40) and named after them. *See also* parasitism (p. 110).

結合體(名)　生物體之間存在的，有利於其中一個生物體或全部生物體的任何關係。就植物而言，頂極(第221頁)羣落(第217頁)是由一個種(第40頁)或少數幾個種佔絕對優勢的，從而也就以這些物種來命名。參見"寄生現象"(第110頁)。

symbiosis (*n*) an association (p. 227) between two or more species (p. 40) of organisms to their mutual benefit, such as the association of the mycorrhiza (p. 49) of certain fungi (p. 46) with the roots of trees whereby the tree provides nutrients (p. 92) for the fungus which helps the tree to take up water and supplies nitrates to the roots. **symbiotic** (*adj*).

共生（名）　兩個或兩個以上的種（第40頁）的生物之間存在的一種相互有益的結合體（第227頁），例如某些眞菌（第46頁）的菌根（第49頁）與一些樹木的根之間形成的結合體，樹木可利用該結合體爲眞菌提供養份（第92頁），而眞菌則可幫助樹木吸收水份，並將硝鹽供給樹根。（形容詞形式爲 symbiotic）

ectotrophic mycorrhiza 外生菌根

symbiosis 共生
e.g. mycorrhizae
圖例：菌根

樹木爲眞菌 tree provides
提供光合 fungus with
作用的 organic products
有機產物 of photosynthesis

眞菌爲 fungus
樹木提供 provides tree
來自土壤的 with inorganic
無機營養素 nutrients from soil

fungal hyphae
眞菌菌絲

endotrophic mycorrhiza
L.S. root
內生菌根
的縱切面

commensalism (*n*) an association (p. 227) in which one species (p. 40) of organism, the commensal, benefits while the other species is neither harmed nor gains benefit. The bacteria (p. 42) in the gut (p. 80) of mammals (p. 80) are commensals.

片利共生（名）　一種結合體（第227頁），在這種結合體中，一種（第40頁）生物即共棲體受益，而另一種生物旣不受害，也不得益。哺乳動物（第80頁）胃腸道（第98頁）中的細菌（第42頁）就是共棲體。

mutualism (*n*) an association (p. 227) between two or more species (p. 40) in which both benefit. In some cases of mutualism, neither species may be able to survive without the other while, in others, both species may be able to survive independently. It is a form of symbiosis (↑). For example, a species of sea anemone lives on the back of the hermit crab and benefits from being transported to new feeding sites where it feeds on debris from the crab's meals while the crab is protected from predation (p. 220) by the stinging tentacles of the anemone.

互利共生（名）　兩個或兩個以上的種（第40頁）的生物之間存在的一種結合體（第227頁），在這種結合體中，各方均獲得益處。在有些互利共生的場合，兩個種都不可能在沒有另一個種的情況下生存，而在另外一些互利共生的場合，兩個種都可能獨立地生存下來。互利共生是共生（↑）的一種形式。例如，海葵有一個體棲息在寄居蟹的背上，由於其被帶到一些新的攝食地，而得到好處，在那裏，海葵以寄居蟹的食物碎屑爲食，而寄居蟹則利用海葵帶刺的觸手保護自己免遭捕食（第220頁）。

epiphyte (*n*) any plant, such as some ferns or lichens (p. 49), which grows on another plant, in a commensal (↑) association (p. 227), using it only for support and not involved in any parasitism (p. 110)

附生植物（名）　在一個共棲（↑）結合體（第227頁）中，附生於別的植物體上的任何植物，例如某些蕨類植物或地衣（第49頁）。這些植物之所以附生於別的植物體上，僅在於利用它作爲支撐，而與任何寄生（第110頁）無關。

epizoite (*n*) any animal, such as the remora fish which is attached to a shark by a powerful sucker, which lives permanently on another animal, using it for transport etc. and not involved in any parasitism (p. 110).

附生動物（名）　永久地棲息在別的動物體上的任何動物，例如可藉强有力的吸盤吸附在鯊魚身上的鮣魚。這些動物之所以附生在別的動物體上，是爲了遷移等的目的，而與任何寄生（第110頁）無關。

farming (*n*) the process in which humans exploit naturally occurring plants and animals to provide food for their own needs either by deliberately cultivating wild species (p. 40) or by developing new types of organisms and then sowing, planting, tending and protecting them.

fishery (*n*) the process in which humans catch fish or other aquatic animals for food, and exploit the natural processes of population (p. 214) control to increase the size of the catch.

maximum sustainable yield the maximum size of catch of, for example, fish that can be obtained and sustained over years from a given area of water which is fished in such a way that the stocks are larger than they would be if they were unfished. The adult fish are removed from the water for food so that the young do not have to compete with the adults to the same degree for food and the biomass (p. 226) of the water is increased by their survival.

agriculture (*n*) all of the processes associated with the growing of food in a systematic way, including cultivation of land, tending of stock, development of new types, and destruction of competing (p. 220) species (p. 40), so that the yield from a given area can be increased to cope with the increasing demands of a growing human population (p. 214).

pest (*n*) any species (p. 40) of animal or plant which, in the light of modern methods of agriculture (↑) where vast tracts of land are given over to one species, is not subject to the controls of a natural ecosystem (p. 217) and may increase rapidly in numbers to destroy the crop.

weed (*n*) any species (p. 40) of plant which may be able to grow in an area which has been given over to the cultivation of food plants and which will compete with those food plants for space, light, water and nutrients (p. 92).

biological control a method of reducing the numbers of weeds (↑) or pests (↑) by introducing a natural predator (p. 220) of the pest species (p. 40). If the predator is also able to feed on species which are not regarded as pests, then its numbers will not be reduced when the numbers of pests have fallen. This attempts to maintain a natural equilibrium between the pest and the predator.

biological control
e.g. ladybirds are introduced to control aphids (pests)
生物防治
圖例：引進瓢蟲防治蚜蟲（害蟲）

aphid (pest)
蚜蟲（害蟲）

ladybird
瓢蟲

耕作（名） 人類爲提供自身需要的食物而對天然存在的動植物資源進行開發的過程。具體作法是對野生種（第40頁）生物進行精心培育、或者是培育出新品種的生物，然後再進行播種、栽植並加管理和保護。

漁業（名） 人類爲其食物需要而捕捉魚類和其他一些水生動物的過程。在此過程中，人類還開發、利用種羣（第214頁）控制方面的一些自然過程，來增加捕獲量。

最大可持續收獲量（名） 可從一個特定水域持續多年獲得的最大捕獲量如魚的最大捕獲量。魚的最大捕獲量是以這樣一種方式取得的，即要使現有魚類資源大於未曾進行捕魚時的魚類資源。當成魚爲覓食而從該水域離去時，幼魚就不必以同樣的程度和成魚爭奪食物，該水域的生物量（第226頁）也會由於這些幼魚的生存而增加。

農業（名） 與以系統的方式栽培糧食有關的全部過程，其中包括土地的耕作、牲畜的照料、新品種的培育和競爭（第220頁）種（第40頁）的消除。因此，可以增加一定區域內的糧食產量，以適應人口（第214頁）增長所需增加的糧食。

有害生物（名） 從採用以大片土地種植作物的現代農業（↑）方法來看，不受自然生態系統（第217頁）支配、其數量迅速增加並能毀壞作物的任何種（第40頁）的動物和植物，均屬有害生物。

雜草（名） 在一個以經栽培食用植物的區域內可以生長、並與食用植物爭奪空間、光照、水份和養份（第92頁）的任何種（第40頁）植物。

生物防治 藉引進有害生物種（第40頁）的天然捕食者（第220頁），以減少雜草（↑）或有害生物（↑）數量的一種方法。如果捕食者能以不被視爲有害生物的物種爲食，那麼，當有害生物的數量已減少時，捕食者本身的數量也不致減少。這就有助於保持有害生物與捕食者之間的自然平衡。

pesticide (*n*) any agent, usually chemical, which is used to control and destroy pests (p. 229).

herbicide (*n*) any agent, usually chemical, which is used to destroy or control weeds (p. 229).

water purification all of the processes, including storage, straining, filtering and sterilizing which are used by the water authorities to maintain drinking water fit for human consumption. Since drinking water is drawn from rivers, lakes and underground wells, it is also important to ensure that pollutants (↓) from industry or agriculture (p. 229) do not enter the supplies to unacceptable levels.

sewage treatment all of the processes, including the removal of sludge by sedimentation, screening to remove large particles of waste, biological oxidation (p. 32), removal of grit, filtering etc, to ensure that the effluent, which would otherwise contain human waste etc, can be returned to the water cycle (p. 226) without the risk of spreading diseases etc.

conservation (*n*) the use of the natural resources in such a way that they are not despoiled. It is usually taken to include the act of study, management and protection of ecosystems (p. 217), habitats (p. 217) or species (p. 40) of organisms in order to maintain the natural balance of wildlife and its environment (p. 218).

endangered species any species (p. 40) of anima or plant which, by changes in the natural environment (p. 218) or by human intervention, are threatened with death and extinction.

over-exploitation the use of natural resources in such a way that natural ecosystems (p. 217) may be irreversibly disturbed, habitats (p. 217) destroyed, or organisms threatened with extinction.

pollution (*n*) the act of introducing into the natural environment (p. 218) any substance or agent which may harm that environment and which is added more quickly than the environment is able to render it safe. **pollutant** (*n*), **pollute** (*v*).

water pollution the pollution (↑) of marine and freshwater habitats (p. 217) by the unthinking introduction of human, agricultural (p. 229), and industrial waste into rivers, lakes, and oceans.

殺蟲劑（名） 用於防治和消滅有害生物（第229頁）的任何製劑，通常都爲化學藥劑。

除草劑、除葵劑（名） 用於消滅或防治雜草（第229頁）的任何製劑，通常都爲化學藥劑。

水淨化作用 爲使飲用水適合人們的消費需要，由水的管理機構進行全部水處理過程，其中包括貯水、粗濾，細濾和消毒。由於飲用水取自河流、湖泊和地下井，所以，確保進入供水的來自工、農業（第229頁）的各種污染物（↓）低於規定含量也是很重要的。

污水處理 爲確保污水能夠返回水循環（第226頁）中又不致有傳播疾病危險，而對其進行的全部處理過程。其中，包括用沉積法除去污泥、篩除大顆粒廢物、進行生物氧化（第32頁）、除去砂粒、進行過濾等。如果不經過這些處理過程，污水中就會含有人類的廢棄物等等。

自然資源保護（名） 以不採取掠奪方式對自然資源進行利用。爲維持野生物及其環境（第218頁）的自然平衡，通常採取的措施是對生態系統（第217頁）、生境（第217頁）和生物種（第40頁）進行研究、管理和保護。

瀕於滅種危險的物種 由於自然環境（第218頁）的種種變化或人爲干涉，而可能發生死亡和滅絕危險的任何動物或植物種（第40頁）。

過度開發 利用自然資源的一種方式。這種方式對自然生態系統（第217頁）的擾亂可能達到不可逆的程度，破壞生境（第217頁），並且使一些生物受到滅絕威脅。

污染（名） 指這樣的一種行爲：將任何可能是有害的物質或藥劑放進自然環境（第218頁）中，而且放進的速度較環境所能使其變得安全無害的速度爲快。（名詞形式爲 pollutant，動詞形式爲 pollute）

水污染 不加思考地將人類的廢棄物及工農業（第229頁）廢物排入河流、湖泊和海洋，造成對海洋和淡水生境（第217頁）的污染（↑）。

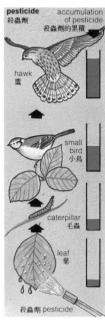

pesticide
殺蟲劑

accumulation of pesticide
殺蟲劑的累積

hawk
鷹

small bird
小鳥

caterpillar
毛蟲

leaf
葉

殺蟲劑 pesticide

principle natural resources exploited by man
人類開發的主要自然資源

renewable resources 可再生的資源

solar energy
太陽能

air
空氣

rain
雨

potentially renewable resources
潛在的可再生的資源

plant communities
植物羣落

utilizable water
可利用的水

animal communities
動物羣落

土壤 soil
養份 nutrients

human population
人羣

土壤 soil

生態系統 ecosystems

non-renewable resources
非再生資源

underground water 地下水

rare plants and animals
稀有動植物

ecosystems
生態系統

oil 石油

marine fisheries
海洋水產

coal 煤

oxygen demand the condition which exists in aquatic environments (p. 218) into which pollutants (↑) have been introduced which promote the growth of aerobic (p. 32) bacteria (p. 42) causing a depletion of the levels of oxygen in the water. Thus, the natural plant life of the environment is reduced and, with it, the animal life that depends upon the plants.

eutrophication (*n*) the situation which occurs when an excess of nutrients (p. 92) is introduced into a freshwater habitat (p. 217) causing a dramatic growth in certain kinds of algae (p. 44). When the nutrients have been used up, the algae die, and the bacterial (p. 42) decomposers (p. 223) which feed on the dead algae use up the oxygen in the water giving rise to an oxygen demand (↑).

需氧量　已輸入污染物(↑)的水生環境(第218頁)中所存在的狀況。這些污染物能促使需氧(第32頁)細菌(第42頁)生長，降低水的含氧量。因而減少該環境中的自然植物，與此同時，也減少依賴這些植物爲生的動物。

富營養化(作用)（名）　當過量營養素(第92頁)被輸入淡水生境(第217頁)中，致使某些藻類(第44頁)驚人生長時所發生的情況。在這些營養素耗盡時，藻類就會死亡，而作爲分解者(第223頁)的那些以死藻體爲食的各種細菌(第42頁)則耗盡水中的氧，從而造成需氧污染(↑)的情況。

algal bloom the dramatically increased population (p. 214) of algae which occurs in an aquatic environment (p. 218) which occurs as a result of eutrophication (p. 231).

air pollution pollution (p. 230) of the atmosphere which results from burning fossil fuels, such as coal and oil, with the introduction into the air of organic (p. 15) and inorganic (p. 15) compounds, such as carbon dioxide, carbon monoxide, sulphur dioxide, etc.

smog (n) fog which has been polluted (p. 230). A mixture of smoke and fog.

marine pollution the pollution (p. 230) of the marine environment (p. 218) primarily by crude oil as a result of the illegal washing of tanks at sea or by accidental loss. The damage to seabird populations (p. 214) is great and well known but there is also poisoning of marine plankton (p. 227) which thereby affects the whole marine food web (p. 226).

radioactive pollution the pollution (p. 230) of the environment (p. 218) by accidental leakage from sites of nuclear energy production or from the dumping of nuclear waste products. The radioactive materials which find their way into the environment can lead to chromosome (p. 13) damage and mutations (p. 206).

terrestrial pollution the pollution (p. 230) of the land environments (p. 218) by the dumping of waste materials from mining industries, for example, or by pesticides (p. 230).

birth control the attempt by humans artificially to limit the rapid growth which has taken place in the world human population (p. 214) which may otherwise place possibly disastrous strains on food supplies and other non-replaceable resources. It involves methods of preventing conception with the use of contraceptives such as the Pill, vasectomy, etc.

hygiene (n) the science which deals with the preservation of human health by such means as improvements of sanitation to prevent the spread of disease. It is thought that hygiene improvements are among the most important factors in the increase in human life expectancy.

藻類過量繁殖　水生環境（第218頁）中發生藻類種羣（第214頁）驚人增殖。這種增殖是由於富營養化作用（第231頁）所致的。

空氣污染　由於燃燒諸如煤和石油之類化石燃料，而將像二氧化碳、一氧化碳和二氧化硫等之類一些有機（第15頁）和無機（第15頁）化合物排入空氣中而造成大氣污染（第230頁）。

霧、烟霧（名）　受到污染（第230頁）的霧，即煙和霧的混合物。

海洋污染　主要是原油造成的海洋環境（第218頁）污染（第230頁）。產生這種污染的原因是，在海上非法清洗油艙，或發生意外溢油事故。海洋污染對海鳥種羣（第214頁）的危害很大，這已爲人所周知，但是，它對海洋浮游生物（第227頁）也有毒害，從而對整個海洋食物網（第226頁）造成損害。

放射性污染　由於核能生產廠地發生意外泄漏事故，或者由於傾倒核廢物而造成環境（第218頁）污染（第230頁）。通過各種途經進入環境的這些放射性物質，可導致染色體（第13頁）損傷和突變（第206頁）。

陸地的污染　因傾倒例如礦業廢物，或者由於農藥（第230頁）的使用而造成對陸地環境（第218頁）污染（第230頁）。

控制生育、節育　人類在人爲限制世界人口（第214頁）迅速增長方面所作出的努力。如果不作出這種努力，在食物供應和其他一些不可恢復資源方面就可能造成災難性的緊張狀態。控制生育所採取的預防措施，包括利用如服避孕丸、作輸精管切除等避孕手段。

衛生學（名）　探討採取改善衛生條件之類的手段，防止疾病傳播，保護人類健康的科學。據認爲，改善衛生在提高人的估計壽命方面是重要的因素。

air pollution smog 空氣污染燜霧

pollution of water 水的污染

disease (*n*) any disorder or illness of the body or organ.

infectious (*adj*) of a disease caused by viruses (p. 43) or other parasitic (p. 92) organisms, such as certain bacteria (p. 42), which can be passed from one individual to another.

contagious (*adj*) of a disease which can be passed from one individual to another by contact, which may be direct touching of the individuals or through objects which have been contaminated by the diseased individual and then handled by another individual.

antiseptic (*adj*) of any agent which destroys the microorganisms which invade the body leading to disease.

aseptic (*adj*) of conditions in which disease-causing microorganisms are not present.

antibiotic (*n*) any substance, produced by a living organism, for example, the fungus (p. 46) *Penicillium*, which is poisonous to other living organisms. Antibiotic substances are used in medicine to destroy disease-causing microorganisms.

antibody (*n*) a protein (p. 21) which is produced by an organism following the invasion of the body fluids by a substance which is not normally present and which may be harmful. The antibody combines with the invading substance thereby removing it from the body.

immunity (*n*) the state in which organisms are protected from the invasion of disease which mainly involves the production of antibodies (↑).

active immunity immunity (↑) in which the body's defensive mechanisms are stimulated by the invasion of foreign microorganisms to produce antibodies (↑).

passive immunity immunity (↑) in which the body's own defensive mechanisms are not stimulated by the invasion of foreign micro-organisms but in which antibodies (↑) have been transferred to it from another animal in which active immunity (↑) has been stimulated.

inherited immunity passive immunity (↑) in which the resistance to certain diseases is inherited genetically (p. 196) from the parents.

the production of antibodies 抗體的產生

harmful substances invade animal
有害物質侵入動物

defence mechanisms of animal produce antibodies
動物的防禦機制產生相應的抗體

antibodies combine with harmful substances
抗體與有害物質相結合

combinations of antibody and now harmless substances removed from the body
抗體的結合物，
即無害物質便可從機體內除去

疾病（名）　身體或器官發生的任何失調或不健康情況。

傳染的、侵染性的（形）　指由病毒（第43頁）或其他一些寄生（第92頁）生物如某些細菌（第42頁）引起的某種疾病而言。這種疾病能從一個個體傳給另一個個體。

接觸傳染的（形）　指經接觸能從一個個體傳給另一個個體的某種疾病而言，這種疾病可能由個體之間直接接觸而傳染或者可能通過接觸患病個體污染的物體而傳染給其他個體。

防腐的、抗菌的（形）　指可對侵入身體、導致疾病的種種微生物起破壞作用的任何制劑而言。

無菌的（形）　指不存在致病微生物環境而言。

抗菌素、抗生素（名）　由一種生物例如青霉屬眞菌（第46頁）產生的，對其他一些生物有毒的任何物質。醫學上利用抗菌物質來破壞各種致病微生物。

抗體（名）　一種蛋白質（第21頁）。在生物體內通常不存在這種蛋白質、且可能是有害的一種物質侵入該生物體體液之後，由該生物體所產生的。抗體與侵入的物質相結合，從而將其從體內除掉。

免疫、免疫性（名）　生物體能夠免受疾病侵害的狀況。它主要涉及抗體（↑）的產生。

自動免疫、主動免疫　機體的防禦機制在受到外來微生物入侵的刺激下，產生抗體（↑）而建立的免疫（↑）。

被動免疫　機體自身的防禦機制並未受到外來微生物入侵的刺激，其抗體（↑）是從另一個動物轉移到其身上的，而在該動物體內，自動免疫（↑）此前已被激活。這樣建立起來的免疫（↑）稱爲被動免疫。

遺傳免疫　機體通過遺傳（第196頁）從親代獲得對某些疾病的抵抗力。這樣建立起來的免疫稱爲遺傳免疫。遺傳免疫屬於被動免疫（↑）。

acquired immunity active immunity (p. 233) by exposure to an infectious (p. 233) disease which is too restricted to cause the symptoms of the disease or passive immunity (p. 233) by the transfer of antibodies (p. 233) from the mother to the offspring across the placenta (p. 192).

vaccination (*n*) the injection into the body of an animal modified forms of the microorganisms which will cause a particular disease so that the body produces antibodies (p. 233) that will resist any possible invasion of the disease itself. The animal gains acquired immunity (↑).

vaccine (*n*) any substance containing antigens (↓) which is injected into an animal's body to produce antibodies (p. 233) and give the animal acquired immunity (↑) to specific diseases.

antigen (*n*) any substance, produced by a micro-organism, which will stimulate the production of antibodies (p. 233).

epidemic (*adj*) of a disease which is not normally present in a population (p. 214) and which, therefore, will spread rapidly from individual to individual and infect (p. 233) a large number of the population because there is no natural immunity (p. 233) to the infection.

endemic (*adj*) of a disease which occurs naturally in particular, geographically restricted populations (p. 214).

pandemic (*adj*) of a disease which occurs throughout the population (p. 214) of a whole continent or even the world.

allergy (*n*) the condition in which certain individuals may be particularly sensitive to substances which are quite harmless to other individuals. For example, asthmatic attacks may be stimulated by breathing dust or pollen (p. 181). Allergic reactions may include inflammation or swelling.

symptom (*n*) a sign or condition of the presence of, e.g. a disease.

contract (*v*) (1) *of diseases* to get or to catch. (2) to become smaller or shorter e.g. muscles (p. 143) contract. **contraction** (*n*), **contractile** (*adj*).

後天免疫、獲得性免疫　機體因接觸一種傳染(第233頁)病而獲得的自動免疫(第233頁)，由於已對該病採取抑制措施，以使機體未能產生該病的症狀；或因抗體(第233頁)從母體經胎盤(第192頁)轉移到後代上而獲得的被動免疫(第233頁)。

接種(名)　將會引起一種特定疾病的微生物的變異型注射在某一動物體內，使該動物體產生抗體(第233頁)，抵抗該病本身的任何可能的侵襲。該動物爲此獲得後天免疫(↑)。

疫苗(名)　含有抗原(↓)的任何物質。將疫苗注射到某一動物體內，即可產生抗體(第233頁)，並使該動物對一些特定疾病具有後天免疫(↑)能力。

抗原(名)　由微生物產生並可刺激抗體(第233頁)形成的任何物質。

流行性的(形)　指一種疾病而言，這種疾病通常並不出現於某一種羣(第214頁)之中，但因對該疾病的傳染沒有天然免疫(第233頁)能力而致在該種羣的許多個體之間迅速傳染(第233頁)蔓延。

地方性的(形)　指一種疾病而言，這種疾病自然發生於一些特殊的、在地理上受到限制的種羣(第214頁)之中。

廣泛流行、大流行的(形)　指一種疾病而言，這種疾病遍播於整個大陸甚至全世界的種羣(第214頁)之中。

過敏反應(名)　某些個體對一些物質可能特別敏感時所表現的狀態，而這些物質對其他一些個體來說，則是完全無害的。例如，由於吸入塵埃或花粉(第181頁)，可能會刺激哮喘病發作。過敏反應也包括炎症和腫脹。

症狀(名)　某種疾病存在的病症或狀態。

感染；收縮(動)　(1)指患病或感染病而言。(2)變得較小或縮短，例如肌肉(第143頁)收縮。(名詞形式爲 contraction，形容詞形式爲 contractile)

acquired immunity 後天免疫

animal with disease-carrying microorganisms
有帶病微生物的動物

a few microorganisms injected into another animal
將一些微生物注射到另一個動物體內

antigens produced by the microorganisms cause animal to produce antibodies
由這些微生物產生的抗原使動物產生抗體

animal has gained an acquired immunity to the disease through vaccination
動物通過接種獲得了對此病的後天免疫

scientific method a means of gaining knowledge of the environment (p. 218) by observation (↓) which leads to the development of a hypothesis (↓). From the hypothesis, predictions (↓) are made which are tested by experiments (↓) that include controls (↓).

observation (*n*) (1) a natural event or phenomenon which is viewed or learned; (2) that which is viewed or learned.

hypothesis (*n*) an idea which has been put forward to explain the occurrence of a natural event or events noted by observation (↑). **hypotheses** (*pl.*).

prediction (*n*) the process of foretelling likely events or phenomena in a given system from those already noted by observation (↑). Predictions follow from the hypothesis.(↑).

experiment (*n*) a means of examining a hypothesis (↑) by testing a prediction (↑) made on the basis of the hypothesis.

control (*n*) an experiment (↑) performed at the same time as the main experiment which differs from it in one factor only. Controls are a means of testing those factors which affect a phenomenon.

theory (*n*) an idea or set of ideas resulting from the scientific method (↑) used as principles (↓) to explain natural phenomena which have been noted by observation (↑).

phenomenon (*n*) any observable fact that can be described scientifically. **phenomena** (*pl.*).

principle (*n*) a general truth or law at the centre of other laws.

adaptation (*n*) a change in structure, function etc, which fits a new use. A particular adaptation may make an organism better fitted to survive (p. 209) in its environment (p. 218). **adapt** (*v*).

structure (*n*) the way in which all the parts of an object, or organism or part of an organism are arranged. The structure of anything is closely related to the function it performs.

function (*n*) the normal action of an object or part of an organism, for example, the function of the ear (p. 157) is to hear (p. 159).

adjacent (*adj*) near by, next to or close to.

amorphous (*adj*) without shape or form, e.g. cells which have not been differentiated.

科學方法　經觀察（↓）獲得有關周圍環境（第218頁）的知識的一種手段。觀察可導致建立假說（↓），根據假說則可作出種種預言（↓），而這些預言皆需經過實驗（↓）包括對照實驗（↓）驗證。

觀察（名）　（1）對一種自然事件或自然現象進行的考察或學習；（2）對某一對象進行的考察或學習。

假說（名）　爲了解釋觀察（↑）中引起注意的某一自然事件或某一自然事件發生的原因而提出的一種概念。（複數形式爲 hypotheses）。

預言（名）　以觀察（↑）中引起注意的那些迹象爲依據，對一個給定系統中某些可能事件或現象作出預測的過程。預言是在假說（↑）的指引下作出的。

實驗（名）　對根據假說所作出的預言（↑）進行試驗，以驗證假說（↑）的一種手段。

對照實驗（名）　在作主實驗的同時所進行的一種實驗（↑），它與主實驗僅在一個因素上有所不同。對照實驗是對影響某一現象的那些因素進行檢驗的一種手段。

學說、理論（名）　由科學方法（↑）得出的一種概念或一組概念、可作爲原理（↓），以解釋在觀察（↑）中引起注意的那些自然現象。

現象（名）　能用科學加以描述的任何可觀察到的事實。（複數形式爲 phenomena）。

原理（名）　各有關定律之核心部份的定律或眞理。

適應（名）　在結構或功能等方面所發生的一種變化。此種變化能適合於新用途的需要。經歷特定的適應之後，生物體能夠更適合於其所處的環境（第218頁）中生存（第209頁）下去。（動詞形式爲 adapt）。

結構（名）　一個物體或一個生物體的所有組成部份或一個生物體的某一部份的排列方式。任何物體或生物體的結構皆與其所發揮的功能密切相關。

功能（名）　一個物體，或一個生物體的一部份所發揮的正常作用，例如耳（第157頁）的功能是聽覺（第159頁）。

鄰近的（名）　靠近的、貼近的或接近的。

無定形的（形）　不具備一定形狀或形態的，例如尚未分化時的細胞。

anterior (*adj*) at, near or towards the front (or head) end of an animal, usually the end directed forward when the animal is moving (in humans the anterior is the ventral (p. 75) part).

articulation (*n*) the movable or non-movable connection or joint between two objects.

axis (*n*) a real or imaginary straight line about which an object rotates e.g. the axis of symmetry (p. 60).

cavity (*n*) a hole or space e.g. the buccal cavity (p. 99).

comatose (*adj*) inactive and in a deep sleep as, for example, in a hibernating (p. 132) animal.

comparable (*adj*) of two or more objects, of similar quality. **compare** (*v*).

concentration (*n*) the strength or quantity of a substance in, for example, a solution (p. 118).

constituent (*n*) a part of the whole. **constituent** (*adj*).

constrict (*v*) to make thinner, for example, a narrowing of the blood vessels (p. 127). **constriction** (*n*).

dilate (*v*) to make wider, for example, in blood vessels (p. 127). **dilation** (*n*).

convoluted (*adj*) rolled or twisted into a spiral or coil, for example, the convoluted tubules in the kidney (p. 136). **convolute** (*v*).

co-ordinate (*v*) to cause two or more things, e.g. limbs, to work together for the same purpose. **co-ordination** (*n*).

crystallize (*v*) to form crystals (regular shapes).

deficiency (*n*) a shortage or lack of something. For example, vitamin deficiency (*see* p. 238).

development (*n*) a stage in growth which includes changes in structures and the appearance of new organs and tissues (p. 83).

duct (*n*) a tube formed of cells.

equilibrium (*n*) the state in which an object is steady or stable because the forces acting upon it are equal.

essential (*adj*) very necessary.

external (*adj*) of the outside.

internal (*adj*) of the inside.

extract (*v*) to remove or draw out one substance from a particular material.

filter (*n*) an instrument used to take solids and other substances out of liquids. **filtration** (*n*).

前端的、前面的(形)　指位置接近或朝向動物的前端(或頭部)而言，通常指的是動物在運動時朝着前進方向的那一端(對人而言，是指腹(第75頁)部)。

關節(名)　位於兩個物體之間的可動的或不可動的連接或接合物。

軸(名)　一個物體圍繞其旋轉的一條眞的直線或虛的直線，例如對稱(第60頁)軸。

腔、洞(名)　孔或間隙，例如口腔(第99頁)。

昏迷的(形)　猶如冬眠(第132頁)動物所呈現出的狀態那樣，不活動的和沉睡的。

可比較的(形)　指具有類似性質的兩個或兩個以上的物體而言。(動詞形式爲 compare)。

濃度(名)　一種物質在一種溶液(第118頁)中的含量或數量。

成份、組份(名)　整體的一部份。(形容詞形式爲 constituent)。

收縮、縮窄(動)　使變細、例如血管(第127頁)收縮。(名詞形式爲 constriction)。

擴張、膨脹(動)　使變粗、例如血管(第127頁)擴張。(名詞形式爲 dilation)。

回旋狀的、盤旋的(形)　捲繞或盤繞成螺旋或盤狀物，例如腎臟(第136頁)的曲小管。(動詞形式爲 convolute)。

協調、配合(動)　使兩個或兩個以上的物體如肢體爲同一目的而一起動作。(名詞形式爲 co-ordination)

結晶(動)　形成晶體(具有規則的形狀)。

缺乏(名)　不足或短缺某種物質。例如，缺乏維生素(第238頁)。

發育(名)　生長的一個階段，在這個階段，結構發生變化，新的器官和組織(第83頁)的外表也發生變化。

管、導管(名)　由細胞構成的管道。

平衡(名)　物體由於作用於其上的各個力相等而處於穩定的狀態。

必需的(形)　非常需要的。

外部的(形)　指外部而言。

內部的(名)　指內部而言。

抽出(動)　從一特殊物質中取出或抽出一種物質。

濾器、濾機(名)　用來使固體和其他物質從液體中分離出來的一種器械。(名詞形式 filtration 意爲過濾)。

flex (*v*) *of a joint* to bend, *of a muscle* (p. 143) to contract.

gradient (*n*) the increase or decrease in a substance over a distance.

increase (*v*) to become or to make greater in some way, for example, in size, value, concentration etc. **increase** (*n*).

decrease (*v*) to become or to make less or fewer in some way, for example, in size, value, concentration etc. **decrease** (*n*).

insulation (*n*) any material used to prevent the passage of heat (or electricity), for example, hair insulates the bodies of mammals (p. 80) and feathers (p. 147) the bodies of birds.

intermediate (*adj*) of an object in the middle, e.g. an intermediate stage in metabolism (p. 26).

lubricate (*v*) to make smooth or slippery in order to make the movement of parts of a machine or organism easier. **lubrication** (*n*).

offspring (*n*) = progeny (p. 200).

parallel (*adj*) of lines or planes which run in the same direction and never meet.

permeable (*adj*) of, for example, a membrane (p. 14) which allows a substance to pass through. *See also* semipermeable membrane (p. 118).

posterior (*adj*) at, near or towards the back or hind end of an animal, usually the end directed backwards when the animal is moving.

product (*n*) a substance that is produced.

byproduct (*n*) a substance that is produced in the course of producing another substance.

protuberance (*n*) a part or thing which swells or sticks out, for example, a pseudopodium (p. 44) of *Amoeba* (p. 44).

sedentary (*adj*) of an animal that remains attached to a surface and does not carry out locomotion (p. 143), for example, a coral polyp (p. 61).

synthesize (*v*) to make a substance from its parts.

tensile (*adj*) of a material which is able to be stretched.

transparent (*adj*) of a material that lets light pass through and through which objects can be clearly seen.

viscous (*adj*) of a fluid that will not flow i.e. is rather solid.

屈曲（動）　指關節的彎曲而言，也指肌肉（第143頁）的收縮而言。

梯度（名）　一種物質在一定間隔中的增加量或減少量。

增大、增加（動）　大小、數值、濃度等方面變或使之變大。（名詞形式爲 increase）

減小、減少（動）　大小數值、濃度等方面變或使之變小（少）。（名詞形式爲 decrease）

絕熱體、絕緣體（名）　可用於阻止熱（或電流）傳導的任何物質。例如，毛髮可以使哺乳動物（第80頁）的身體絕熱，而羽毛（第147頁）則可使鳥類的身體絕熱。

中間的、居間的（形）　指居於中間的事物而言，例如新陳代謝（第26頁）的一個中間階段。

潤滑（動）　爲使機器或生物體的各組成部份易於運動，而對它們的表面進行光滑或滑溜處理。（名詞形式爲 lubrication）

子孫（名）　同後代（第200頁）。

平行的（形）　指朝同一方向延伸而永遠不會相交的一些直線或平面而言。

可透過的、可滲透的（形）　指可容許一種物質通過的東西例如膜而言。參見"半透性膜"（第118頁）。

後面的、後端的（形）　指位置在，接近或朝着動物的後面或後端而言，通常是指動物在運動時，朝向後方的那一端。

產品（名）　生產出來的物質。

副產品（名）在生產一種物質的過程中產生出來的另一種物質。

突起、突出物（名）　隆起或突出的部份或東西，例如變形蟲（第44頁）的僞足（第44頁）。

固着的（形）　指某種動物而言，該動物始終附着於某一表面，而不進行運動（第143頁），例如珊瑚蟲（第61頁）。

合成（動）　用某種物質的各組份來製造該物質。

拉伸的（形）　指某種材料而言，該材料能夠予以拉伸。

透明的（形）　指某種材料而言，該材料可讓光線通過，且透過該材料可清楚地看見其他一些物件。

黏滯的（形）　指某種液體而言，該液體不會流動，也就是說，它是頗爲稠密的。

Vitamins

NAME	LETTER	MAIN SOURCES	FUNCTION	EFFECTS OF DEFICIENCY	FAT (F) OR WATER (W) SOLUBLE
retinol	A	liver, milk, vegetables containing yellow and orange pigments e.g. carrots	light perception, healthy growth, resistance to disease	night blindness, poor growth, infection, drying and degeneration of the cornea	F
calciferol	D	fish liver, eggs, cheese, action of sunlight on the skin	absorption of calcium and phosphorus and their incorporation into bone	bone disorders e.g. rickets	F
tocopherol	E	many plants, such as wheatgerm and green vegetables	cell respiration, conservation of other vitamins	in humans, no proved effect, may cause sterility, muscular dystrophy in rats	F
phylloquinone	K	green vegetables, egg yolk, liver	synthesis of blood clotting agents	haemorrhage, prolonged blood clotting times	F
thiamin	B_1	most meats and vegetables, especially cereals and yeast	coenzyme in energy metabolism	beri-beri, loss of apetite and weakness	W
riboflavine	B_2	milk, eggs, fish, green vegetables	coenzyme in energy metabolism	ulceration of the mouth, eyes and skin	W

維生素

名　稱	字母代稱	主要來源	功　能	缺乏時的症狀	脂(F)溶性或水 (W)溶性
視黃醇	A	肝、奶、含黃色和橙色色素的蔬菜如胡蘿蔔	感知光線、促進健康生長、抗病	夜盲、發育不良、易感染、角膜乾燥和變性	F
鈣化醇	D	魚肝、蛋、乳酪和日光對皮膚的照射作用	促進鈣和磷的吸收及其對骨內的滲入	骨頭病徵如佝僂病	F
生育酚	E	多種植物諸如麥芽和綠色蔬菜	細胞呼吸、貯存其他維生素	未能證實對人的副作用。可導致老鼠不育和肌肉萎縮。	F
葉綠醌	K	綠色蔬菜、蛋黃、肝	合成凝血因子	出血、延長血液凝固時間	F
硫胺素	B_1	大多數肉類、蔬菜，尤其是穀類和酵母	作能量代謝中的輔酶	脚氣病、喪失食慾、身體虛弱	W
核黃素	B_2	奶、蛋、魚、綠色蔬菜	作能量代謝中的輔酶	口腔、眼睛和皮膚潰瘍	W

niacin	B complex (B$_2$)	fish, meat, green vegetables, wheatgerm	coenzyme in energy metabolism	pellagra: skin infections, weakness; mental illness	W
pantothenic acid	B$_5$	most foods, especially yeast, eggs, cereals	coenzyme in energy metabolism	headache, tiredness, poor muscle co-ordination	W
pyridoxine	B$_6$	most foods, especially meat, cabbage, potatoes	release of energy, formation of amino acids	nausea, diarrhoea, weight loss	W
biotin	B complex (H)	most foods, especially milk, yeast, liver, egg yolk	coenzyme in energy metabolism	dermatitis	W
folic acid	B$_c$	green vegetables, liver, kidneys	similar to vitamin B$_{12}$	a form of anaemia	W
cobalamin	B$_{12}$	meats e.g. liver, heart, herrings, yeast, some green plants	maturing red blood cells, growth, metabolism	a form of anaemia	W
ascorbic acid	C	citrus fruits, green vegetables	collagen formation	scurvy: tooth loss, weakness susceptibility to disease, weight loss	W

烟 酸	B 複合物 (B₂)	魚、肉、綠色蔬菜、麥芽	作能量代謝中的輔酶	糙皮病、皮膚感染、身體虛弱、精神病	W
泛 酸	B₅	大多數食物，尤其是酵母、蛋、穀類	作能量代謝中的輔酶	頭痛、疲倦、肌肉協調機能不良	W
吡哆醇	B₆	大多數食物，尤其是肉、包心菜、馬鈴薯	釋放能量，形成氨基酸	噁心、腹瀉、體重下降	W
生物素	B 複合物 (H)	大多數食物，尤其是奶、酵母、肝、蛋黃	作能量代謝中的輔酶	皮炎	W
葉 酸	BC	綠色蔬菜、肝、腎	與維生素 B₁₂ 類似	某種類型的貧血	W
鈷胺素	B₁₂	肉類例如肝、心、以及鯡魚、酵母、某些綠色植物	促使紅血細胞成熟、促進生長和新陳代謝	某種類型的貧血	W
抗壞血酸	C	柑橘、綠色蔬菜	形成膠原蛋白	壞血病、牙齒脫落、易感染疾病、體重下降	W

Nutrients

carbon dioxide a colourless, odourless gas at normal temperature and pressure with the chemical formula CO_2. It is denser than oxygen and occurs in the atmosphere at lower levels. It is absorbed by plants and is used to make complex organic compounds especially by photosynthesis. It is a waste product of respiration.

oxygen a colourless, odourless gas at normal temperature and pressure with the chemical formula O_2. It is a vital element of the inorganic and organic compounds, such as carbohydrates, proteins and fats, which make up all living organisms. It is taken in by plants as gaseous oxygen in the dark and as carbon dioxide and water and released as a gas from photosynthesis. It is essential for respiration in aerobic organisms.

water a colourless, tasteless liquid, at normal temperatures and pressures, with the chemical formula H_2O. Most nutrients are soluble in water. Water takes part in many of the chemical reactions involved in nutrition and is also an essential fluid in the transport of materials throughout the body of an organism. It is a waste product of respiration and is essential in photosynthesis.

PLANT NUTRIENTS

macronutrients

potassium a macronutrient which is absorbed by plants in the form of potassium salts and which is required as a component of enzymes and amino acids. Potassium deficiency will eventually lead to the plant's death and is indicated by yellow edges to the leaves.

calcium a macronutrient which is absorbed by plants in the form of calcium salts and which is required in cell walls. Calcium deficiency will cause a plant to have stunted roots and shoots because the growing points die.

nitrogen a macronutrient present in the atmosphere as a colourless, odourless gas at normal temperatures and pressures but absorbed by plants in the form of nitrates. It is an essential part of proteins and amino acids etc. Nitrogen deficiency causes the plant to show stunted growth with yellowing of the leaves.

phosphorus a macronutrient which is absorbed by plants as H_2PO_4 and is found in proteins, ATP and nucleic acids. Phosphorus deficiency causes the plant to show stunted growth with dull dark green leaves.

magnesium a macronutrient which is absorbed by plants in the form of magnesium salts and is found in chlorophyll. Magnesium deficiency causes yellowing of the leaves.

sulphur a macronutrient which is absorbed by plants as sulphates and is found in certain proteins. Sulphur deficiency causes roots to develop poorly as well as yellowing of the leaves.

iron a macronutrient which is absorbed by plants as iron salts and is found in cytochromes. Iron deficiency causes yellowing of the leaves.

營養素

二氧化碳　一種在常溫、常壓下為無色、無臭的氣體，化學式為 CO_2。二氧化碳的密度大於氧，在大氣中的含量較低。植物可吸收二氧化碳，特別是經過光合作用，將它用來製造複雜的有機化合物。二氧化碳是呼吸作用產生的一種廢物。

氧　一種在常溫、常壓下為無色、無臭的氣體，其化學式為 O_2。氧是構成所有生物體的無機化合物和諸如碳水化合物、蛋白質、脂肪的有機化合物的一種重要元素。植物在黑暗中吸入氧態氧，而在光合作用中，則是以二氧化碳和水的形式吸入氧，再以氧態氧的形式釋放出來。氧對需氧生物的呼吸作用是必不可少的。

水　一種在常溫、常壓下無色、無味的液體，其化學式為 H_2O。大多數營養素可溶解於水。水參與許多涉及營養的化學反應，也是為生物體全身輸送物質必不可少的一種液體。水是呼吸作用產生的一種廢物，而在光合作用中則是不可缺少的。

植物營養素

主要營養素

鉀　植物以鉀鹽的形式吸收的一種大量營養素。它是酶和氨基酸的一種必需成份。缺鉀最終都將導致植物死亡，其症狀是葉子的邊緣變黃。

鈣　植物以鈣鹽的形式吸收的一種大量營養素。它是細胞壁的必需成份。缺鈣將使植物根部和枝條的生長受阻，原因是其生長點死亡。

氮　在常溫、常壓下作為無色、無臭的氣體存在於大氣中的一種大量營養素，但是它以硝酸鹽的形式被植物吸收。氮是蛋白質和氨基酸等不可缺少的組成部份。缺氮將使植物生長受阻，葉子發黃。

磷　植物以 H_2PO_4 的形式吸收的一種大量營養素。蛋白質、ATP 和核酸中含有磷。缺磷將使植物生長受阻，葉子呈暗綠色。

鎂　植物以鎂鹽的形式吸收的一種大量營養素。葉綠素中含有鎂。缺鎂導致葉子發黃。

硫　植物以硫酸鹽的形式吸收的一種主要營養素。某些蛋白質中含硫。缺硫使根部發育不良，葉子發黃。

鐵　植物以鐵鹽的形式吸收的一種大量營養素。細胞色素中含鐵。缺鐵導致葉子發黃。

micronutrients

boron a micronutrient absorbed by plants in the form of borates. It is important after pollination in the stimulation of germination of the pollen grains as well as in the absorption of calcium through the roots. Boron deficiency results in certain diseases of plants, such as internal cork in apples.

zinc a micronutrient which is absorbed by plants in the form of zinc salts. It is important in the activation of certain enzymes and in the production of leaves. Zinc deficiency results in the abnormal growth of leaves.

copper a micronutrient which is absorbed by plants in the form of copper salts. It is required by some enzymes. Copper deficiency results in the growth of the plant showing certain kinds of abnormality.

molybdenum a micronutrient absorbed by plants in the form of molybdenum salts. It is important in the function of certain enzymes for the reduction of nitrogen. Molybdenum deficiency results in the overall growth of the plants being reduced.

chlorine a micronutrient which is absorbed by plants in the form of chlorides. It is important in osmosis etc. although it cannot easily be shown to have effects if there is a deficiency.

manganese a micronutrient which is absorbed by plants in the form of manganese salts. It is an important activator of certain enzymes. Manganese deficiency results in the yellowing of the leaves as well as grey mottling.

ANIMAL NUTRIENTS

minerals

calcium a mineral, present in milk products, fish, hard water and in bread, which is required for healthy bones and teeth to aid in the clotting of blood and in muscles. The average adult human requires 1.1 grams per day and the total body content is about 1000 grams.

phosphorus a mineral, present in most foods but especially cheese and yeast extract, which is required for healthy bones and teeth, and takes part in the DNA, RNA and ATP metabolism. The average human adult requires 1.4 grams per day and the total body content is about 780 grams.

sulphur a mineral, present in foods containing proteins, such as peas, beans and milk products. It is required as a constituent of certain proteins, such as keratin and vitamins, such as thiamine. The average human adult requires 0.85 grams per day and the total body content is about 140 grams.

potassium a mineral present in a variety of foods, such as potatoes, mushrooms, meats and cauliflower, which is required for nerve transmission acid-base balance. The human requires 3.3 grams per day and the total body content is about 140 grams.

微量營養素

硼 植物以硼酸鹽形式吸收的一種微量營養素。在傳粉以後，它對促進花粉粒的萌發以及根部對鈣的吸收起重要作用。缺硼導致植物的某些病害，例如蘋果內部栓化病。

鋅 植物以鋅鹽的形式吸收的一種微量營養素。它對某些酶的活化和葉子的生長起重要作用。缺鋅可導致葉子異常生長。

銅 植物以銅鹽形式吸收的一種微量營素。它為某些酶所必需。缺銅導致植物發生異常生長。

鉬 植物以鉬鹽形式吸收的一種微量營養素。它對某些酶還原氮的功能起重要作用。缺鉬導致整個植物的生長減慢。

氯 植物以氯化物形式吸收的一種微量營養素。它在滲透作用等方面起重要作用，但還難以顯示缺氯的影響。

錳 植物以錳鹽形式吸收的一種微量營養素。它是某些酶的一種重要激活劑。缺錳導致葉子發黃，並出現灰色斑紋。

動物營養素
礦物質

鈣 乳製品、魚、硬水和面包含有的一種礦物質。它為健康的骨骼和牙齒所必需，對於凝血和肌肉收縮也很有幫助。平均每個成人每天需鈣1.1克，整個身體的含鈣量約為1000克。

磷 大多數食物尤其是乳酪和酵母抽提物中含有的一種礦物質。它為健康的骨骼和牙齒所必需，參與 DNA、RNA 和 ATP 的新陳代謝。平均每個成人每天需磷1.4克，整個身體的磷量約為780克。

硫 豌豆及其他豆類和乳製品類含蛋白質食物中含的一種礦物質。它是某些蛋白質如角蛋白以及維生素如硫胺素的一種成份。平均每個成人每天需硫0.85克，整個身體的含硫量約為140克。

鉀 多種食物諸如馬鈴薯、蘑菇、肉類和花椰菜中含有的一種礦物質它為神經傳遞和酸鹼平衡所必需。每人每天需鉀3.3克，整個身體的含鉀量約為140克。

sodium a mineral present in a variety of 'salty' foods but especially table salt (sodium chloride), cheese and bacon, and which is required for nerve transmission and acid-base balance. The average human requires about 4.4 grams per day and the total body content is about 100 grams.

chlorine as chloride ions, a mineral found with sodium in table salt and in meats, which is required for acid-base balance and for osmoregulation. The average human adult requires 5.2 grams per day and the total body content is about 95 grams.

magnesium a mineral present in most foods, but especially cheese and green vegetables, which is required to activate enzymes in metabolism. The average human adult requires about 0.34 grams per day and the total body content is about 19 grams.

iron a mineral present in liver, eggs, beef and some drinking water, which is an essential constituent of haemoglobin and catalase. The average human adult requires 16 milligrams per day and the total body content is about 4.2 grams.

fluorine as fluoride, a mineral found in sea water and sea foods and sometimes added to drinking water. It is a constituent of bones and teeth and prevents tooth decay. The average human requires 1.8 milligrams per day and the total body content is about 2.6 grams.

zinc a mineral found in most foods, but especially meat and beans, which is required as a constituent of many enzymes. It is also thought to promote healing. The average human adult requires 13 milligrams per day and the total body content is about 2.3 grams.

copper a mineral found in most foods, but especially in liver, peas and beans, which is required for the formation of haemoglobin and certain enzymes. The average human adult requires 3.5 milligrams per day and the total body content is about 0.07 grams.

iodine a mineral found in sea foods and some drinking water and vegetables, which is required as a constituent of thyroxine. The average adult human requires 0.2 milligrams per day and the total body content is only about 0.01 grams.

manganese a mineral found in most foods, but especially tea and cereals, which is required in bones and to activate certain enzymes in amino acid metabolism. The average adult human requires 3.7 milligrams per day and the total body content is only 0.01 grams.

chromium a mineral found in meat and cereals.

cobalt a mineral found in most foods, but especially meat and yeast products, which is an essential constituent of vitamin B_{12}. The average adult human requires 0.3 milligrams per day and the total body content is as little as 0.001 grams.

鈉　多種 "含鹽" 食物、尤其是食鹽（氯化鈉），乳酪和鹹肉中含的一種礦物質。它爲神經傳遞和酸鹼平衡所必需。平均每人每天需鈉4.4克，整個身體的含鈉量約爲100克。

氯　以氯離子形式和鈉一起存在於食鹽和肉類中的一種礦物質。它爲酸鹼平衡和滲透調節所必需。平均每個成人每天需氯5.2克，整個身體的含氯量約爲95克。

鎂　大多數食物、尤其是乳酪和綠色蔬菜中含的一種礦物質。在新陳代謝中，需要它來激活某些酶。平均每個成人每天需鎂0.34克，整個身體的含鎂量約爲19克。

鐵　肝、蛋、牛肉和某些飲用水中含有的一種礦物質。它是血紅蛋白和過氧化氫酶的一種必不可少的成份。平均每個成人每天需鐵約16毫克，整個身體的含鐵量約爲4.2克。

氟　以氟化物形式存在於海水、海味中和有時被添加於飲用水中的一種礦物質。它是骨骼和牙齒的一種成份，可防止牙腐蝕。平均每人每天需氟1.8毫克，整個身體的含氟量約爲2.6克。

鋅　大多數食物、尤其是肉和豆類中含有的一種礦物質。它是許多酶的一種成份。它也可促進愈合。平均每個成人每天需鋅13毫克，整個身體的含鋅量約爲2.3克。

銅　大多食物、尤其是肝、豌豆和其他豆類中含有的一種礦物質。它爲形成血紅蛋白和其他酶所必需。平均每個成人每天需銅3.5毫克，整個身體的含銅量約爲0.07克。

碘　海味、某些飲用水和蔬菜中含有的一種礦物質。它爲甲狀腺素的一種成份。平均每個成人每天需碘0.2毫克，整個身體的含碘量僅約0.01克。

錳　大多數食物、尤其是茶葉和谷類中含有的一種礦物質。它爲骨骼以及在氨基酸代謝中激活某些酶所必需。平均每個成人每天需錳3.7毫克，整個身體的含錳量僅爲0.01克。

鉻　肉和谷類中含的一種礦物質。

鈷　大多數食物尤其是肉和酵母製品中含的一種礦物質。它是維生素 B_{12} 不可缺少的一種成份。平均每個成人每天需鈷0.3毫克，整個身體的含鈷量僅爲0.001克。

International System of Units (SI)
國際單位制

PREFIXES 前綴

PREFIX 前綴		FACTOR 因數	SIGN 記號	PREFIX 前綴		FACTOR 因數	SIGN 記號
milli-	毫	x 10^{-3}	m	kilo-	千	x 10^{3}	k
micro-	微	x 10^{-6}	μ	mega-	兆	x 10^{6}	M
nano-	納（毫微）	x 10^{-9}	n	giga-	吉	x 10^{9}	G
pico-	皮（微微）	x 10^{-12}	p	tera-	太	x 10^{12}	T

BASIC UNITS 基本單位

UNIT	單位	SYMBOL 符號	MEASUREMENT	量
metre	米	m	length	長度
kilogram	千克（公斤）	kg	mass	質量
second	秒	s	time	時間
ampere	安培	A	electric current	電流
kelvin	開爾文	K	temperature	溫度
mole	摩爾	mol	amount of substance	物質的量

DERIVED UNITS 導出單位

UNIT	單位	SYMBOL 符號	MEASUREMENT	量
newton	牛頓	N	force	力
joule	焦耳	J	energy, work	能量、功
hertz	赫茲	Hz	frequency	頻率
pascal	帕斯卡	Pa	pressure	壓強、壓力
coulomb	庫倫	C	quantity of electric charge	電荷量
volt	伏特	V	electrical potential	電位
ohm	歐姆	Ω	electrical resistance	電阻

Index 索引

Bilingual Edition Publisher : Willie Shen
雙語版出版人：　　　　　沈維賢

Author :　　　　　　　　Neil Curtis
原著者：　　　　　　　　尼爾・寇蒂斯

Managing Editor :　　　　Aman Chiu
策劃編輯：　　　　　　　趙嘉文

Translator:　　　　　　　Zhang Jingjing
翻譯：　　　　　　　　　張靜靜

Reviser :　　　　　　　　Xu Ke Zhung
審訂：　　　　　　　　　徐克莊

Editor :　　　　　　　　Chan Kai Kan
編輯：　　　　　　　　　陳繼勤

Glycolysis
糖酵解

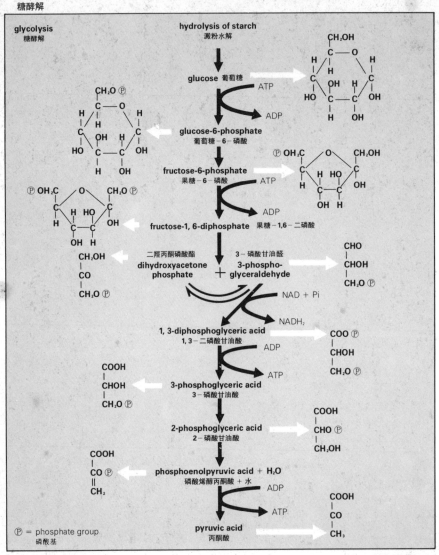